Maxwell Equation

Inverse Scattering in Electromagnetism

Maxwell Equation

Inverse Scattering in Electromagnetism

Hiroshi Isozaki

University of Tsukuba, Japan

World Scientific

NEW JERSEY · LONDON · SINGAPORE · BEIJING · SHANGHAI · HONG KONG · TAIPEI · CHENNAI · TOKYO

Published by

World Scientific Publishing Co. Pte. Ltd.
5 Toh Tuck Link, Singapore 596224
USA office: 27 Warren Street, Suite 401-402, Hackensack, NJ 07601
UK office: 57 Shelton Street, Covent Garden, London WC2H 9HE

Library of Congress Cataloging-in-Publication Data
Names: Isozaki, Hiroshi, 1950–   author.
Title: Maxwell equation : inverse scattering in electromagnetism / by Hiroshi Isozaki
     (University of Tsukuba, Japan).
Description: New Jersey : World Scientific, 2018. | Includes bibliographical references and index.
Identifiers: LCCN 2018016259 | ISBN 9789813232693 (hardcover : alk. paper)
Subjects: LCSH: Maxwell equations. | Electromagnetic theory.
Classification: LCC QC670 .I75 2018 | DDC 530.14/1--dc23
LC record available at https://lccn.loc.gov/2018016259

British Library Cataloguing-in-Publication Data
A catalogue record for this book is available from the British Library.

For any available supplementary material, please visit
http://www.worldscientific.com/worldscibooks/10.1142/10782#t=suppl

Printed in Singapore

Preface

The Maxwell equation is a fundamental equation in electromagnetism and constitutes the core of classical physics as well as quantum physics. Viewed from the mathematical side, it is a source of a wide variety of important problems such as partial differential equations in analysis and cohomology theory in geometry. The main purpose of this book is to give a mathematical foundation for wave propagations associated with the Maxwell equation, especially the scattering phenomenon. Suppose that the free electromagnetic waves are given at infinity. They will be scattered by the matter or the boundary of the domain and reflected back to infinity. How can we describe the propagation of waves, and what can be known about the media or the domain by observing the waves at infinity? The scattering theory aims at answering these questions. It is a classical subject, and a lot of nice ideas and powerful techniques have been developed. Therefore, it is worthwhile to summarize recent achievements in the inverse problem for the Maxwell equation on the firm and easily accesible basis of the forward problem.

We will discuss the following subjects in this monograph. Chapter 1 is devoted to the elementary notions and formulas of vector calculus. Basic knowledge of surfaces is also summarized. In Chapter 2, we study fundamental solutions for the Laplace equation, the Helmholtz equation and the wave equation. In Chapter 3, we study the potential theory. The Fredholm theory for integral equations is the main theme. We apply them to solve the boundary value problem for the Laplace equation. The radiation problem is also introduced. The study of the Maxwell equation begins from Chapter 4. The solution formulas and linking numbers are given there, and some problems for electric conductors are solved. Chapter 5 is devoted to the cohomology theory on domains in $\mathbf{R}^n$, which is related to decompositions of electromagnetic fields in the physical side, and to topological

properties of domains in the mathematical side. The advanced study of
the Maxwell equation starts from Chapter 6. We prepare the theory of
Fourier transforms and use functional analysis to solve the initial-boundary
value problem for the Maxwell equation. We then enter into the spectral
theory in Chapter 7. Introducing the spaces $H(\mathrm{div}\,;\,\Omega)$ and $H(\mathrm{curl}\,;\,\Omega)$,
which are also fundamental in fluid dynamics, we study the basic spectral
properties of the Maxwell operator. The main purpose of Chapter 7 is the
asymptotic completeness of wave operators, which characterizes the asymptotic behavior of solutions to the Maxwell equation in an exterior domain.
The eigenfunction expansion theory is studied in Chapter 8. We construct
a complete system of generalized eigenfunctions of the Maxwell operator
and observe their far field patterns. In Chapters 9 and 10, we study the
inverse scattering problem. In Chapter 9, we show that the S-matrix, obtained from the far field patterns of solutions to the Maxwell equation,
determines the electromagnetic map for the boundary value problem in a
suitable interior domain. This reduces the inverse scattering problem to
the inverse boundary value problem on a bounded domain. Chapter 10 is
devoted to the study of inverse problem for the Maxwell equation. We explain the Faddeev theory of direction dependent Green operators as well as
the method of complex geometrical optics solutions to the interior boundary value problem. We then elucidate the inverse boundary value problem.
Other interesting problems such as anisotropic media, domain identification
problem, linear sampling method, transmission eigenvalue problem are also
discussed. We finally prove that the Betti number of the boundary of the
domain is determined from the S-matrix.

In summary, Chapters 1 and 2 cover the basic undergraduate course of
vector calculus and introductory partial differential equations. Chapters 3
and 5 are fairly advanced for undergraduate students, however, essential
in studying the mathematical aspect of electromagnetism. Chapters 4 and
6 are the standard graduate school course for the Maxwell equation and
linear partial differential equations. The remaining chapters are devoted
to the scattering theory, Chapters 7 and 8 to the forward problem, and
Chapetrs 9 and 10 to the inverse problem. Graduate students as well as
young researchers can find an overview on this field.

Because of its mathematical and physical importance, lots of excellent
books concerning the mathematical theory of the Maxwell equation and
scattering theory have been published. Friedrichs [33] deals with the basis
of electromagnetic waves and scattering theory. A beautiful theory of scattering based on the idea of outgoing-incoming subspaces was developed by

Lax-Phillips [65]. Petkov [82] deals with non-selfadjoint cases and Petkov-Stoyanov [83] studies the propagation of singularities. Colton-Kress [22] is a treatise of the boundary value problem and scattering theory for the Helmholtz equation based on the potential theory, and [23] is an excellent exposition of stationary scattering theory for the Maxwell equation. Cessenat [19] is written in the same spirit with more emphasis on mathematical side. Isakov [43] deals with inverse problems from wide view points. Recently, the computational aspect of scattering has been more developed. In Kirsch [56], Monk [74], Cakoni-Colton [14], Ammari-Kang [6], one can see both of the theory and numerical results of scattering. As a good expository article of scattering theory, we refer to Chadan-Colton-Päivärinta-Rundell [20]. The boundary control method, which is a recently established method of inversion procedure from the spectral data and makes it possible to reconstruct not only the coefficients of the differential equation but also the underlying manifold, is also applied to the inverse scattering problem. See Katchalov-Kurylev-Lassas [49], Isozaki-Kurylev [47].

We use the standard notation. Let $\mathbf{R}$ and $\mathbf{C}$ be the set of the real and complex numbers, respectively, and $\mathbf{Z}_+$ the set of all non-negative integers. An element $\alpha = (\alpha_1, \cdots, \alpha_n) \in \mathbf{Z}_+^n$ is said to be a multi-index. We put

$$|\alpha| = \alpha_1 + \cdots + \alpha_n, \quad \alpha \in \mathbf{Z}_+^n,$$

$$x^\alpha = x_1^{\alpha_1} \cdots x_n^{\alpha_n}, \quad x = (x_1, \cdots, x_n) \in \mathbf{R}^n.$$

We use the notation $\partial_{x_i} = \partial/\partial_{x_i}$ and

$$\partial_x^\alpha = (\partial_{x_1})^{\alpha_1} \cdots (\partial_{x_n})^{\alpha_n}.$$

For an open set U in $\mathbf{R}^n$, $C(U; \mathbf{C}^m)$ is the set of $\mathbf{C}^m$-valued continuous functions on U. When $m = 1$, $\mathbf{C}^m$ is often omitted. $C^\infty(U)$ denotes the set of all infinitely many times differentiable functions on U, while $C^\infty(\overline{U})$ is the set of functions which are infinitely differentiable on an open set containing $\overline{U}$, the closure of U. $C_0(U)$ is the set of compactly supported continuous functios on U, and $C_0^\infty(U) = C_0(U) \cap C^\infty(U)$. For open sets U_1 and U_2, $U_1 \subset\subset U_2$ means that $U_1 \subset \overline{U_1} \subset U_2$ and $\overline{U_1}$ is compact. The Euclidean inner product of $x, y \in \mathbf{R}^n$ is denoted by

$$x \cdot y = \langle x, y \rangle = \sum_{i=1}^n x_i y_i.$$

From the knowledge of Lebesgue integral, we only use the following theorem: Let Ω be an open set in $\mathbf{R}^m$. Suppose that a sequence of functions $\{f_n(x)\}_{n=1}^{\infty}$ converges to $f(x)$ on Ω, and that there exists $g(x)$ such that

$$\int_{\Omega} |g(x)|dx < \infty, \quad |f_n(x)| \le g(x), \quad \forall x \in \Omega, \quad \forall n \ge 1.$$

Then

$$\lim_{n \to \infty} \int_{\Omega} f_n(x)dx = \int_{\Omega} f(x)dx.$$

Let us recall basic notions from functional analysis. Given a complex vector space $\mathcal{X}$, a *norm* on $\mathcal{X}$ is a function $p(\cdot)$, usually denoted as $\| \cdot \|$, from $\mathcal{X}$ to $\mathbf{R}$ having the following properties:

- $\|x\| \ge 0, \quad \forall x \in \mathcal{X}$.
- $\|x\| = 0 \Longleftrightarrow x = 0$.
- $\|\alpha x\| = |\alpha|\|x\|, \quad \forall \alpha \in \mathbf{C}, \forall x \in \mathcal{X}$.
- $\|x + y\| \le \|x\| + \|y\|, \quad \forall x, y \in \mathcal{X}$.

The space $\mathcal{X}$ is said to be a *Banach space*, if it is complete with respect to the norm, i.e. for any sequence $\{x_n\}_{n=1}^{\infty}$ satisfying $\|x_n - x_m\| \to 0$ as $m, n \to \infty$, there exists an $x \in \mathcal{X}$ such that $\|x_n - x\| \to 0$. An *inner product* on a complex vector space $\mathcal{H}$ is a function $(\cdot, \cdot) : \mathcal{H} \times \mathcal{H} \to \mathbf{C}$ having the following properties

- $(x, x) \ge 0, \quad \forall x \in \mathcal{H}$.
- $(x, x) = 0 \Longleftrightarrow x = 0$.
- $(x, y) = \overline{(y, x)}, \quad \forall x, y \in \mathcal{H}$.
- $(\alpha x + \beta y, z) = \alpha(x, z) + \beta(y, z), \quad \forall x, y, z \in \mathcal{H}, \quad \forall \alpha, \beta \in \mathbf{C}$.

Using these definitions, one can derive the Cauchy-Schwartz inequality

$$|(x, y)| \le \|x\|\|y\|, \quad x, y \in \mathcal{H},$$

and define a norm on $\mathcal{H}$ by

$$\|x\| = \sqrt{(x, x)}.$$

The space $\mathcal{H}$ is said to be a *Hilbert space* if it is complete with respect to this norm. Let Ω be an open set in $\mathbf{R}^n$. Then the set of all square integrable functions $f(x)$ on Ω, i.e. $\int_{\Omega} |f(x)|^2 dx < \infty$, is a Hilbert space equipped with the inner product

$$(f, g)_{L^2(\Omega)} = \int_{\Omega} f(x)\overline{g(x)}dx.$$

For $1 < p < \infty$, $L^p(\Omega)$ is the set of functions $f(x)$ on Ω satisfying

$$\|f\|_{L^p(\Omega)} = \left(\int_\Omega |f(x)|^p dx \right)^{1/p} < \infty.$$

This is a Banach space. Given two Banach spaces $\mathcal{X}$ and $\mathcal{Y}$, a linear operator A from $\mathcal{X}$ to $\mathcal{Y}$ is said to be *bounded* if there is a constant $C > 0$ such that

$$\|Ax\| \leq C\|x\|, \quad \forall x \in \mathcal{X}.$$

The *norm* (operator norm) of A is then defined by

$$\|A\| = \sup_{x \in \mathcal{X}, \|x\|=1} \|Ax\|.$$

The set of all bounded linear operators from $\mathcal{X}$ to $\mathcal{Y}$ is denoted by $\mathbf{B}(\mathcal{X}; \mathcal{Y})$. If $\mathcal{X} = \mathcal{Y}$, it is denoted by $\mathbf{B}(\mathcal{X})$. The other necessary notions will be explained in the text.

The author is indebted to Dr. Hisashi Morioka for his help in drawing the figures.

This work is supported by Grants-in-Aid for Scientific Research (S) 15H05740, (B) 16H0394, Japan Society for the Promotion of Science. The author expresses his gratitude to JSPS.

Kyoto
September, 2017

Hiroshi Isozaki

Contents

Preface   v

1. Vector calculus   1

    1.1   Implicit function theorem   1
    1.2   Local definition of surfaces   4
    1.3   Tangent space   6
    1.4   Induced metric   9
    1.5   Volume element   11
    1.6   Global definition of surfaces   13
          1.6.1   Submanifold of $\mathbf{R}^N$ 13
          1.6.2   Tangent space   14
          1.6.3   Partition of unity   14
          1.6.4   Integral on M   15
          1.6.5   Convergence of volume integral to line integral . .   16
    1.7   Integral formulas   17
          1.7.1   Integration by parts   17
          1.7.2   Theorem of Gauss-Ostrogradsky   18
          1.7.3   Theorem of Green   19
          1.7.4   Surfaces with boundaries   19
          1.7.5   Stokes' Theorem   20
          1.7.6   Integral of 1-form   21
    1.8   A short course on differential forms   22
          1.8.1   The case of two variables   23
          1.8.2   The case of three variables   26

2. Fundamental solutions   31

  2.1   Fundamental solution on $\mathbf{R}^n$ 31
  2.2   Harmonic functions 36
  2.3   Uniqueness theorem 38
  2.4   Wave equation in $\mathbf{R}^n$ 39
     2.4.1   d'Alembert's formula 39
     2.4.2   Kirchhoff's formula 40
     2.4.3   Poisson's formula 41
     2.4.4   Huygens' principle 42
     2.4.5   Duhamel's principle 42
     2.4.6   Energy inequality 44

3. Potential theory   47

  3.1   Dirac measure and principal value 47
  3.2   Function spaces on S 52
  3.3   Continuity of tangential derivatives 53
  3.4   Single and double layer potentials 59
  3.5   Dual space of a Banach space 65
  3.6   Mapping properties 67
  3.7   Fredholm's alternative theorem 71
     3.7.1   Theory of Riesz-Schauder 71
     3.7.2   Dual system 73
  3.8   Boundary value problems for the Laplace equation 75
     3.8.1   Boundary components 75
     3.8.2   Uniqueness 76
     3.8.3   Boundary integral equations on a single boundary 78
     3.8.4   Boundary integral equations on multiple boundaries 79
  3.9   Potential theory for the Helmholtz equation 86
     3.9.1   Radiation condition 86
     3.9.2   Uniqueness 87
     3.9.3   Boundary integral equations 89

4. Maxwell equation   93

  4.1   Equations in the vacuum 93
  4.2   Gauge transformation 95
     4.2.1   Existence of the potential 95
     4.2.2   Coulomb gauge 97
     4.2.3   Lorentz gauge 97

4.3   Biot-Savart's law 99

4.4   Linking number 99

4.5   Maxwell equation in medium 103

4.6   Conductor 104

       4.6.1   Electrostatic capacity 104

       4.6.2   Electrostatic shield 108

5. Cohomology   111

5.1   Bounded planar domain 111

5.2   Closed 2-dimensional surface 113

5.3   Handle body 115

5.4   Divergence free fields 116

5.5   Harmonic fields 121

5.6   Gradient fields 124

5.7   Hodge decomposition theorem 126

6. Initial value problem for the Maxwell equation   129

6.1   Fourier transform 129

       6.1.1   Rapidly decreasing functions 129

       6.1.2   Fourier transforms of L^2-functions 130

       6.1.3   Tempered distributions 131

6.2   Distributions with support on surfaces 132

6.3   Sobolev spaces 136

6.4   Fourier transformation and the Maxwell equation 137

6.5   Hilbert space approach 140

       6.5.1   Self-adjoint operator and unitary group 140

       6.5.2   Maxwell equation in the vacuum 141

       6.5.3   Maxwell equation in the medium 142

7. Boundary value problem for the Maxwell equation   143

7.1   Curl and divergence spaces 143

       7.1.1   Trace operator 143

       7.1.2   Spaces $H(\mathrm{div}\,;\Omega)$ and $H(\mathrm{curl}\,;\Omega)$ 144

7.2   Local compactness 148

7.3   Self-adjointness 152

7.4   Spectral properties of self-adjoint operators 154

       7.4.1   Spectral measure 154

       7.4.2   Classification of the spectrum 156

       7.4.3   Local decay of scattering solutions 157
  7.5   Spectrum of the Maxwell operator 159
       7.5.1   Interior problem 159
       7.5.2   Free Maxwell operator 159
       7.5.3   Exterior problem 161
  7.6   Time-dependent scattering theory 163
  7.7   Tools from scattering theory 167
       7.7.1   Functional calculus 167
       7.7.2   Pseudo-differential calculus 170

8. Stationary scattering theory   177

  8.1   Resolvent estimates for the free Laplacian 177
  8.2   Resolvent estimates for the free Maxwell operator 180
  8.3   Radiation condition 182
  8.4   Resolvent estimates in an exterior domain 184
  8.5   Spectral representation 190
  8.6   The surjectivity of $\mathcal{F}_\pm$ 194
  8.7   S-matrix 196
  8.8   Homogeneous equation 198

9. Scattering and interior boundary value problem   203

  9.1   Mapping properties of the resolvent 204
  9.2   Layer potentials 205
  9.3   Mapping properties on Sobolev spaces 211
       9.3.1   Function spaces 211
       9.3.2   Layer potentials on Sobolev spaces 213
       9.3.3   Perturbed layer potentials 216
  9.4 E-M map 218
  9.5   S-matrix determines E-M map 219

10. Inverse scattering   223

  10.1   Green operators and scattering amplitudes 224
       10.1.1   Abstract theory 224
       10.1.2   Faddeev's Green operator 226
       10.1.3   Green operator of Eskin-Raston 229
       10.1.4   Inverse boundary value problem 230
  10.2   Direction dependent Green operators for the free Maxwell
       equation 232

10.3   Perturbed direction dependent Green operators 234
10.4   From physical scattering amplitude to Faddeev
       scattering amplitude 246
10.5   Determination of medium 248
10.6   Anisotropic medium 250
10.7   Identifiability of the domain 250
10.8   Linear sampling method 252
10.9   Transmission eigenvalue 254
10.10 Determination of Betti number 256

Appendix A   Formulae in vector calculus   259

Appendix B   1-forms and vector fields   261

Appendix C   Stationary phase method   267

Appendix D   Interpolation theorem   271

Appendix E   Theory of quadratic forms   273

Bibliography   275

Index   281

Chapter 1

Vector calculus

Throughout this chapter, n denotes an integer ≥ 2. A point $x \in \mathbf{R}^n$ is represented by a column vector $x = {}^t(x_1, \cdots, x_n)$, however, following the convention in Riemannian geometry, we sometimes use the upper indices: $x = {}^t(x^1, \cdots, x^n)$, and also omit superscript ^{t} for the sake of simplicity. In the case $n = 2$ or 3, we often denote the coordinates by (x, y) or (x, y, z). Given an $\mathbf{R}^N$-valued function $f(x) = {}^t(f_1(x), \cdots, f_N(x))$ defined on $\mathbf{R}^n$, its differential or Jacobi-matrix is defined by

$$f'(x) = \frac{\partial f}{\partial x} = \left(\frac{\partial f}{\partial x_1}, \cdots, \frac{\partial f}{\partial x_n} \right) = \begin{pmatrix} \dfrac{\partial f_1}{\partial x_1} & \dfrac{\partial f_1}{\partial x_2} & \cdots & \dfrac{\partial f_1}{\partial x_n} \\ \dfrac{\partial f_2}{\partial x_1} & \dfrac{\partial f_2}{\partial x_2} & \cdots & \dfrac{\partial f_2}{\partial x_n} \\ \cdots & \cdots & \cdots & \cdots \\ \dfrac{\partial f_N}{\partial x_1} & \dfrac{\partial f_N}{\partial x_2} & \cdots & \dfrac{\partial f_N}{\partial x_n} \end{pmatrix}. \tag{1.1}$$

For a point $p \in \mathbf{R}^n$ and a constant $r > 0$, $B(p, r)$ denotes the ball of radius r centered at p:

$$B(p, r) = \{ x \in \mathbf{R}^n \,;\, |x - p| < r \}. \tag{1.2}$$

For an open set U in $\mathbf{R}^n$, $C^\infty(U; \mathbf{R}^N)$ denotes the set of all $\mathbf{R}^N$-valued C^∞-functions defined on U. Let U and V be open sets in $\mathbf{R}^n$. By a *diffeomorphism* $\varphi : U \to V$, we mean a function in $C^\infty(U; \mathbf{R}^n)$ which is 1 to 1 and onto from U to V with smooth inverse $\varphi^{-1} : V \to U$.

1.1 Implicit function theorem

Suppose we are given a curve on the plane

$$C = \{ (x, y) \,;\, f(x, y) = 0 \}. \tag{1.3}$$

1

If at a point $p_0 = (x_0, y_0) \in C$, the condition

$$f_y(x_0, y_0) = \frac{\partial f}{\partial y}(x_0, y_0) \neq 0 \tag{1.4}$$

holds, then C is represented as $y = g(x)$ near p_0, where $g(x)$ is a function defined near x_0. A simple example is $f(x, y) = ax + by$. If $b \neq 0$, the line $f(x, y) = 0$ is represented as $y = -ax/b$. As will be explained in §2, $\nu = (f_x(x_0, y_0), f_y(x_0, y_0))$ is orthogonal to the curve C. The condition (1.4) means that ν is not parallel to the x-axis. This is the reason why C is represented as $y = g(x)$.

Example 1.1. (*Lemniscate*) Put $f(x, y) = (x^2 + y^2)^2 - (x^2 - y^2)$, and consider a curve $C = \{(x, y) ; f(x, y) = 0\}$.
(1) On the curve C,

$$f_x(x, y) = 0 \iff (x, y) = (0, 0), \pm(\sqrt{3/8}, \pm\sqrt{1/8}),$$

$$f_y(x, y) = 0 \iff (x, y) = (0, 0), \pm(1, 0).$$

(2) C is written as

$$\begin{cases} y = \pm\sqrt{(\sqrt{1 + 8x^2} - 1 - 2x^2)/2}, & -1 < x < 1, \\ x = \pm\sqrt{(\sqrt{1 - 8y^2} + 1 - 2y^2)/2}, & -\sqrt{1/8} < y < \sqrt{1/8}. \end{cases}$$

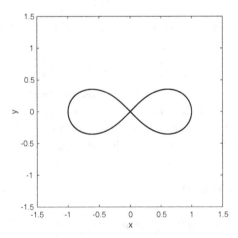

Fig. 1.1   Lemniscate

Before entering into the general case of implicit function theorem, let us consider a linear map $f : \mathbf{R}^m \times \mathbf{R}^n \to \mathbf{R}^n$ defined by

$$f(x, y) = {}^t(f_1(x, y), \cdots, f_n(x, y)),$$

$$f_i(x, y) = a_{i1}x_1 + \cdots + a_{im}x_m + b_{i1}y_1 + \cdots + b_{in}y_n, \quad 1 \le i \le n.$$

This is rewritten as

$$f(x, y) = Ax + By,$$

where $A = (a_{ij}), B = (b_{ik})$. The implicit function theorem for f asserts that from the equation $f(x, y) = 0$ we can represent y as a function of x. This is possible only when $\det B \neq 0$. Obviously, we have $B = \dfrac{\partial f}{\partial y}$. Therefore, for the linear case, the implicit function theorem holds if $\det \left(\dfrac{\partial f}{\partial y} \right) \neq 0$. This suggests the theorem for the general case.

Theorem 1.1. *Let $\mathcal{O}$ be an open set in $\mathbf{R}^m \times \mathbf{R}^n$, and $f \in C^\infty(\mathcal{O}; \mathbf{R}^n)$. Suppose there exists $(x_0, y_0) \in \mathcal{O}$ such that*

$$\det \left(\frac{\partial f}{\partial y} \right)(x_0, y_0) \neq 0. \tag{1.5}$$

Then there exist open sets $U_0 \subset \mathbf{R}^m, V_0 \subset \mathbf{R}^n$ and $g \in C^\infty(U_0; \mathbf{R}^n)$ such that $(x_0, y_0) \in U_0 \times V_0 \subset W_0$, and

$$f(x, y) = 0, \ (x, y) \in U_0 \times V_0 \Longrightarrow y = g(x).$$

Differentiating $f_i(x, g(x)) = 0$ with respect to x_j, we have

$$\left(\frac{\partial f_i}{\partial x_j} \right)(x, g(x)) + \sum_{k=1}^n \left(\frac{\partial f_i}{\partial y_k} \right)(x, g(x)) \left(\frac{\partial g_k}{\partial x_j} \right)(x) = 0,$$

from which we can compute the differential of $g(x)$ as follows

$$\frac{\partial g}{\partial x}(x) = - \left(\frac{\partial f}{\partial y}(x, g(x)) \right)^{-1} \frac{\partial f}{\partial x}(x, g(x)). \tag{1.6}$$

Example 1.2. Let $f(x, y, z, t) = tx + p(t)y + z + q(t)$, $g(x, y, z, t) = x + p'(t)y + q'(t)$. Assume that $p''(t)y + q''(t) \neq 0$. Then from the equation $f = g = 0$, one can represent z and t as functions of x, y, i.e. $z = z(x, y), t = t(x, y)$, and $(z_{xy})^2 = z_{xx}z_{yy}$.

In fact, if $f(x, y, z, t) = g(x, y, z, t) = 0$, we have

$$\left(\frac{\partial}{\partial z} \begin{pmatrix} f \\ g \end{pmatrix}, \frac{\partial}{\partial t} \begin{pmatrix} f \\ g \end{pmatrix} \right) = \begin{pmatrix} 1 & g \\ 0 & p''(t)y + q''(t) \end{pmatrix}.$$

Then one can apply Theorem 1.1 to obtain $z = z(x, y), t = t(x, y)$. Differentiating $t(x, y)x + p(t(x, y))y + z(x, y) + q(t(x, y)) = 0$ with respect to x, y and using $x + p'(t(x, y))y + q'(t(x, y)) = 0$, we obtain $z_x = -t$, $z_y = -p(t)$. This yields $z_{xy} = -t_y = -p'(t)t_x$. We also have $z_{xx} = -t_x, z_{yy} = -p'(t)t_y$, which proves $(z_{xy})^2 = z_{xx}z_{yy}$.

Another fundamental tool is the following inverse function theorem.

Theorem 1.2. *Let U be an open set in $\mathbf{R}^n$ and $f \in C^\infty(U; \mathbf{R}^n)$. Suppose* $\det\left(\frac{\partial f}{\partial x}\right)(x_0) \neq 0$ *at some $x_0 \in U$. Then there exist open sets $U_0, V_0 \subset \mathbf{R}^n$ such that $x_0 \in U_0 \subset U$, $f(x_0) \in V_0$ and $f : U_0 \to V_0$ is a diffeomorphism.*

Example 1.3. Let $f(x_1, x_2) = {}^t(x_1 + x_2, x_1 x_2)$ on $B(0, 1)$. Then $\det \frac{\partial f}{\partial x} = 0$ if and only if $x_1 = x_2$. Let $U_\pm = \{(x_1, x_2) \in B(0, 1) ; \pm(x_1 - x_1) > 0\}$, $V = \{(y_1, y_2) ; 2(y_1^2 - 1) < 4y_2 < y_1^2\}$. Then each of $f : U_+ \to V$ and $f : U_- \to V$ is a diffeomorphism.

Note that x_1, x_2 are the roots of the equation $t^2 - y_1 t + y_2 = 0$, where $y_1 = x_1 + x_2, y_2 = x_1 x_2$, and the equality $x_1{}^2 + x_2{}^2 = y_1{}^2 - 2y_2$ holds.

1.2 Local definition of surfaces

Let us study m-dimensional surfaces in $\mathbf{R}^n$, where $0 < m < n - 1$. We begin with the case of $m = n - 1$, and the *local* definition, i.e. a small part of the surface. There are three ways to define it.

(1) *Definition 1 (Graph of a function).* Given a function $g(x^1, \cdots, x^{n-1}) \in C^\infty(U; \mathbf{R})$, U being an open set in $\mathbf{R}^{n-1}$, we define the surface S by

$$S \ni x \Longleftrightarrow x^n = g(x^1, \cdots, x^{n-1}), \quad (x^1, \cdots, x^{n-1}) \in U. \qquad (1.7)$$

For example, the upper hemisphere is defined by $x^n = \left(1 - \sum_{i=1}^{n-1}(x^i)^2\right)^{1/2}$.

(2) *Definition 2 (Zeros of a function).* Given a function $f(x) \in C^\infty(\mathcal{O}; \mathbf{R})$, $\mathcal{O}$ being an open set in $\mathbf{R}^n$, we define the surface S by

$$S \ni x \Longleftrightarrow f(x) = 0, \quad x \in \mathcal{O}. \qquad (1.8)$$

Here we need to impose the condition

$$\frac{\partial f}{\partial x} \neq 0 \quad \text{on} \quad S. \tag{1.9}$$

For example, the unit sphere in $\mathbf{R}^n$ is defined by $\sum_{i=1}^{n} (x^i)^2 - 1 = 0$.

To show that Definition 1 implies Definition 2, we have only to put $f(x) = x^n - g(x^1, \cdots, x^{n-1})$. Since

$$\frac{\partial f}{\partial x} = \left(-\frac{\partial g}{\partial x^1}, \cdots, -\frac{\partial g}{\partial x^{n-1}}, 1 \right) \neq 0,$$

the condition (1.9) is satisfied. Conversely, suppose S is as in Definition 2. Take $p \in S$ arbitrarily. By (1.9), we can assume without loss of generality that $\partial f / \partial x^n \neq 0$ at p. By the implicit function theorem, the points in S near p are represented as $x^n = g(x^1, \cdots, x^{n-1})$ by some function g.

For an $n \times m$-matrix $A = (A_1, \cdots, A_m)$, where $A_i \in \mathbf{R}^n$, the rank of A is defined to be the dimension of the range of A. This is equal to the number of linearly independent vectors in $\{A_1, \cdots, A_m\}$. Recall the following well-known fact.

Lemma 1.1. *Let A be an $n \times m$ matrix of rank m with $0 < m \leq n$. Then by interchanging the row vectors of A suitably, one can assume that*

$$\det \begin{pmatrix} a_{11} & \cdots & a_{1m} \\ \cdots & \cdots & \cdots \\ a_{m1} & \cdots & a_{mm} \end{pmatrix} \neq 0.$$

(3) *Definition 3 (Representation by parameters).* Given an open set $U \subset \mathbf{R}^{n-1}$, and $F = (F^1, \cdots, F^n) \in C^\infty(U; \mathbf{R}^n)$, we define the surface S by

$$S = \{ F(u) \, ; \, u \in U \}. \tag{1.10}$$

Here we impose the condition

$$\text{rank} \, \frac{\partial F}{\partial u} = n - 1, \quad \text{on} \quad U. \tag{1.11}$$

It is easy to see that Definition 1 implies Definition 3. In fact, we take $u = (x^1, \cdots, x^{n-1})$, and put $F(u) = {}^t(u, g(u))$, i.e. $F^i = u^i, 1 \leq i \leq n-1, F^n = g(u)$. Then we have

$$\frac{\partial F}{\partial u} = \begin{pmatrix} I_{n-1} \\ a \end{pmatrix}, \quad a = \frac{\partial g}{\partial u},$$

where I_{n-1} is the $(n-1) \times (n-1)$ identity matrix. Therefore, the condition (1.11) is satisfied. Conversely, suppose S is given by (1.10). Take a point

$u_0 \in U$, and let $G(u) = {}^t(F^1(u), \cdots, F^{n-1}(u))$. In view of Lemma 1.1, one can assume that $\det \dfrac{\partial G}{\partial u} \neq 0$ near u_0. Then by the inverse function theorem (Theorem 1.2), $G(u)$ is a diffeomorphism between neighborhoods of u_0 and $G(u_0)$. Let $u(x^1, \cdots, x^{n-1})$ be the inverse of $G(u)$. Then the points x on S near $F(u_0)$ satisfies $x^n = F^n(u(x^1, \cdots, x^{n-1}))$. Therefore Definition 3 implies Definition 1.

Example 1.4. The unit sphere $S^{n-1} = \{x \in \mathbf{R}^n \,;\, |x| = 1\}$ is parameterized by polar coordinates as follows:

$$\begin{cases} x^n = \cos\theta^{n-1}, \\ x^{n-1} = \sin\theta^{n-1}\cos\theta^{n-2}, \\ \quad\cdots\cdots\,, \\ \quad\cdots\cdots\,, \\ x^2 = \sin\theta^{n-1}\sin\theta^{n-2}\cdots\cdots\sin\theta^2\cos\theta^1, \\ x^1 = \sin\theta^{n-1}\sin\theta^{n-2}\cdots\cdots\sin\theta^2\sin\theta^1, \end{cases} \tag{1.12}$$

$$0 \leq \theta^i \leq \pi, \quad (2 \leq i \leq n-1), \quad 0 \leq \theta^1 \leq 2\pi.$$

Example 1.5. The 2-dimensional torus in $\mathbf{R}^3$ is the surface obtained by rotating the circle $(x-a)^2 + z^2 = b^2$ in the x-z plane around the z-axis. It is represented as follows:

$$x = (a + b\cos\theta)\cos\phi, \quad y = (a + b\cos\theta)\sin\phi, \quad z = b\sin\theta, \tag{1.13}$$

where $0 < b < a$, and

$$0 \leq \theta \leq 2\pi, \quad 0 \leq \phi \leq 2\pi.$$

1.3   Tangent space

Take a point p from a surface S. The *tangent space* of S at p, denoted by $T_p(S)$, is the 1st order approximation of S at p. We compute it in three ways by using the above definitions of the surface.

(1) If S is defined by the graph: $x^n = g(x') = g(x^1, \cdots, x^{n-1})$ and $p = (p', p^n)$, $T_p(S)$ is defined by the equation

$$x^n - p^n = \sum_{i=1}^{n-1} \frac{\partial g}{\partial x^i}(p')(x^i - p'^i). \tag{1.14}$$

(2) If S is defined by the equation $f(x) = 0$, $T_p(S)$ is defined by the equation

$$\sum_{i=1}^{n} \frac{\partial f}{\partial x^i}(p)(x^i - p^i) = 0. \tag{1.15}$$

We show that (1.14) and (1.15) define the same plane. Taking $f(x) = x^n - g(x')$, we obtain (1.14) from (1.15). To show the converse, without loss of generality, we assume that $\frac{\partial f}{\partial x^n}(p) \neq 0$. By the implicit function theorem, there exists $g(x')$ such that near p, the zeros of $f(x)$ satisfy $x^n = g(x')$. Hence $f(x', g(x')) = 0$. Differentiating it with respect to x^i, we have

$$\frac{\partial f}{\partial x^i}(p', p^n) = -\frac{\partial f}{\partial x^n}(p', p^n)\frac{\partial g}{\partial x^i}(p'), \quad 1 \leq i \leq n-1.$$

We replace $\frac{\partial f}{\partial x^i}(p)$ in (1.15) by the right-hand side to obtain (1.14).

The following lemma is an obvious consequence of the formula (1.15).

Lemma 1.2. *If a surface S is defined by $f(x) = 0$, the vector $\dfrac{\partial f}{\partial x}(p)$ is orthogonal to S at $p \in S$.*

(3) Suppose S is represented as (1.10). Take $p = x(u) \in S$, and consider the following curve on S passing through p:

$$(u^i - \epsilon, u^i + \epsilon) \ni t \to x(u^1, \cdots, u^{i-1}, t, u^{i+1}, \cdots, u^{n-1}) \in S.$$

Differentiating it with respect to t and letting $t = u^i$, we get a vector tangent to S at p. Therefore, we have the following $n - 1$ vectors tangent to S at p:

$$\frac{\partial x}{\partial u^1}(u), \cdots, \frac{\partial x}{\partial u^{n-1}}(u). \tag{1.16}$$

They are linearly independent because of the condition (1.11). We define the tangent space $T_p(S)$ by

$$T_p(S) = \Big\{ p + \sum_{i=1}^{n-1} c^i \frac{\partial x}{\partial u^i}(u); \ c^i \in \mathbf{R} \Big\}. \tag{1.17}$$

This space does not depend on the choice of parameters. In fact, if S is represented by other parameters $v \in V$, V being an open set in $\mathbf{R}^{n-1}$, there is a diffeomorphism $V \ni v \to u(v) \in U$. We put $y(v) = x(u(v))$. Then

$$\sum_{j=1}^{n-1} c^j \frac{\partial x}{\partial u^j} = \sum_{i=1}^{n-1} \Big(\sum_{j=1}^{n-1} \frac{\partial v^i}{\partial u^j} c^j \Big) \frac{\partial y}{\partial v^i}.$$

Since $\left(\frac{\partial v^i}{\partial u^j}\right)$ is a non-singular matrix, we have

$$\left\{\sum_{i=1}^{n-1} c^i \frac{\partial x}{\partial u^i}(u); \ c^i \in \mathbf{R}\right\} = \left\{\sum_{i=1}^{n-1} b^i \frac{\partial y}{\partial v^i}(v); \ b^i \in \mathbf{R}\right\}.$$

If S is represented as $x^n = g(x')$, (1.17) agrees with (1.14). In fact, in this case we can take $u = x'$, hence the points on $T_p(S)$ are written as

$$x^i = p^i + c^i, \ (i = 1, \cdots, n-1), \quad x^n = p^n + \sum_{i=1}^{n-1} c^i \frac{\partial g}{\partial x^i}(p').$$

In the above arguments, the tangent space $T_p(S)$ is defined as the $(n-1)$-dimensional affine space passing through p. This agrees with our geometrical picture. However, we usually define $T_p(S)$ by translating p to the origin of $\mathbf{R}^n$ so that it is the $(n-1)$-dimensional vector subspace of $\mathbf{R}^n$.

Definition 1.1. If S is represented as (1.10), (1.11), the tangent space of S at $p = x(u) \in S$ is defined by

$$T_p(S) = \left\{\sum_{i=1}^{n-1} c^i \frac{\partial x}{\partial u^i}(u); \ c^i \in \mathbf{R}\right\}. \tag{1.18}$$

We compute the normal vector of S. Consider the following $(n-1) \times n$-matrix

$$\begin{pmatrix} \frac{\partial x^1}{\partial u^1} & \frac{\partial x^1}{\partial u^2} & \cdots & \frac{\partial x^1}{\partial u^{n-1}} \\ \frac{\partial x^2}{\partial u^1} & \frac{\partial x^2}{\partial u^2} & \cdots & \frac{\partial x^2}{\partial u^{n-1}} \\ \vdots & \vdots & \vdots & \vdots \\ \frac{\partial x^n}{\partial u^1} & \frac{\partial x^n}{\partial u^2} & \cdots & \frac{\partial x^n}{\partial u^{n-1}} \end{pmatrix}. \tag{1.19}$$

Remove the ith row from this matrix, and take the determinant, which we denote by $\frac{\partial(x^1, \cdots, \widehat{x^i}, \cdots, x^n)}{\partial(u^1, \cdots, u^{n-1})}$. Then the vector

$$N = (N^1, \cdots, N^n), \quad N^i = (-1)^{i-1} \frac{\partial(x^1, \cdots, \widehat{x^i}, \cdots, x^n)}{\partial(u^1, \cdots, u^{n-1})} \tag{1.20}$$

is normal to S. In fact, enlarge the matrix (1.19) by inserting the column vector $(\frac{\partial x^1}{\partial u^i}, \cdots, \frac{\partial x^n}{\partial u^i})$ to the right of the last column, and take the determinant of the resulting $n \times n$-matrix. Expanding this determinant with respect to the last column, we have $N \times \frac{\partial x}{\partial u^i} = 0$, $i = 1, \cdots, n-1$. Hence $N \perp T_p(S)$.

Note that for $n = 3$, N is written by the vector product:

$$N = \frac{\partial x}{\partial u^1} \times \frac{\partial x}{\partial u^2}. \tag{1.21}$$

The properties of vector product are summarized in Lemma A.1 in Appendix A.

1.4 Induced metric

For $x = (x^1, \cdots, x^n), y = (y^1, \cdots, y^n) \in \mathbf{R}^n$, let

$$\langle x, y \rangle = x^1 y^1 + \cdots + x^n y^n$$

be their inner product. For $p = x(u) \in S$, we put

$$g_{ij}(u) = \left\langle \frac{\partial x}{\partial u^i}(u), \frac{\partial x}{\partial u^j}(u) \right\rangle, \tag{1.22}$$

and consider the bilinear form defined on $T_p(S)$ by

$$G_p(v, w) = \sum_{i,j=1}^{n-1} g_{ij}(u) v^i w^j, \tag{1.23}$$

where $v, w \in T_p(S)$ written as

$$v = \sum_{i=1}^{n-1} v^i \frac{\partial x}{\partial u^i}(u), \quad w = \sum_{i=1}^{n-1} w^i \frac{\partial x}{\partial u^i}(u).$$

Lemma 1.3. *(1) For any $v \in T_p(S)$, $\left(G_p(v, v) \right)^{1/2}$ is equal to the length of v in $\mathbf{R}^n$.*
(2) $G_p(\cdot, \cdot)$ is a positive definite bilinear form on $T_p(S)$.

Proof. The Euclidean length of v is computed as

$$|v|^2 = \left\langle \sum_{i=1}^{n-1} v^i \frac{\partial x}{\partial u^i}(u), \sum_{j=1}^{n-1} v^j \frac{\partial x}{\partial u^j}(u) \right\rangle.$$

Expanding the right-hand side, we obtain (1). By (1), $G_p(v, v) \geq 0$, and if $G_p(v, v) = 0$, we have $v = 0$, which proves the positive definiteness. $\square$

We call $G_p(\cdot, \cdot)$ the *Riemannian metric* induced from the Euclidean metric of $\mathbf{R}^n$. It is also called *induced metric*. The traditional notation is

$$ds^2 = \sum_{i,j=1}^{n-1} g_{ij}(u) du^i du^j. \tag{1.24}$$

Its meaning is easily understood by the formal computation. Given the surface S parametrized by $u = (u^1, \cdots, u^{n-1})$, we take the total differential

$$dx^k = \sum_{i=1}^{n-1} \frac{\partial x^k}{\partial u^i} du^i,$$

and sum up $\sum_{k=1}^{n}(dx^k)^2$, which is denoted by ds^2.

For example, if S is represented as $x_n = f(x_1, \cdots, x_{n-1})$, one has

$$ds^2 = \sum_{i=1}^{n-1}\left(1 + \left(\frac{\partial f}{\partial x^i}\right)^2\right)(dx^i)^2 + 2\sum_{1 \le i < j \le n-1}\frac{\partial f}{\partial x^i}\frac{\partial f}{\partial x^j}dx^idx^j. \qquad (1.25)$$

By the polar coordinates $x = \sin\theta\cos\varphi$, $y = \sin\theta\sin\varphi$, $z = \cos\theta$, the induced metric on S^2 is written as

$$ds^2 = (d\theta)^2 + \sin^2\theta(d\varphi)^2. \qquad (1.26)$$

By the parameters (1.13), the induced metric on T^2 is computed as

$$ds^2 = b^2(d\theta)^2 + (a + b\cos\theta)^2(d\phi)^2. \qquad (1.27)$$

We use Einstein's summation convention: for a sum containing both of upper and lower indices i ranging over $\{1, \cdots, n-1\}$, we omit the symbol $\sum_{i=1}^{n-1}$ and write as follows

$$a_ib^i = \sum_{i=1}^{n-1}a_ib^i.$$

The metric (1.24) is then written as

$$ds^2 = g_{ij}(u)du^idu^j.$$

The expression of the Riemannian metric ds^2 does not depend on the choice of parameter u. In fact, suppose we have another open set $\overline{U} \subset \mathbf{R}^n$ (note that this is not the closure of U) and a diffeomorphism : $U \ni u \to \overline{u} \in \overline{U}$. Suppose the surface S is also parametrized by $\overline{u}$: $S = \{\overline{x}(\overline{u})\,;\,\overline{u} \in \overline{U}\}$. Since

$$\frac{\partial\overline{x}}{\partial\overline{u}^i} = \frac{\partial u^k}{\partial\overline{u}^i}\frac{\partial x}{\partial u^k},$$

we have

$$\overline{g}_{ij} := \left\langle\frac{\partial\overline{x}}{\partial\overline{u}^i}, \frac{\partial\overline{x}}{\partial\overline{u}^j}\right\rangle = \frac{\partial u^k}{\partial\overline{u}^i}\frac{\partial u^\ell}{\partial\overline{u}^j}g_{k\ell}. \qquad (1.28)$$

Therefore, if a vector $v \in T_p(S)$ is represented as

$$v = a^i\frac{\partial x}{\partial u^i} = \overline{a}^i\frac{\partial\overline{x}}{\partial\overline{u}^i},$$

we have

$$a^i\frac{\partial x}{\partial u^i} = \overline{a}^i\frac{\partial u^k}{\partial\overline{u}^i}\frac{\partial x}{\partial u^k} = \overline{a}^k\frac{\partial u^i}{\partial\overline{u}^k}\frac{\partial x}{\partial u^i}.$$

It implies

$$a^i = \frac{\partial u^i}{\partial \overline{u}^k} \overline{a}^k. \tag{1.29}$$

Let us also represent $w \in T_p(S)$ as

$$w = b^i \frac{\partial x}{\partial u^i} = \overline{b}^i \frac{\partial \overline{x}}{\partial \overline{u}^i}.$$

Using (1.28) and (1.29), one then obtains

$$g_{ij} a^i b^j = \overline{g}_{ij} \overline{a}^i \overline{b}^j, \tag{1.30}$$

which shows the invariance of $ds^2 = g_{ij} du^i du^j$.

Note the following by-product of the above computation. By (1.28), we have

$$(\overline{g}_{ij}) = \left(\frac{\partial u^k}{\partial \overline{u}^i} \right) (g_{k\ell}) \,^t \left(\frac{\partial u^\ell}{\partial \overline{u}^j} \right).$$

Therefore, letting

$$\frac{\partial(u^1, \cdots, u^{n-1})}{\partial(\overline{u}^1, \cdots, \overline{u}^{n-1})} = \det \left(\frac{\partial u^k}{\partial \overline{u}^i} \right), \tag{1.31}$$

we have proven the following formula

$$\sqrt{\det(\overline{g}_{ij})} = \left| \frac{\partial(u^1, \cdots, u^{n-1})}{\partial(\overline{u}^1, \cdots, \overline{u}^{n-1})} \right| \sqrt{\det(g_{ij})}. \tag{1.32}$$

1.5 Volume element

A function on S actually depends on $u \in U$ through the parametrization $x(u)$. By definition, $f(x) \in C_0(S)$ if and only if $f(x(u)) \in C_0(U)$, similarly $f(x) \in C_0^\infty(S)$ if and only if $f(x(u)) \in C_0^\infty(U)$. The integral of $f(x) \in C_0(S)$ is defined by

$$\int_S f(x) dS = \int_U f(x(u)) \sqrt{g} \, du, \tag{1.33}$$

where $du = du^1 \cdots du^{n-1}$ and

$$g = \det(g_{ij}). \tag{1.34}$$

The integral (1.33) is independent of the choice of the parameter u, since for another parameter $\overline{u} \in \overline{U}$ such that $x(u) = \overline{x}(\overline{u})$, we have

$$\int_{\overline{U}} f(\overline{x}(\overline{u})) \sqrt{\overline{g}} \, d\overline{u} = \int_U f(x(u)) \sqrt{\overline{g}} \left| \frac{\partial(\overline{u}^1, \cdots, \overline{u}^{n-1})}{\partial(u^1, \cdots, u^{n-1})} \right| du,$$

which coincides with (1.33) by virtue of (1.32). We put

$$dS = \sqrt{g(u)}\, du^1 \cdots du^{n-1}, \tag{1.35}$$

and call it the *volume element* or *surface element* of S.

Example 1.6. For an $(n-1)$-dimensional surface S in $\mathbf{R}^n$ and $a > 0$, let $S(a) = aS = \{ax\,;\, x \in S\}$. Then $dS(a) = a^{n-1}dS$.

The g_{ij} for $S(a)$ is a^2-times that for S. Therefore the assertion follows from (1.34) and (1.35).

Example 1.7. By the polar coordinates, the surface element of S^{n-1} is written as

$$dS^{n-1} = (\sin\theta^{n-1})^{n-2} \cdots (\sin\theta^2)d\theta^1 \cdots d\theta^{n-1}.$$

For the case $n = 2$, this is well-known. Suppose it is true for $n - 1$. Then

$$\int_{S^{n-1}} f(x)dS^{n-1} = \int_0^\pi \left(\int_S f(x)dS \right) d\theta^{n-1}, \tag{1.36}$$

where $S = (\sin\theta^{n-1})S^{n-2}$. By Example 1.6, $dS = (\sin\theta^{n-1})^{n-2}dS^{n-2}$. Therefore $dS^{n-1} = (\sin\theta^{n-1})^{n-2}dS^{n-2}d\theta^{n-1}$.

Example 1.8. $\left| S^{n-1} \right| = \dfrac{2\pi^{n/2}}{\Gamma(n/2)}$, where $\left| S^{n-1} \right|$ denotes the surface area of S^{n-1}.

For $n = 2$, this is true. Assuming it for $n - 1$, we have by (1.36)

$$\left| S^{n-1} \right| = \int_0^\pi (\sin\theta^{n-1})^{n-2} \left| S^{n-2} \right| d\theta^{n-1}.$$

The assertion then follows from the well-known formula for the Gamma function

$$\int_0^{\pi/2} \sin^m \theta\, d\theta = \frac{\sqrt{\pi}}{2} \frac{\Gamma(\frac{m+1}{2})}{\Gamma(\frac{m}{2} + 1)}.$$

Example 1.9. When the surface S is defined by $x^n = f(x^1, \cdots, x^{n-1})$,

$$dS = \left(1 + \sum_{i=1}^{n-1} \left(\frac{\partial f}{\partial x^i} \right)^2 \right)^{1/2} dx^1 \cdots dx^{n-1}.$$

Since $g_{ij} = \delta_{ij} + a_i a_j$ with $a_i = \dfrac{\partial f}{\partial x^i}$, we have only to use the following lemma.

Lemma 1.4.
$$\det\left(\delta_{ij} + a_i a_j\right) = 1 + |a|^2.$$

Proof. Assume $a \neq 0$. Letting $A = (a_i a_i)$ and $\omega = a/|a|$, we have
$$Ax = |a|^2 (\omega \cdot x)\omega.$$
The eigenvalues of A are then $|a|^2, 0, \cdots, 0$, which proves the lemma. $\qquad\square$

Example 1.10. If a 2-dimensional surface S in $\mathbf{R}^3$ is parametrized as $S = \{x(u)\,;\, u \in U\}$, its surface element is written as
$$dS = \left| \frac{\partial x}{\partial u^1} \times \frac{\partial x}{\partial u^2} \right| du^1 du^2.$$

This is a consequence of *Lagrange's identity*:
$$|a|^2 |b|^2 = (a \cdot b)^2 + |a \times b|^2, \quad a, b \in \mathbf{R}^3.$$

1.6 Global definition of surfaces

1.6.1 *Submanifold of* $\mathbf{R}^N$

So far we have considered $(n-1)$-dimensional surfaces in $\mathbf{R}^n$. Let us discuss m-dimensional surfaces in $\mathbf{R}^n$ with $1 \leq m \leq n$. Namely, given an open set U in $\mathbf{R}^m$ and $\varphi(u) \in C^\infty(U; \mathbf{R}^n)$, which is 1 to 1 and
$$\text{rank}\left(\frac{\partial \varphi}{\partial u^1}, \cdots, \frac{\partial \varphi}{\partial u^m} \right) = m, \tag{1.37}$$
we define a surface $S = \varphi(U)$ of dimension m. The tangent space and the induced metric are defined in the same way as above. The *global* surface is defined by gluing these *local* surfaces. We use the terminologies *submanifolds* and *local coordinates* instead of surfaces and parameters.

Suppose we are given several open subsets $U_\alpha \subset \mathbf{R}^m$, α varies over some index set A, and $\varphi_\alpha \in C^\infty(U_\alpha; \mathbf{R}^n)$, which is assumed to be 1 to 1 and satisfy the rank condition (1.37) on U_α. Then the set
$$M = \bigcup_{\alpha \in A} \varphi_\alpha(U_\alpha) \tag{1.38}$$
is said to be an m-dimensional submanifold of $\mathbf{R}^n$.

Let us first note the following lemma.

Lemma 1.5. *If $\varphi_\alpha(U_\alpha) \cap \varphi_\beta(U_\beta) \neq \emptyset$, there is a diffeomorphism : $U_{\alpha,\beta} \to U_{\beta,\alpha}$, where $U_{\alpha,\beta} = \{u_\alpha \in U_\alpha \,;\, \varphi_\alpha(u_\alpha) \in \varphi_\beta(U_\beta)\}$.*

Proof.   Proof. Since φ_α's are 1 to 1, there is a 1 to 1 onto map $U_{\alpha,\beta} \ni u_\alpha \to u_\beta \in U_{\beta,\alpha}$ satisfying $\varphi_\alpha(u_\alpha) = \varphi_\beta(u_\beta)$. By the rank condition (1.37), one can assume that $\det\left(\partial\varphi_\alpha^i / \partial u_\alpha^j\right)_{1 \le i,j \le m} \neq 0$. The implicit function theorem then implies that $u_\alpha \in U_{\alpha,\beta}$ is a C^∞-function of $x^1, \cdots, x^m$, which are also C^∞-functions of $u_\beta \in U_{\beta,\alpha}$. Similarly, $u_\beta \in U_{\beta,\alpha}$ is a C^∞-function of $u_\alpha \in U_{\alpha,\beta}$. □

1.6.2   Tangent space

For $p \in M$, there is α such that $p = \varphi_\alpha(u_\alpha^0) \in \varphi_\alpha(U_\alpha)$. The tangent space $T_p(M)$ is defined by

$$T_p(M) = \left\{ \sum_{i=1}^m c^i \frac{\partial\varphi_\alpha}{\partial u_\alpha^i}(u_\alpha^0)\,;\, c^i \in \mathbf{R} \right\}. \tag{1.39}$$

This is an m-dimensional subspace of $\mathbf{R}^n$. As has been discussed before, as a geometrical object we should think of the affine subspace

$$\left\{ p + \sum_{i=1}^m c^i \frac{\partial\varphi_\alpha}{\partial u_\alpha^i}(u_\alpha^0)\,;\, c^i \in \mathbf{R} \right\}.$$

There may be another $\beta \in A$ such that $p \in \varphi_\beta(U_\beta)$. However, (1.39) does not depend on the choice of α. In fact, suppose $p \in \varphi_\alpha(U_\alpha) \cap \varphi_\beta(U_\beta)$. Then, near p, $x \in M$ is represented as $x = \varphi_\alpha(u_\alpha) = \varphi_\beta(u_\beta)$. Therefore

$$\frac{\partial\varphi_\alpha}{\partial u_\alpha^i} = \frac{\partial u_\beta^j}{\partial u_\alpha^i} \frac{\partial\varphi_\beta}{\partial u_\beta^j}.$$

Since $\left(\partial u_\beta^j / \partial u_\alpha^i\right)$ is a non-singular matrix by Lemma 1.5, this means that the set (1.39) is invariant if we replace α by β. Moreover, it also implies that if $v \in T_p(M)$ is represented as

$$v = c_\alpha^i \frac{\partial\varphi_\alpha}{\partial u_\alpha^i} = c_\beta^j \frac{\partial\varphi_\beta}{\partial u_\beta^j}, \tag{1.40}$$

we have

$$c_\beta^i = c_\alpha^j \frac{\partial u_\beta^i}{\partial u_\alpha^j}. \tag{1.41}$$

1.6.3   Partition of unity

By definition, a function $f(x)$ is in $C^\infty(M)$ if $f(\varphi_\alpha(u)) \in C^\infty(U_\alpha)$ for all $\alpha \in A$. In the following, we consider the case where M is compact, and the index set A is finite. The set of $\{\rho_\alpha(x)\}_{\alpha \in A}$ in the following lemma is called a *partition of unity* on M.

Lemma 1.6. *There exists* $\rho_\alpha(x) \in C^\infty(M)$ *such that*

$$\text{supp} \, \rho_\alpha \subset \varphi_\alpha(U_\alpha), \quad \forall \alpha \in A, \tag{1.42}$$

$$\sum_{\alpha \in A} \rho_\alpha(x) = 1, \quad on \quad M. \tag{1.43}$$

Proof. Let $B(c\,;r)$ be an open ball in $\mathbf{R}^m$ with center c and radius $r > 0$. One can then find a finite number of $c_\alpha^{(n)} \in U_\alpha$ and $r_\alpha^{(n)} > 0$ such that $B(c_\alpha^{(n)}; r^{(n)}) \subset U_\alpha$ and

$$\underset{\alpha,n}{\cup} \varphi_\alpha\big(B(c_\alpha^{(n)}; r^{(\alpha)}/2)\big) = M. \tag{1.44}$$

We then construct $f_{\alpha,n}(u_\alpha) \in C^\infty(U_\alpha)$ such that

$$f_{\alpha,n}(u_\alpha) = \begin{cases} 1 & \text{on} \quad B(c_\alpha^{(n)}; r_\alpha^{(n)}/2), \\ 0 & \text{outside} \quad B(c_\alpha^{(n)}; r_\alpha^{(n)}), \end{cases} \tag{1.45}$$

and put

$$f_\alpha(x) = \sum_n f_{\alpha,n}(u_\alpha), \quad \text{for} \quad x = \varphi_\alpha(u_\alpha).$$

Then, $\text{supp} \, f_\alpha(x) \subset \varphi_\alpha(U_\alpha)$ and by (1.44), (1.45), $f(x) = \sum_\alpha f_\alpha(x) > 0$ on M. Therefore, we have only to put $\rho_\alpha(x) = f_\alpha(x)/f(x)$. $\square$

1.6.4 *Integral on M*

We now put

$$g_\alpha(u_\alpha) = \det\big(g_{ij,\alpha}(u_\alpha)\big), \quad g_{ij,\alpha} = \frac{\partial \varphi_\alpha}{\partial u_\alpha^i} \cdot \frac{\partial \varphi_\alpha}{\partial u_\alpha^j}, \tag{1.46}$$

$$dM_\alpha = \sqrt{g_\alpha(u_\alpha)} \, du_\alpha^1 \cdots du_\alpha^m. \tag{1.47}$$

Then the volume element

$$dM = dM_\alpha, \quad \text{on} \quad \varphi_\alpha(U_\alpha), \tag{1.48}$$

is well-defined, since $dM_\alpha = dM_\beta$ on $\varphi_\alpha(U_\alpha) \cap \varphi_\beta(U_\beta)$. The integral of a function $f(x)$ on M is then defined by

$$\int_M f(x) \, dM = \sum_{\alpha \in A} \int_{\varphi_\alpha(U_\alpha)} f(x)\rho_\alpha(x) \, dM_\alpha. \tag{1.49}$$

This is independent of the choice of partition of unity, since for two partitions of unity $\{\rho_\alpha\}$, $\{\rho'_{\alpha'}\}$, their product $\{\rho_\alpha \rho'_{\alpha'}\}$ is again a partition of unity.

When $m = 1$, the above definition coincides with the usual line integral. A curve C in $\mathbf{R}^n$ is a 1 to 1 C^∞ map : $[a, b] \ni t \to x(t) \in \mathbf{R}^n$,

$$C = \{x(t) \, ; \, a \leq t \leq b\}, \tag{1.50}$$

satisfying $x(t) \neq x(s)$ for $t \neq s$, and $\dot{x}(t) \neq 0$. The *line element* then becomes

$$dC = |\dot{x}(t)| dt, \tag{1.51}$$

and the *line integral* is defined by

$$\int_C f(x) \, dC = \int_a^b f(x(t)) |\dot{x}(t)| \, dt. \tag{1.52}$$

1.6.5   *Convergence of volume integral to line integral*

The line integral $\int_C f(x) \, dC$ can be derived from a volume integral $\int_V f(x) dx$ by shrinking the domain $V \subset \mathbf{R}^n$ to the curve C. We consider the case $n = 3$, and let $C = \{c(s) \, ; \, , a \leq s \leq b\}$ be a C^∞-curve such that $|\dot{c}(s)| = 1$. Let $v_1(s) = \dot{c}(s)$, and take $v_2(s)$, $v_3(s)$ so that $v_1(s), v_2(s), v_3(s)$ form an orthormormal basis of $\mathbf{R}^3$. Using the equation

$$x = c(s_1) + s_2 v_2(s_1) + s_3 v_3(s_1),$$

we make the change of variables $x \to (s_1, s_2, s_3)$. Letting

$$g = \det \left(\frac{\partial x}{\partial s_i} \cdot \frac{\partial x}{\partial s_j} \right),$$

and

$$V_\epsilon = \{x \, ; \, a \leq s_1 \leq b, \ s_2^2 + s_3^2 \leq \epsilon^2\},$$

we have

$$\int_{V_\epsilon} f(x) dx = \int_a^b \left(\int_{s_2^2 + s_3^2 \leq \epsilon^2} f(x) \sqrt{g} ds_2 ds_3 \right) ds_1.$$

Noting that $g = 1$ if $s_2 = s_3 = 0$, we then have

$$\lim_{\epsilon \to 0} \frac{1}{\pi \epsilon^2} \int_{V_\epsilon} f(x) dx = \int_a^b f(x(s)) ds = \int_C f(x) \, dC. \tag{1.53}$$

1.7 Integral formulas

1.7.1 *Integration by parts*

Let D be a bounded open set in $\mathbf{R}^n$ ($n \geq 2$), whose boundary $S = \partial D$ is assumed to be a C^∞-submanifold of $\mathbf{R}^n$. Then for any $p \in S$, there is a unit normal $\nu(p) \in S^{n-1}$ such that

$$\nu(p) \cdot v = 0, \quad \forall v \in T_p(S), \tag{1.54}$$

where $T_p(S)$ is defined by (1.18). We also assume that it is *outward*, i.e. there exists an $\epsilon > 0$ such that

$$p + s\nu(p) \notin D, \quad 0 < \forall s < \epsilon. \tag{1.55}$$

For a function $f(x)$ in D, by definition, $f(x) \in C^\infty(\overline{D})$ means that there exists an open set Ω containing $\overline{D}$ such that $f(x) \in C^\infty(\Omega)$. This is equivalent to that all the derivatives of $f(x)$ are uniformly continuous near S and have continuous boundary values on S. The same remark is true for C^m-functions.

Theorem 1.3. *For $u(x), v(x) \in C^\infty(\overline{D})$, we have*

$$\int_D \frac{\partial u}{\partial x^j}(x) v(x) dx = \int_S \nu^j(x) u(x) v(x) dS - \int_D u(x) \frac{\partial v}{\partial x^j}(x) dx,$$

where $\nu(x) = (\nu^1(x), \cdots, \nu^n(x))$.

Proof. We consider the case in which $j = n$ and D is defined by

$$D = \{x = (x', x^n)\,;\, x' \in D',\ f_-(x') < x^n < f_+(x')\},$$

where D' is an open set in $\mathbf{R}^{n-1}$. A suitable division of D reduces the proof to this case. Take $w(x) \in C^\infty(\overline{D})$. Then

$$\int_{f_-(x')}^{f_+(x')} \frac{\partial w}{\partial x^n}(x', y) dy = w(x', f_+(x')) - w(x', f_-(x')).$$

Integrating over D', we have

$$\int_D \frac{\partial w}{\partial x^n}(x) dx = \int_{D'} w(x', f_+(x')) dx' - \int_{D'} w(x', f_-(x')) dx'.$$

The surface elements of $S_\pm = \{(x', f_\pm(x'))\,;\, x' \in D'\}$ are written as (see Example 1.9)

$$dS_\pm = \left(1 + |\nabla_{x'} f_\pm(x')|^2\right)^{1/2} dx^1 \cdots dx^{n-1},$$

where $\nabla_{x'} = (\partial/\partial x^1, \cdots, \partial/\partial x^{n-1})$, and the outer unit normals are

$$\nu_{\pm}(x) = \pm \frac{(-\nabla_{x'} f_{\pm}(x'), 1)}{(1 + |\nabla_{x'} f_{\pm}(x')|^2)^{1/2}}.$$

Therefore

$$dx' = \pm \nu_{\pm}^n(x') dS_{\pm},$$

which implies

$$\int_D \frac{\partial w}{\partial x^n} dx = \int_{S_+} w\nu_+^n dS_+ + \int_{S_-} w\nu_-^n dS_- = \int_S w\nu^n dS,$$

since $\nu^n = 0$ on the lateral sides. Taking $w = uv$, we get the theorem.   □

1.7.2   *Theorem of Gauss-Ostrogradsky*

By using the notation *nabla*,

$$\nabla = \left(\frac{\partial}{\partial x^1}, \cdots, \frac{\partial}{\partial x^n} \right), \tag{1.56}$$

we define for a real-valued $f(x)$ the *gradient* of f by

$$\text{grad } f(x) = \nabla f(x) = \left(\frac{\partial f}{\partial x^1}(x), \cdots, \frac{\partial f}{\partial x^n}(x) \right), \tag{1.57}$$

and for a vector-valued function $v(x) = (v^1(x), \cdots, v^n(x))$ the *divergence* of v by

$$\text{div } v(x) = \nabla \cdot v(x) = \frac{\partial v^1}{\partial x^1}(x) + \cdots + \frac{\partial v^n}{\partial x^n}(x). \tag{1.58}$$

The Laplacian on $\mathbf{R}^n$ is defined by

$$\Delta = \sum_{i=1}^n \left(\frac{\partial}{\partial x^i} \right)^2. \tag{1.59}$$

Note that

$$\Delta = \text{div} \cdot \text{grad} = \nabla \cdot \nabla. \tag{1.60}$$

The following *divergence theorem*, whose precursor appeared in [106], was proven by [109] and [107].

Theorem 1.4. *Let D be a bounded open set in $\mathbf{R}^n$ with C^{∞} boundary S, and $\nu(x)$ the outer unit normal to S. Then for $v(x) \in C^{\infty}(\overline{D}; \mathbf{R}^n)$*

$$\int_D \text{div } v(x)\, dx = \int_S \nu(x) \cdot v(x)\, dS.$$

The proof is immediate from Theorem 1.3.

1.7.3   *Theorem of Green*

As above, let D be a bounded open set in $\mathbf{R}^n$ with C^∞ boundary S, and $\nu(x)$ the outer unit normal to S. The normal derivative at the boundary S is defined by

$$\frac{\partial}{\partial \nu} = \nu(x) \cdot \nabla, \quad \text{on} \quad S. \tag{1.61}$$

The following theorem is due to [108].

Theorem 1.5. *For* $u(x), v(x) \in C^\infty(\overline{D}; \mathbf{R})$,

$$\int_D (\Delta u(x))\, v(x)\, dx = \int_S \frac{\partial u}{\partial \nu}(x) v(x)\, dS - \int_D \nabla u(x) \cdot \nabla v(x)\, dx. \tag{1.62}$$

$$\int_D \Big((\Delta u(x))\, v(x) - u(x)\Delta v(x) \Big)\, dx = \int_S \Big(\frac{\partial u}{\partial \nu}(x)v(x) - u(x)\frac{\partial v}{\partial \nu}(x) \Big)\, dS. \tag{1.63}$$

Proof.   We apply Theorem 1.3 to the integrals $\int_D \left(\frac{\partial}{\partial x^i} \frac{\partial u}{\partial x^i} \right) v\, dx$ and sum them to obtain (1.62). Exchanging u and v and subtracting them, we obtain (1.63). $\qquad\square$

1.7.4   *Surfaces with boundaries*

The theorem of Stokes reduces integrals over 2-dimensional surfaces to those on their boundaries. Since we must be careful about the sign of the integral, we start with explaining the orientation of the boundary.

Let D be an open set in $\mathbf{R}^2$ whose boundary ∂D is a 1-dimensional C^∞-submanifold of $\mathbf{R}^2$ in the sense of Subsection 1.6.1. Intuitively, ∂D is said to be *positively oriented* if you walk along ∂D you always see D on your left. More precisely, it is defined as follows. Assume that ∂D consists of C^∞-curves $C_i = \{x_i(t)\,;\, 0 \le t \le 1\}$, $\dot{x}_i(t) \ne 0$, $x_i(0) = x_i(1)$, $i = 1, \cdots, k$. For $p = x_i(t) \in C_i$, we put $v_{tan}(p) = \dot{x}_i(t)/|\dot{x}_i(t)| = (b^1(p), b^2(p))$. Let $n_{nor}(p) = (a^1(p), a^2(p))$ be the unit vector orthogonal to $v_{tan}(p)$ with direction toward the exterior of D. Then ∂D is said to be positively oriented if

$$\det \begin{pmatrix} a^1(p) & a^2(p) \\ b^1(p) & b^2(p) \end{pmatrix} > 0, \quad \forall p \in \partial D. \tag{1.64}$$

Note that if D is an annulus : $D = \{0 < r < |x| < R < \infty\}$, the outer boundary $|x| = R$ is oriented counterclockwise, and the inner boundary $|x| = r$ is oriented clockwise.

A subset S of $\mathbf{R}^3$ is said to be a *surface with boundary* if the following condition (C) is satisfied : Let $U = \{u = (u^1, u^2) \in \mathbf{R}^2 ; |u| < 1\}$ and $U_+ = \{u = (u^1, u^2) \in U ; u^2 \geq 0\}$.

(C) For any $p \in S$, there exists an open set $\mathcal{O}_p$ in $\mathbf{R}^3$ such that $p \in \mathcal{O}_p$ and $\varphi \in C^\infty(U ; \mathbf{R}^3)$, which is 1 to 1 on U, satisfying

$$\mathrm{rank}\left(\frac{\partial\varphi}{\partial u^1}, \frac{\partial\varphi}{\partial u^2}\right) = 2 \quad \text{on} \quad U, \tag{1.65}$$

and

$$\text{either} \quad (i) \ \varphi(U) = S \cap \mathcal{O}_p,$$

$$\text{or} \quad (ii) \ \varphi(U_+) = S \cap \mathcal{O}_p, \ p \in \varphi(U_+ \cap \{u^2 = 0\})$$

is satisfied.

In the case (i), p is in the interior of S, while in the case (ii), p is on the boundary ∂S. As above, we assume that ∂S consists of non-intersecting closed curves $C_1, \cdots, C_k$: $C_i = \{x_i(t) ; 0 \leq t \leq 1\}$, $\dot{x}_i(t) \neq 0$. By (1.65), $T_p(S)$ is well-defined for $p \in \partial S$, and $T_p(\partial S)$ is regarded as a subspace of $T_p(S)$. For $p \in \partial S$, let $n_{out}(p) = (a^1(p), a^2(p), a^3(p)) \in T_p(S)$ be the unit vector orthogonal to $T_p(\partial S)$ with direction toward the exterior to S, $v_{tan}(p) = \dot{x}_i(t)/|\dot{x}_i(t)| = (b^1(p), b^2(p), b^3(p))$ the unit tangent vector to ∂S. We assume that there exists a C^∞-unit normal field $\nu_S(p) = (c^1(p), c^2(p), c^3(p))$ on S, i.e. $\nu_S(p) \perp T_p(S)$. Then we say that ∂S is positively oriented if $n_{nor}(p) \times v_{tan}(p) = \nu_S(p)$ for all $p \in \partial S$, i.e.

$$\det\begin{pmatrix} a^1(p) & a^2(p) & a^3(p) \\ b^1(p) & b^2(p) & b^3(p) \\ c^1(p) & c^2(p) & c^3(p) \end{pmatrix} > 0, \quad \forall p \in \partial S. \tag{1.66}$$

We also say that S has an *orientation compatible with ∂S*.

1.7.5 *Stokes' Theorem*

For a vector-valued function $V(x) = (V^1(x), V^2(x), V^3(x))$, we define

$$\begin{aligned}\mathrm{curl}\, V(x) &= \nabla \times V(x) \\ &= (\partial_2 V^3 - \partial_3 V^2, \partial_3 V^1 - \partial_1 V^3, \partial_1 V^2 - \partial_2 V^1),\end{aligned} \tag{1.67}$$

where $\partial_i = \partial/\partial x^i$. It is also written as $\mathrm{rot}\, V(x)$. In the classical vector form, Stokes' theorem is stated as follows.

Theorem 1.6. *Let S be a 2-dimensional surface with boundary in $\mathbf{R}^3$, and suppose that its boundary ∂S is positively oriented. Let ν_S be the unit normal field on S, and v_{tan} the unit tangent vector field along ∂S. Let Ω be an open set in $\mathbf{R}^3$ such that $S \subset \Omega$. Then we have for any $V(x) \in C^\infty(\Omega\,;\,\mathbf{R}^3)$*

$$\int_S \langle \operatorname{curl} V, \nu_S \rangle dS = \int_{\partial S} \langle V, v_{tan} \rangle dC, \qquad (1.68)$$

where $\langle\,,\,\rangle$ denotes the standard inner product of $\mathbf{R}^3$ and dC is the line element of ∂S.

1.7.6 Integral of 1-form

Theorem 1.6 is reduced to a simpler and more beautiful formula. For functions $a_i(x) \in C^\infty(\mathbf{R}^n\,;\,\mathbf{R})$, the expression

$$\omega = \sum_{i=1}^n a_i(x) dx^i \qquad (1.69)$$

is called a *1-form*. Let $C = \{x(t)\,;\, a \le t \le b\}$ be a C^∞-curve in $\mathbf{R}^n$. The integral of ω along C is defined by

$$\int_C \omega = \int_a^b \sum_{i=1}^n a_i(x(t))\,\dot{x}^i(t)\,dt. \qquad (1.70)$$

Let us first prove the 2-dimensional version of Stokes' Theorem. We write (x, y) instead of (x^1, x^2).

Theorem 1.7. *Let D be a bounded open set in $\mathbf{R}^2$ with C^∞-boundary ∂D, which is positively oriented. Then for any $a(x,y), b(x,y) \in C^\infty(\overline{D}\,;\,\mathbf{R})$, we have*

$$\int_D \left(\frac{\partial b}{\partial x} - \frac{\partial a}{\partial y} \right) dxdy = \int_{\partial D} adx + bdy. \qquad (1.71)$$

Proof. Suppose $\partial D = C$ is parametrized as $C = \{(x(t), y(t))\,;\, a \le t \le b\}$. Then the outer unit normal at $(x(t), y(t)) \in \partial D$ is

$$\nu = (\nu_x, \nu_y) = \left(\dot{y}(t)/\sqrt{\dot{x}(t)^2 + \dot{y}(t)^2}, -\dot{x}(t)/\sqrt{\dot{x}(t)^2 + \dot{y}(t)^2} \right). \qquad (1.72)$$

By Theorem 1.3

$$\int_D \left(\frac{\partial b}{\partial x} - \frac{\partial a}{\partial y} \right) dxdy = \int_C (\nu_x b - \nu_y a)\, dC.$$

The formulas (1.51) and (1.72) imply

$$\nu_x dC = \dot{y}(t)dt, \quad \nu_y dC = -\dot{x}(t)dt,$$

which prove (1.71). $\qquad\square$

We now prove Theorem 1.6 in the case where S is represented as $S = \varphi(U)$, U being an open set in $\mathbf{R}^2$. Then we have *global* coordinates $u = (u^1, u^2)$ on $S : S = \{x(u)\,;\, u \in U\}$. Therefore in view of Example 1.10

$$\nu_S dS = \frac{\partial x}{\partial u^1} \times \frac{\partial x}{\partial u^2}\, du^1 du^2. \tag{1.73}$$

On the other hand, putting

$$W(u) = (W^1(u), W^2(u)) = \left(V(x(u)) \cdot \frac{\partial x}{\partial u^1}, V(x(u)) \cdot \frac{\partial x}{\partial u^2} \right),$$

we have

$$
\begin{aligned}
\frac{\partial W^2}{\partial u^1} - \frac{\partial W^1}{\partial u^2} &= \frac{\partial V}{\partial u^1} \cdot \frac{\partial x}{\partial u^2} - \frac{\partial V}{\partial u^2} \cdot \frac{\partial x}{\partial u^1} \\
&= \left(V'(x) \frac{\partial x}{\partial u^1} \right) \cdot \frac{\partial x}{\partial u^2} - \left(V'(x) \frac{\partial x}{\partial u^2} \right) \cdot \frac{\partial x}{\partial u^1} \\
&= (\nabla \times V) \cdot \left(\frac{\partial x}{\partial u^1} \times \frac{\partial x}{\partial u^2} \right),
\end{aligned}
$$

where we have used (A.16) in Appendix A. In view of (1.73) and Theorem 1.7, we then have

$$
\begin{aligned}
\int_S (\nabla \times V) \cdot \nu_S dS &= \int_U \left(\frac{\partial W^2}{\partial u^1} - \frac{\partial W^1}{\partial u^2} \right) du^1 du^2 \\
&= \int_{\partial U} W^1 du^1 + W^2 du^2 \\
&= \int_{\partial S} V \cdot dx,
\end{aligned}
$$

which proves the theorem.

To show the theorem in the general case, we split S into small pieces on which we can apply the above result. Here we note that Theorem 1.7 also holds when the boundary is piecewise smooth. Summing up the resulting formulas and taking note that the integrals on the inner boundaries cancel, we complete the proof.

Stokes' theorem first appeared in a letter from Thomson to Stokes in 1850 ([112]). See also [113], [104].

1.8   A short course on differential forms

The above theorems of Gauss-Ostrogradsky-Green-Stokes can be summarized in one simple formula in terms of differential forms. It is based on the notion of *differential*, dx, which has been used from the beginning of the history of integral and differential calculus as *infinitesimal*. Its precise meaning is given in Appendix B.

1.8.1 *The case of two variables*

1.8.1.1 *Exterior product and exterior differential*

A function $f(x, y)$ on $\mathbf{R}^2$ is called a *0-form*. As usual, the total differential of $f(x, y)$ is denoted by

$$df(x, y) = f_x(x, y)dx + f_y(x, y)dy. \tag{1.74}$$

In view of this formula, let us call the following symbol

$$a(x, y)dx + b(x, y)dy$$

a *1-form* on $\mathbf{R}^2$, where $a(x, y), b(x, y)$ are functions on $\mathbf{R}^2$. We also call the symbol

$$a(x, y)dx \wedge dy$$

a *2-form* on $\mathbf{R}^2$. Here the product $\wedge$, called *exterior product*, is assumed to have the following properties (i), (ii), (iii). The first two properties are standard:

(i) For functions a, b, p, q, f,

$$(adx + bdy) + (pdx + qdy) = (a + p)\, dx + (b + q)\, dy,$$
$$f(adx + bdy) = (fa)dx + (fb)dy,$$
$$adx \wedge dy + bdx \wedge dy = (a + b)dx \wedge dy,$$
$$f(adx \wedge dy) = (fa)\, dx \wedge dy = dx \wedge (fady).$$

(ii) For any 1-forms $\omega_1, \omega_2, \omega_3$,

$$(\omega_1 \wedge \omega_2) \wedge \omega_3 = \omega_1 \wedge (\omega_2 \wedge \omega_3).$$

The feature of the exterior product is the following property.

(iii) $dx \wedge dy = -dy \wedge dx.$

Due to the property (iii), we have

$$dx \wedge dx = 0, \quad dy \wedge dy = 0, \tag{1.75}$$

since $dx \wedge dx = -dx \wedge dx$. Therefore, we have

$$(adx + bdy) \wedge (pdx + qdy) = (aq - bp)\, dx \wedge dy. \tag{1.76}$$

The property (ii) implies that the triple product $\omega_1 \wedge \omega_2 \wedge \omega_3$ is well-defined. However, in the 2-dimensional case, we have

$$\omega_1 \wedge \omega_2 \wedge \omega_3 = 0. \tag{1.77}$$

In fact, by (1.76), $\omega_1 \wedge \omega_2 \wedge \omega_3$ is of the form

$$a dx \wedge dy \wedge (b dx + c dy).$$

Using (ii) and (1.75), we have

$$dx \wedge dy \wedge dx = -(dx \wedge dx) \wedge dy = 0, \quad dx \wedge dy \wedge dy = dx \wedge (dy \wedge dy) = 0,$$

which proves that $\omega_1 \wedge \omega_2 \wedge \omega_3 = 0$.

The other important notion is the *exterior differential* defined as follows. For a 0-form $f(x, y)$, df is the total differential (1.74). For a 1-form $\omega = a(x, y)dx + b(x, y)dy$, $d\omega$ is defined to be the 2-form

$$d(a dx + b dy) = (da) \wedge dx + (db) \wedge dy. \tag{1.78}$$

By (1.75), this is computed as

$$d(a dx + b dy) = (b_x - a_y) dx \wedge dy.$$

For a 2-form $\omega = a(x, y)dx \wedge dy$, we define

$$d(a dx \wedge dy) = (da) \wedge dx \wedge dy.$$

However, by (1.77), this is 0.

Example 1.11. For any k-form ω on $\mathbf{R}^2$, $d^2\omega = d(d\omega) = 0$.

We have only to consider a 0-form $\omega = f(x, y)$. Since $df = f_x dx + f_y dy$, we then have $d^2 f = (f_{yx} - f_{xy}) dx \wedge dy = 0$.

1.8.1.2 *Integration of differential forms*

Given a curve $C = \{(x(t), y(t)) \, ; \, t_0 \leq t \leq t_1\}$ in $\mathbf{R}^2$, the integral of a 1-form $a(x, y)dx + b(x, y)dy$ on C is defined by

$$\int_C a(x, y)dx + b(x, y)dy = \int_{t_0}^{t_1} \left(a(x(t), y(t))\dot{x}(t) + b(x(t), y(t))\dot{y}(t) \right) dt.$$

The integral of a 2-form $a(x, y)dx \wedge dy$ on a domain $D \subset \mathbf{R}^2$ is defined by

$$\int_D a(x, y)dx \wedge dy = \int_D a(x, y)dx dy,$$

where the right-hand side means the standard integral on D. Then one can rewrite Theorem 1.7 as follows.

Theorem 1.8. *Let ω be a 1-form on $\mathbf{R}^2$ and D a bounded open set in $\mathbf{R}^2$ with boundary ∂D which is positively oriented. Then we have*

$$\int_D d\omega = \int_{\partial D} \omega.$$

1.8.1.3 *Hodge $*$ operator*

The $*$-operator by Hodge is defined as follows

$$*1 = dx \wedge dy,$$
$$*dx = dy,$$
$$*dy = -dx,$$
$$*dx \wedge dy = 1,$$

and then extended linearly to any p-form. Note that, letting ω be any of $1, dx, dy, dx \wedge dy$, we have

$$\omega \wedge *\omega = dx \wedge dy.$$

The δ-operator is defined as follows

$$\delta = -*d*,$$

which assigns a $(p-1)$-form $\delta\omega$ to a given p-form ω. Note that

$$\delta f = 0, \quad \text{for } 0-\text{form } f,$$
$$\delta(adx + bdy) = -a_x - b_y,$$
$$\delta(adx \wedge dy) = -a_x dy + a_y dx.$$

Let us compare these notation with those of vector calculus. For a function f and a vector field $A = (a, b)$, we put

$$\omega_A^{(1)} = adx + bdy,$$
$$\omega_f^{(2)} = fdx \wedge dy.$$

Then we have the following correspondence:

$$df = \omega_{\nabla f}^{(1)},$$
$$d\omega_A^{(1)} = \omega_{\nabla \times A}^{(2)},$$
$$\delta\omega_A^{(1)} = -\nabla \cdot A,$$

where $\nabla \times A = b_y - a_x$.

We define the inner product of p-forms ω_1, ω_2 as follows.

(0) For $p = 0$, and $\omega_i = f_i(x, y)$, $i = 1, 2$,

$$(\omega_1, \omega_2) = \int_{\mathbf{R}^2} f_1(x, y)\overline{f_2(x, y)}dxdy.$$

(1) For $p = 1$, $\omega_i = a_i(x, y)dx + b_i(x, y)dy$, $i = 1, 2$,

$$(\omega_1, \omega_2) = \int_{\mathbf{R}^2} a_1(x, y)\overline{a_2(x, y)}dxdy + \int_{\mathbf{R}^2} b_1(x, y)\overline{b_2(x, y)}dxdy.$$

(2) For $p = 2$, $\omega_i = a_i(x, y)dx \wedge dy$, $i = 1, 2$,

$$(\omega_1, \omega_2) = \int_{\mathbf{R}^2} a_1(x, y)\overline{a_2(x, y)}dxdy.$$

We have for any p-form ω_1 and $p + 1$-form ω_2 with compact support,

$$(d\omega_1, \omega_2) = (\omega_1, \delta\omega_2).$$

We also have for any p-form

$$d\delta + \delta d = -\Delta,$$

where for a p-form ω, $\Delta\omega$ means that we apply Δ to each coefficient of ω. A p-form satisfying $d\omega = 0$ and $\delta\omega = 0$ is called a *harmonic* p-form. Note the *minus* sign in the right-hand side. In fact, we often denote $d\delta + \delta d = \Delta$ in differential geometry.

1.8.2 *The case of three variables*

A 0-form on $\mathbf{R}^3$ is a function $f(x)$ depending on the variables $x = (x^1, x^2, x^3)$ in $\mathbf{R}^3$. A k-form $\omega^{(k)}$ on $\mathbf{R}^3$ is defined by

$$\omega^{(1)} = \sum_{i=1}^{3} a_i(x)dx^i, \ \omega^{(2)} = \sum_{1 \leq i,j \leq 3} a_{ij}(x)dx^i \wedge dx^j, \ \omega^{(3)} = a(x)dx^1 \wedge dx^2 \wedge dx^3.$$

As above, they obey the following rule of computation:

$$dx^i \wedge dx^j = -dx^j \wedge dx^i.$$

Then the 2-form is represented as

$$\omega^{(2)} = a_1 dx^2 \wedge dx^3 + a_2 dx^3 \wedge dx^1 + a_3 dx^1 \wedge dx^2.$$

For 1-forms $\omega_j^{(1)}$, $j = 1, \cdots, 4$, we have

$$\omega_1^{(1)} \wedge \omega_2^{(1)} \wedge \omega_3^{(1)} \wedge \omega_4^{(1)} = 0.$$

The exterior differential d is defined in the same way as in the case of 2-variables: for instance, for the 0-form

$$df = \sum_{i=1}^{3} \frac{\partial f}{\partial x^i} dx^i,$$

and for the 1-form $\omega^{(1)}$

$$d\omega^{(1)} = \sum_{i=1}^{3} (da_i) \wedge dx^i.$$

For a vector field $A(x) = (A_1(x), A_2(x), A_3(x))$ and a function $f(x)$, we put

$$\omega_A^{(1)} = A_1 dx^1 + A_2 dx^2 + A_3 dx^3,$$
$$\omega_A^{(2)} = A_1 dx^2 \wedge dx^3 + A_2 dx^3 \wedge dx^1 + A_3 dx^1 \wedge dx^2,$$
$$\omega_f^{(3)} = f\, dx^1 \wedge dx^2 \wedge dx^3.$$

Then we have the following correspondence to the formulas in vector calculus

$$df = \omega_{\nabla f}^{(1)},$$
$$d\omega_A^{(1)} = \omega_{\nabla \times A}^{(2)}, \tag{1.79}$$
$$d\omega_A^{(2)} = \omega_{\nabla \cdot A}^{(3)}.$$

Lemma 1.7. *For any k-form ω on $\mathbf{R}^3$, we have $d^2\omega = 0$.*

Proof. For a 0-form, we have

$$d^2 f = d\omega_{\nabla f}^{(1)} = \omega_{\nabla \times \nabla f}^{(2)} = 0,$$

since $\nabla \times \nabla = 0$. For a 1-form, we have

$$d^2 \omega_A^{(1)} = d\omega_{\nabla \times A}^{(2)} = \omega_{\nabla \cdot \nabla \times A}^{(3)} = 0,$$

since $\nabla \cdot (\nabla \times) = 0$. $\qquad \square$

The integration of a 1-form $\omega^{(1)}$ along a curve $C = \{x(t)\,;\, a \le t \le b\}$ is defined by

$$\int_C \omega^{(1)} = \int_a^b \sum_{i=1}^3 a_i(x(t))\dot{x}^i(t)dt.$$

To define the integral of a 2-form on a surface S, we need to assume that S is orientable. (As a matter of fact, we should have done it in Subsection 1.6.4.) Recall that a 2-dimensional surface S in $\mathbf{R}^3$ is written as $S = \cup_\alpha S_\alpha$, where $S_\alpha = \{x_\alpha(u) = (x_\alpha^1(u_\alpha), x_\alpha^2(u_\alpha), x_\alpha^3(u_\alpha))\,;\, u_\alpha = (u_\alpha^1, u_\alpha^2) \in U_\alpha\}$, where U_α is an open set in $\mathbf{R}^2$, and the rank condition

$$\text{rank}\left(\frac{\partial x_\alpha(u_\alpha)}{\partial u_\alpha^1}, \frac{\partial x_\alpha(u_\alpha)}{\partial u_\alpha^2}\right) = 2, \quad \text{on} \quad U_\alpha$$

is assumed. By definition, S is *orientable* if the local coordinates u_α are chosen so that when $S_\alpha \cap S_\beta \neq \emptyset$,

$$\frac{\partial(u_\alpha)}{\partial(u_\beta)} = \det \begin{pmatrix} \dfrac{\partial u_\alpha^1}{\partial u_\beta^1} & \dfrac{\partial u_\alpha^1}{\partial u_\beta^2} \\[2mm] \dfrac{\partial u_\alpha^2}{\partial u_\beta^1} & \dfrac{\partial u_\alpha^2}{\partial u_\beta^2} \end{pmatrix} > 0.$$

This is equivalent to the fact that a normal vector field is defined globally on S, since on $S_\alpha \cap S_\beta$

$$\frac{\partial x_\alpha}{\partial u_\alpha^1} \times \frac{\partial x_\alpha}{\partial u_\alpha^2} = \frac{\partial(u_\beta)}{\partial(u_\alpha)} \frac{\partial x_\beta}{\partial u_\beta^1} \times \frac{\partial x_\beta}{\partial u_\beta^2}.$$

Then the integral of the 2-form $\omega^{(2)}$ on S_α is defined by

$$\int_{S_\alpha} \omega^{(2)} = \sum_{i,j} \int_{U_\alpha} a_{ij}(x_\alpha(u_\alpha)) \frac{\partial(x_\alpha^1, x_\alpha^j)}{\partial(u_\alpha^1, u_\alpha^2)} du_\alpha^1 du_\alpha^2.$$

As in (1.52), the integral of $\omega^{(2)}$ on S is defined by using the partition of unity $\{\rho_\alpha\}$ on S:

$$\int_S \omega^{(2)} = \sum_\alpha \int_{S_\alpha} \rho_\alpha(x)\omega^{(2)}.$$

The integral of the 3-form $\omega^{(3)}$ on a domain D in $\mathbf{R}^3$ is defined by

$$\int_D \omega^{(3)} = \int_D a(x)\, dx^1 dx^2 dx^3. \tag{1.80}$$

Then the theorem of Gauss-Ostrogradsky is rewritten as follows.

Theorem 1.9. *Let ω be a 2-form on $\mathbf{R}^3$ and D a bounded open set in $\mathbf{R}^3$ with boundary ∂D which is positively oriented. Then we have*

$$\int_D d\omega = \int_{\partial D} \omega.$$

Proof. By Theorem 1.4 and (1.79), the left-hand side is equal to $\int_{\partial D} \nu(x) \cdot A(x) dS$. Since $\nu(x) dS = \frac{\partial x}{\partial u^1} \times \frac{\partial x}{\partial u^2} du^1 du^2$, where $u = (u^1, u^2)$ is the local coordinate of S, we have

$$A(x) \cdot \nu(x) dS$$

$$= A(x) \cdot \left(\frac{\partial x}{\partial u^1} \times \frac{\partial x}{\partial u^2} \right) du^1 du^2 \tag{1.81}$$

$$= \left(A_1(x) \frac{\partial(x^2, x^3)}{\partial(u^1, u^2)} + A_2(x) \frac{\partial(x^3, x^1)}{\partial(u^1, u^2)} + A_3(x) \frac{\partial(x^1, x^2)}{\partial(u^1, u^2)} \right) du^1 du^2,$$

which proves the theorem.   □

Stokes' theorem is rewritten as follows.

Theorem 1.10. *Let S be a 2-dimensional surface in $\mathbf{R}^3$ whose boundary ∂S is positively oriented. Then for a 1-form ω*

$$\int_S d\omega = \int_{\partial S} \omega.$$

Proof. We replace $A(x)$ by $\nabla \times A(x)$ in (1.81) to get

$$\int_S (\nabla \times A) \cdot \nu dS = \int_S \omega^{(2)}_{\nabla \times A} = \int_S d\omega^{(1)}_A.$$

This and Theorem 1.6 prove the present theorem. $\qquad\square$

We have thus seen that Theorems 1.4, 1.5, 1.8 are rewritten as the following same form

$$\int_M d\omega = \int_{\partial M} \omega. \qquad (1.82)$$

The formula (1.82) holds for more general k-form ω defined on a k-dimensional manifold M with boundary.

A differential form ω is said to be *closed* if it satisfies $d\omega = 0$. It follows from Theorem 1.10 that if ω is a closed 1-form, the integral $\int_C \omega$ over a closed curve C does not change if we deform C continuously. Let us state this fact more precisely.

For two curves $C_i = \{x_i(t)\,;\, 0 \le t \le 1\}$, $i = 1, 2$, such that $x_1(1) = x_2(0)$, $C_1 + C_2$ means the curve defined by $\{x(t)\,;\, 0 \le t \le 2\}$, where $x(t) = x_1(t)$ for $0 \le t \le 1$ and $x(t) = x_2(t-1)$ for $1 \le t \le 2$. In particular, if $x(t)$ is 1-periodic, i.e. $x(t) = x(t+1)$, for a closed curve $C = \{x(t)\,;\, 0 \le t \le 1\}$, nC is defined by $nC = \{x(t)\,;\, 0 \le t \le n\}$. For a curve $C = \{x(t)\,;\, 0 \le t \le 1\}$, the curve $-C$ is defined by $-C = \{x(t-1)\,;\, 0 \le t \le 1\}$, which is the curve C with opposite direction.

Let $C_i = \{x_i(t)\,;\, 0 \le t \le 1\}$, $i = 0, 1$, be closed curves in a domain D. By $C_0 \simeq C_1$, we mean that C_0 is continuously deformed to C_1 within D, i.e. there exists a function $x(s, t) \in C([0, 1] \times [0, 1]\,;\, D)$ such that $x(0, t) = x_0(t)$, $x(1, t) = x_1(t)$, and for each $0 \le s \le 1$, $x(s, t)$ defines a smooth curve in D.

Theorem 1.11. *Let ω be a closed 1-form in D and C_1, C_2 closed curves in D such that $C_1 \simeq C_2$. Then*

$$\int_{C_1} \omega = \int_{C_2} \omega.$$

Proof. Let S be a surface in D with boundary $C_1 - C_2$. Applying Theorem 1.10, we have

$$0 = \int_S d\omega = \int_{\partial S} \omega = \int_{C_1} \omega - \int_{C_2} \omega.$$

$\qquad\square$

The Hodge $*$-operator is also defined for the 3-dimensional case:

$$*1 = dx^1 \wedge dx^2 \wedge dx^3,$$
$$*dx^i = dx^j \wedge dx^k,$$
$$*(dx^i \wedge dx^j) = dx^k,$$
$$*(dx^1 \wedge dx^2 \wedge dx^3) = 1,$$

where $(i, j, k) = (1, 2, 3), (2, 3, 1), (3, 1, 2)$. For a p-form, we define

$$\delta = (-1)^p * d * .$$

We then have

$$\delta\omega_A^{(1)} = -\nabla \cdot A,$$
$$\delta\omega_A^{(2)} = \omega_{\nabla \times A}^{(1)}, \qquad (1.83)$$
$$\delta\omega_f^{(3)} = -\omega_{\nabla f}^{(2)}.$$

By a direct computation, we have

$$(d\omega_1, \omega_2) = (\omega_1, \delta\omega_2),$$

for any compactly supported p-form ω_1 and $p + 1$-form ω_2. Moreover,

$$d\delta + \delta d = -\Delta.$$

Chapter 2

Fundamental solutions

The *Helmholtz equation* is the following equation

$$(-\Delta - z)u = f, \tag{2.1}$$

where Δ is the *Laplacian*

$$\Delta = \sum_{i=1}^{n} \left(\frac{\partial}{\partial x_i}\right)^2, \tag{2.2}$$

and $z \in \mathbf{C}$ is a parameter. It is called the *Poisson equation* when $z = 0$, and the *Laplace equation* when $z = 0$, $f = 0$. We shall study in this section an explicit formula of solutions and its properties in the case of $\mathbf{R}^n$. We also discuss the solution formula of the wave equation

$$\partial_t^2 u = c^2 \Delta u, \tag{2.3}$$

and its basic properties.

2.1 Fundamental solution on $\mathbf{R}^n$

By the *fundamental solution* of (2.1), we mean a function $K_z(x)$ having the following property: If we put

$$u(x) = \int_{\mathbf{R}^n} K_z(x - y)f(y)dy, \quad f \in C_0^\infty(\mathbf{R}^n), \tag{2.4}$$

it satisfies (2.1) on $\mathbf{R}^n$. The fundamental solution is not unique, since if $K_z(x)$ is a fundamental solution, for any function $\Phi(x)$ satisfying $(-\Delta - z)\Phi = 0$ in $\mathbf{R}^n$, $K_z(x) + \Phi(x)$ is also a fundamental solution. A standard choice is the following one:

$$R_0(x; z) = \frac{i}{4}\left(\frac{\sqrt{z}}{2\pi|x|}\right)^{(n-2)/2} H_{(n-2)/2}^{(1)}(\sqrt{z}|x|), \tag{2.5}$$

where $H_\nu^{(1)}(t)$ is the Hankel function of 1st kind, and

$$z \neq 0, \quad \operatorname{Im}\sqrt{z} \geq 0. \tag{2.6}$$

In particular, for $n = 1$ and 3, it has the following explicit form:

$$R_0(x; z) = \frac{i}{2\sqrt{z}} e^{i\sqrt{z}|x|}, \quad n = 1, \tag{2.7}$$

$$R_0(x; z) = \frac{e^{i\sqrt{z}|x|}}{4\pi|x|}, \quad n = 3. \tag{2.8}$$

For the case $z = 0$, we take

$$R_0(x; 0) = R_0(x) = \begin{cases} \dfrac{1}{(n-2)|S^{n-1}|}|x|^{-(n-2)}, & (n \neq 2), \\ \dfrac{1}{2\pi}\log\dfrac{1}{|x|}, & (n = 2), \end{cases} \tag{2.9}$$

where

$$|S^{n-1}| = \frac{2\pi^{n/2}}{\Gamma(n/2)} \tag{2.10}$$

is the surface measure of the unit sphere S^{n-1}.

Let us recall basic properties of $H_\nu^{(1)}(t)$. It is an analytic function of $t \in \mathbf{C} \setminus (-\infty, 0]$ and satisfies the equation

$$w'' + \frac{1}{t}w' + \left(1 - \frac{\nu^2}{t^2}\right)w = 0. \tag{2.11}$$

It has the following asymptotic expansion

$$H_\nu^{(1)}(t) = \sqrt{\frac{2}{\pi t}} e^{i(t - (2\nu+1)\pi/4)}\left(1 + a(t)\right), \tag{2.12}$$

where $a(t)$ satisfies

$$a(t) = O(t^{-1}), \quad a'(t) = O(t^{-2}), \tag{2.13}$$

as $|t| \to \infty$, $-\pi + \epsilon < \arg t < 2\pi - \epsilon$, $\epsilon > 0$ being an arbitrary small constant (see [94], p. 197) , and also

$$H_{(n-2)/2}^{(1)}(t) = \begin{cases} \pi^{-1/2}\left(\dfrac{t}{2}\right)^{-1/2}(1 + b(t)), & n = 1, \\ 2i\pi^{-1}\log\left(\dfrac{t}{2}\right)(1 + b(t)), & n = 2, \\ -i\pi^{-1}\Gamma\left(\dfrac{n-2}{2}\right)\left(\dfrac{t}{2}\right)^{-(n-2)/2}(1 + b(t)), & n \geq 3, \end{cases} \tag{2.14}$$

where $b(t)$ satisfies as $t \to 0$

$$b(t) = O(t), \quad b'(t) = O(1). \tag{2.15}$$

Note that $R_0(x; z) \to R_0(x)$ as $z \to 0$ when $n \neq 2$. However, $R_0(x; z)$ is not continuous at $z = 0$ when $n = 2$.

We make use of the following properties of $R_0(x; z)$.

Lemma 2.1. *(1) For $x \neq 0$, $R_0(x; z)$ satisfies the equation*

$$(-\Delta - z)R_0(x; z) = 0.$$

(2) For $\sqrt{z} = \sqrt{k^2 \pm i0} = \pm k$ where $k > 0$, we have as $r = |x| \to \infty$

$$R_0(x; z) = C_n^{(\pm)}(k) \frac{e^{\pm ikr}}{r^{(n-1)/2}} (1 + a(r)),$$

$$a(r) = O(r^{-1}), \quad a'(r) = O(r^{-2}),$$

$$C_n^{(\pm)}(k) = \frac{e^{\mp(n-3)\pi i/4}}{2k} \left(\frac{k}{2\pi}\right)^{(n-1)/2}.$$

(3) For $\operatorname{Im}\sqrt{z} \geq 0$, $z \neq 0$, we have as $r = |x| \to 0$,

$$|R_0(x; z)| \leq C_n(z) \begin{cases} 1, & n = 1, \\ |\log r|, & n = 2, \\ r^{2-n}, & n \geq 3, \end{cases} \tag{2.16}$$

where $C_n(z)$ is a constant depending only on n and z.
(4) For $\operatorname{Im}\sqrt{z} \geq 0$, $z \neq 0$, we have as $r = |x| \to 0$,

$$\frac{\partial}{\partial r} R_0(x; z) = -|S^{n-1}|^{-1} r^{1-n} (1 + O(r)). \tag{2.17}$$

Proof. Noting that $R_0(x; z)$ depends only on $r = |x|$, we put $k(r) = R_0(x; z)$, and $w(\sqrt{z}r) = r^{(n-2)/2} k(r)$. Since $w(t)$ satisfies

$$w''(t) + \frac{1}{t}w'(t) + \left(1 - \frac{(n-2)^2}{4t^2}\right)w(t) = 0,$$

we see that $w(\sqrt{z}r)$ satisfies

$$\partial_r^2 w + \frac{1}{r}\partial_r w + \left(z - \frac{(n-2)^2}{4r^2}\right)w = 0.$$

Inserting $w(\sqrt{z}r) = r^{(n-2)/2} k(r)$, we obtain

$$k'' + \frac{(n-1)}{r} k' + zk = 0,$$

which proves (1). The expansion (2) follows from (2.12) and (2.13). The expansions (3) and (4) follow from (2.14) and (2.15). $\qquad\square$

When $z = 0$, $R_0(x)$ has properties similar to $R_0(x; z)$. We omit the proof of the following lemma which follows easily from Lemma 2.1.

Lemma 2.2. *(1) For $x \neq 0$, $R_0(x)$ satisfies the equation*

$$\Delta R_0(x) = 0.$$

(2) As $r = |x| \to 0$,

$$|R_0(x)| \leq C \begin{cases} 1, & n = 1, \\ |\log r|, & n = 2, \\ r^{2-n}, & n \geq 3, \end{cases} \tag{2.18}$$

$$\frac{\partial}{\partial r} R_0(x) = -\frac{1}{|S^{n-1}| r^{n-1}}. \tag{2.19}$$

The basic properties of solutions to the Helmholtz and Poisson equations are derived from the following formula.

Lemma 2.3. *Let $D \subset \Omega \subset \mathbf{R}^n$ be open sets such that $\overline{D} \subset \Omega$. For any $u \in C^\infty(\Omega)$, we put*

$$u_D(x) = \int_D R_0(x - y; z)(-\Delta_y - z)u(y)dy. \tag{2.20}$$

Then we have

$$u_D(x) = w(x) + \int_{\partial D} u(y)\partial_{\nu_y} R_0(x - y; z)dS_y$$

$$- \int_{\partial D} R_0(x - y; z)\partial_{\nu_y} u(y)dS_y, \tag{2.21}$$

where $\partial_{\nu_y} = \nu(y) \cdot \nabla_y$, $\nu(y)$ being the outer unit normal at $y \in \partial D$, and

$$w(x) = \begin{cases} u(x), & x \in D, \\ 0, & x \notin D. \end{cases} \tag{2.22}$$

Proof. If $x \notin D$, there is no singularity in the integrand of $u_D(x)$. Therefore, by integration by parts, using Lemma 2.1 (1),

$$u_D(x) = \int_{\partial D} u(y)\partial_{\nu_y} R_0(x - y; z)dS_y - \int_{\partial D} R_0(x - y; z)\partial_{\nu_y} u(y)dS_y.$$

This proves (2.22) for $x \notin D$.

When $y \in D$, we take $\epsilon > 0$ small enough and put

$$D_{1,\epsilon} = \{y \in D; |x - y| > \epsilon\}, \quad D_{2,\epsilon} = \{y \in D; |x - y| < \epsilon\},$$

and split the right-hand side of (2.20) into two parts

$$\int_{D_{1,\epsilon}} \cdots dy + \int_{D_{2,\epsilon}} \cdots dy =: f_{1,\epsilon}(x) + f_{2,\epsilon}(x).$$

By Lemma 2.1 (3), we can easily check

$$\lim_{\epsilon \to 0} f_{2,\epsilon}(x) = 0.$$

We apply Green's formula to $f_{1,\epsilon}(x)$. Let $S_{1,\epsilon} = \{y \, ; \, |y - x| = \epsilon\}$ and pass to the polar coordinates $y - x = r\omega$, $\omega \in S^{n-1}$. Then

$$f_{1,\epsilon}(x)$$
$$= \int_{\partial D} u(y) \Big(\partial_{\nu_y} R_0(x - y; z) \Big) dS - \int_{\partial D} \Big(\partial_{\nu_y} u(y) \Big) R_0(x - y; z) dS$$
$$- \int_{S_{1,\epsilon}} u(y) \Big(\frac{\partial}{\partial r} R_0(x - y; z) \Big) dS_{1,\epsilon} + \int_{S_{1,\epsilon}} \Big(\frac{\partial}{\partial r} u(y) \Big) R_0(x - y; z) dS_{1,\epsilon}.$$
$$(2.23)$$

Since $dS_{1,\epsilon} = \epsilon^{n-1} d\omega$, where $d\omega$ is the standard measure on S^{n-1}, the 4th term of the right-hand side of (2.23) tends to 0 as $\epsilon \to 0$. By Lemma 2.2 (4), the 3rd term tends to $u(x)$. This proves (2.22) when $x \in D$. □

Theorem 2.1. *For $f \in C_0^\infty(\mathbf{R}^n)$, we put $u_z(x) = \int_{\mathbf{R}^n} R_0(x - y; z) f(y) dy$.*
Then we have
(1) $(-\Delta_x - z)u_z = f$ *on* $\mathbf{R}^n$.
(2) *If* $\operatorname{Im} \sqrt{z} > 0$, $u_z \in L^2(\mathbf{R}^n)$.
(3) *If* $\operatorname{Im} \sqrt{z} = 0$, u_z *satisfies*

$$u_z(x) = O(|x|^{-(n-1)/2}), \quad |x| \to \infty, \tag{2.24}$$

$$\Big(\frac{\partial}{\partial r} - i\sqrt{z} \Big) u_z(x) = O(|x|^{-(n+1)/2}), \quad |x| \to \infty. \tag{2.25}$$

Proof. Let

$$u_z(x) = \int_{\mathbf{R}^n} R_0(x - y; z) f(y) dy = \int_{\mathbf{R}^n} R_0(y; z) f(x - y) dy.$$

Using $\Delta_x f(x - y) = \Delta_y f(x - y)$, we have

$$(-\Delta_x - z)u_z(x) = \int_{\mathbf{R}^n} R_0(y; z)(-\Delta_y - z) f(x - y) dy$$
$$= \int_{\mathbf{R}^n} R_0(x - y; z)(-\Delta_y - z) f(x - y) dy.$$

Since $f(y)$ is compactly supported, we apply Lemma 2.3 with $D = \{|x| < R\}$ and take R large enough to obtain the assertion (1). The assertion (2) follows from (2.12), since $R_0(x; z)$ decays exponentially as $|x| \to \infty$. The assertion (3) follows from Lemma 2.1 (2). □

Letting $z = 0$, we have the following theorem.

Theorem 2.2. *For $f \in C_0^\infty(\mathbf{R}^n)$, we put $u(x) = \int_{\mathbf{R}^n} R_0(x - y)f(y)dy$.*
Then we have
(1) $-\Delta_x u = f$ on $\mathbf{R}^n$.
(2) As $|x| \to \infty$,

$$u(x) = O(|x|^{-(n-2)}), \tag{2.26}$$

$$\frac{\partial}{\partial r}u(x) = O(|x|^{-(n-1)}). \tag{2.27}$$

2.2 Harmonic functions

Let Ω be an open set in $\mathbf{R}^n$. A function $u(x) \in C^\infty(\Omega)$ is said to be *harmonic* in Ω if it satisfies

$$\Delta u(x) = 0, \quad \text{in} \quad \Omega. \tag{2.28}$$

If $n = 1$, $u(x)$ is a linear function: $u(x) = ax + b$, hence its value on the middle point of an interval is equal to the mean of the values of two end points. This is also the case for $n \geq 2$ and is called the *mean value property*.

Theorem 2.3. *Suppose $u(x)$ is harmonic in Ω. Take $x_0 \in \Omega$ and $\delta > 0$ such that $B(x_0, \delta) = \{x \in \mathbf{R}^n \,;\, |x - x_0| \leq \delta\} \subset \Omega$. Then, $u(x)$ has the following properties.*
(1) For $0 < r \leq \delta$,

$$u(x_0) = \frac{1}{|S_r|} \int_{S_r} u(x)dS_r,$$

where $S_r = \{x \in \mathbf{R}^n \,;\, |x - x_0| = r\}$, and $|S_r|$ is the surface measure of S_r.
(2) For $0 < r \leq \delta$,

$$u(x_0) = \frac{1}{|B_r|} \int_{B_r} u(x)dx,$$

where $B_r = \{x \in \mathbf{R}^n \,;\, |x - x_0| < r\}$ and $|B_r|$ is the volume of B_r.

Proof. Take the polar coordinates $x = x_0 + r\omega$, $\omega \in S^{n-1}$. Letting $v(x) = 1$ and applying Green's formula, we then have

$$0 = \int_{B_t} (\Delta u)v dx - \int_{B_t} u\Delta v dx = \int_{S_t} \frac{\partial u}{\partial r}dS_t, \quad 0 < t < \delta.$$

This equality together with Lemma 2.3 with $z = 0$ and (2.19) yields

$$u(x_0) = \frac{1}{t^{n-1}|S^{n-1}|} \int_{S_t} u(x)dS_t,$$

which proves (1). Integrating this equality on $(0, r) \times S^{n-1}$ with respect to $t^{n-1}dtd\omega$, we otain (2). $\qquad\square$

This mean value property implies the following *maximum principle*.

Lemma 2.4. *Let Ω be an open set in $\mathbf{R}^n$, and $u(x)$ a real-valued harmonic function on Ω. Then for any connected open set D such that $\overline{D} \subset \Omega$, we have*

$$\sup_{x \in D} u(x) \leq \max_{x \in \partial D} u(x). \tag{2.29}$$

Moreover, $u(x)$ is constant in D if there exists $x_0 \in D$ such that

$$u(x_0) = \sup_{x \in D} u(x).$$

Proof. Let $M = \max_{x \in \overline{D}} u(x)$, and put

$$D_1 = \{x \in D \,;\, u(x) = M\}, \quad D_2 = \{x \in D \,;\, u(x) < M\}.$$

Then, $D = D_1 \cup D_2$, $D_1 \cap D_2 = \emptyset$. If $D_1 = \emptyset$, the lemma holds. If $D_1 \neq \emptyset$, there exists $x_0 \in D$ such that $u(x_0) = M$. Take $\delta > 0$ small enough so that $B(x_0, \delta) \subset D$. Then since $M - u(x) \geq 0$ in $B(x_0, \delta)$, we have by (2) of Theorem 2.3

$$0 = M - u(x_0) = \frac{1}{|B(x_0, \delta)|} \int_{B(x_0,\delta)} (M - u(x))dx \geq 0.$$

Therefore, $M - u(x) = 0$ on $B(x_0, \delta)$, which implies that D_1 is an open set. Since D_2 is cleary an open set, the connectivity of D yields $D_1 = D$. This proves the lemma. $\qquad\square$

For the Helmholtz equation, the maximum principle does not hold. For example, in $\mathbf{R}^3$, $u(x) = r^{-1}\sin kr$, $r = |x|$, satisfies

$$\Delta u + k^2 u = 0, \quad \text{in} \quad \mathbf{R}^3,$$

however,

$$u = 0 \quad \text{for} \quad r = m\pi/k, \quad m \in \mathbf{Z}.$$

2.3   Uniqueness theorem

The fundamental result for the uniqueness of solutions to the Helmholtz equation

$$(-\Delta - k^2)u(x) = 0 \tag{2.30}$$

in $\mathbf{R}^n$, $n \geq 2$, was given by Rellich [86] and Vekua [91].

Theorem 2.4. *Let $k > 0$, and $u(x)$ satisfy (2.30) on $D = \{x \in \mathbf{R}^n \, ; \, |x| > R_0\}$ for a constant $R_0 > 0$. If*

$$\lim_{r \to \infty} r^{(n-1)/2} u(x) = 0, \tag{2.31}$$

then $u(x) = 0$ on D.

Let us give here the proof for the case $n = 3$. Passing to the polar coordinates, the equation (2.30) is rewritten as

$$\left(\frac{\partial^2}{\partial r^2} + \frac{2}{r} \frac{\partial}{\partial r} + \frac{\Lambda}{r^2} + k^2 \right) u = 0,$$

$$\Lambda = \frac{\partial^2}{\partial \theta^2} + \cot\theta \frac{\partial}{\partial \theta} + \frac{1}{\sin^2\theta} \frac{\partial^2}{\partial \phi^2}.$$

Expanding u by a spherical harmonics $\{Y_n^m(\theta, \phi)\}$:

$$u(x) = \sum_{n=0}^{\infty} \sum_{m=-n}^{n} u_{nm}(r) Y_n^m(\theta),$$

we obtain an equation satisfied by u_{nm}:

$$\left(\frac{d^2}{dr^2} + \frac{2}{r} \frac{d}{dr} + k^2 - \frac{n(n+1)}{r^2} \right) u_{nm} = 0.$$

Therefore, it is written by Hankel functions of the 1st kind and 2nd kind:

$$u_{nm}(r) = a_{nm} H_n^{(1)}(kr) + b_{nm} H_n^{(2)}(kr).$$

Comparing the behavior of Hankel functions (2.12) with (2.31), we have $a_{nm} = b_{nm} = 0$, hence $u(x) = 0$.

The characteristic feature of Theorem 2.4 consists in the fact the equation (2.30) is satisfied only near infinity of $\mathbf{R}^n$. The decay rate (2.31) is optimal. In fact, the fundamental solution to the Helmholtz equation satisfies (2.30) except for the origin, and behaves like $r^{-(n-1)/2} e^{ikr}$ as $r \to \infty$. Note also that when $k = 0$, Theorem 2.4 does not hold. In fact, on $\mathbf{R}^2 = \mathbf{C}$, $u(x) = z^{-m}$, m being any integer, gives a counter example.

Theorem 2.4 has been extended in many directions, e.g. to the Schrödinger operator $-\Delta + V(x)$ ([50]), or to the partial differential operator of constant coefficients ([41]). The decay assumption (2.31) is replaced by integral mean:

$$\lim_{R \to \infty} \frac{1}{R} \int_{R < |x| < 2R} |u(x)|^2 dx = 0. \tag{2.32}$$

For the Laplace equation, the following theorem holds.

Theorem 2.5. *Suppose* $u(x) \in C^\infty(\mathbf{R}^n)$ *satisfies*

$$\Delta u(x) = 0 \quad in \quad \mathbf{R}^n,$$

and

$$u(x) \to 0 \quad as \quad |x| \to \infty.$$

Then, $u(x) = 0$ *in* $\mathbf{R}^n$.

Proof. We have only to consider the case where $u(x)$ is real-valued. Suppose there exists $x_0 \in \mathbf{R}^n$ such that $u(x_0) \neq 0$. We can assume that $u(x_0) > 0$. By the assumption of the theorem, there exists $R > 0$ such that $|x_0| < R$ and $|u(x)| < u(x_0)/2$ on $\{|x| = R\}$. This contradicts the maximum principle. □

2.4 Wave equation in $\mathbf{R}^n$

In this section, we study the initial value problem of the wave equation

$$\partial_t^2 u = c^2 \Delta u, \quad (t, x) \in \mathbf{R} \times \mathbf{R}^n, \tag{2.33}$$

$$u\big|_{t=0} = f(x) \in C^\infty(\mathbf{R}^n), \quad \partial_t u\big|_{t=0} = g(x) \in C^\infty(\mathbf{R}^n), \tag{2.34}$$

where c is a positive constant representing the speed of the wave.

2.4.1 *d'Alembert's formula*

Theorem 2.6. *For* $n = 1$, *the solution of the equation (2.33) with initial data (2.34) is written as*

$$u(t, x) = \frac{1}{2}\big(f(x + ct) + f(x - ct)\big) + \frac{1}{2c} \int_{x-ct}^{x+ct} g(y) dy.$$

Proof. By the change of the variables $p = x + ct$, $q = x - ct$, the equation becomes $\partial^2 u/\partial p \partial q = 0$. Integrating $\partial(\partial u/\partial q)/\partial p = 0$, we see that $\partial u/\partial q$ is a function depending only on q, which we denote by $F_1(q)$. Take $G(q)$ such that $G'(q) = F_1(q)$. Then, $\partial(u - G(q))/\partial q = 0$, hence $u - G(q)$ depends only on p. Therefore,

$$u = F(p) + G(q) = F(x + ct) + G(x - ct).$$

By the initial condition,

$$F(x) + G(x) = f(x), \quad F'(x) - G'(x) = \frac{1}{c}g(x).$$

This implies $F(x) - G(x) = \frac{1}{c}\int_0^x g(y)dy$. Hence, we have

$$F(x) = \frac{1}{2}\left(f(x) + \frac{1}{c}\int_0^x g(y)dy + C\right), \quad G(x) = \frac{1}{2}\left(f(x) - \frac{1}{c}\int_0^x g(y)dy - C\right),$$

which prove the theorem. $\qquad\qquad\qquad\qquad\qquad\qquad\qquad\qquad\qquad\square$

2.4.2 Kirchhoff's formula

Theorem 2.7. *For $n = 3$, the solution of the equation (2.33) with initial data (2.34) is written as*

$$u(t, x) = \frac{\partial}{\partial t}\left(\frac{1}{4\pi c^2 t}\int_{|y - x| = ct} f(y)dS_y\right) + \frac{1}{4\pi c^2 t}\int_{|y - x| = ct} g(y)dS_y.$$

Proof. The uniqueness of the solution is proven later in Theorem 2.12. We show that the above function is a solution by a direct computation. First, let us consider the case $f(x) = 0$. Letting $y = x + ct\omega$, $\omega \in S^2$, we have

$$u(t, x) = \frac{t}{4\pi}\int_{S^2} g(x + ct\omega)d\omega,$$

which yields

$$\partial_t u(t, x) = \frac{1}{4\pi}\int_{S^2} g(x + ct\omega)d\omega + \frac{ct}{4\pi}\int_{S^2} \omega \cdot (\nabla g)(x + ct\omega)d\omega.$$

By the divergence theorem, the 2nd term of the right-hand side is equal to

$$\frac{1}{4\pi ct}\int_{|z| < ct} (\Delta g)(x + z)dz = \frac{1}{4\pi ct}\int_0^{ct}\left(\int_{S^2} (\Delta g)(x + r\omega)d\omega\right) r^2 dr.$$

Hence, we have

$$\partial_t u(t, x) = \frac{1}{4\pi}\int_{S^2} g(x + ct\omega)d\omega + \frac{1}{4\pi ct}\int_0^{ct}\left(\int_{S^2} (\Delta g)(x + r\omega)d\omega\right) r^2 dr.$$

Again differentiating with respect to t, we have

$$\partial_t^2 u(t,x) = \frac{c}{4\pi} \int_{S^2} \omega \cdot (\nabla g)(x + ct\omega) d\omega - \frac{1}{4\pi ct^2} \int_{|z|<ct} (\Delta g)(x+z) dz$$
$$+ \frac{c^2 t}{4\pi} \int_{S^2} (\Delta g)(x + ct\omega) d\omega.$$

Applying the divergence theorem to the 1st term of the right-hand side, we have

$$\partial_t^2 u(t,x) = \frac{c^2 t}{4\pi} \int_{S^2} (\Delta g)(x + ct\omega) d\omega = \frac{1}{4\pi t} \int_{|x-y|=ct} (\Delta g)(y) dS_y.$$

Since

$$\Delta u(t,x) = \frac{t}{4\pi} \int_{S^2} (\Delta g)(x + ct\omega) d\omega = \frac{1}{4\pi c^2 t} \int_{|y-x|=ct} \Delta g(y) dS_y,$$

we have proven the theorem when $f(x) = 0$.

Now, let us consider the operator

$$U(t) : h(x) \to (U(t)h)(x) = \frac{1}{4\pi c^2 t} \int_{|x-y|=ct} h(y) dS_y. \tag{2.35}$$

As has been proven above, $v(t,x) = (U(t)f)(x)$ satisfis $\partial_t^2 v = c^2 \Delta v$, and $v(0,x) = 0$, $v_t(0,x) = f(x)$. Then, $w(t,x) = \partial_t v(t,x)$ satisfies $\partial_t^2 w = \Delta w$, $w(0,x) = v_t(0,x) = f(x)$, and $w_t(0,x) = v_{tt}(0,x) = \Delta v(0,x) = 0$. We have thus seen that $u(t,x) = \partial_t(U(t)f)(x) + (U(t)g)(x)$ is the solution to the problem (2.33), (2.34). $\square$

2.4.3 Poisson's formula

Theorem 2.8. *For $n = 2$, the solution of the equation (2.33) with initial data (2.34) is written as*

$$u(t,x) = \frac{\partial}{\partial t}\left(\frac{1}{2\pi c^2 t} \int_{|x-y|<ct} \frac{f(y)}{\sqrt{c^2 t^2 - |x-y|^2}} dy\right)$$
$$+ \frac{1}{2\pi c^2 t} \int_{|x-y|<ct} \frac{g(y)}{\sqrt{c^2 t^2 - |x-y|^2}} dy.$$

Proof. Put $X = (x,z) \in \mathbf{R}^3$, where $x \in \mathbf{R}^2$. For $g(x) \in C^\infty(\mathbf{R}^2)$. Let $\widetilde{g}(X) = g(x)$. Using (2.35), we put $\widetilde{u}(t,X) = (U(t)\widetilde{g})(X)$. Since $\widetilde{g}(X)$ does not depend on z, we have

$$\partial_z \widetilde{u}(t,X) = (U(t)\partial_z \widetilde{g})(X) = 0.$$

Hence, $\widetilde{u}(t, X)$ is independent of z. Then, putting $u(t, x) = \widetilde{u}(t, x, 0)$, we see that $u(t, x)$ is a solution to the 2-dimensional problem (2.33), (2.34) with $f(x) = 0$. It is rewritten as

$$u(t, x) = \frac{1}{4\pi c^2 t} \int_{|(y,z) - (x,0)| = ct} \widetilde{g}(y, z) dS_{y,z}.$$

We split the sphere $|(y, z) - (x, 0)| = ct$ into two parts: $S_+ \cup S_-$, where

$$S_\pm = \left\{ (x_1 + s_1, x_2 + s_2, \pm\sqrt{c^2 t^2 - s_1^2 - s_2^2}) ;\ s_1^2 + s_2^2 < c^2 t^2 \right\},$$

on which the surface element is written as $dS_{y,z} = (c^2 t^2 - s_1^2 - s_2^2)^{-1/2} ds_1 ds_2$. Therefore, $u(t, x)$ is rewritten as

$$u(t, x) = \frac{1}{2\pi c^2 t} \int_{|x-y| < ct} \frac{g(y)}{\sqrt{c^2 t^2 - |x - y|^2}} dy.$$

The rest of the proof is the same as that of Theorem 2.7. $\qquad\qquad\square$

2.4.4 *Huygens' principle*

The formulas in Theorems 2.6 and 2.7 show that if the support of the initial data is in the ball $|x - x_0| < \epsilon$, the support of the solution $u(t, x)$ is in the annulus $ct - \epsilon \le |x - x_0| \le ct + \epsilon$. Since $\epsilon > 0$ can be chosen arbitrarily small, one can conclude that, when $n = 1, 3$, the influence of the point $x_0 \in \mathbf{R}^n$ at time 0 lies only on the sphere $|x - x_0| = ct$ at time t. Namely, for $n = 1, 3$, the wave propagates exactly with speed c. In other words, in sace-time, the wave emanating from a point x_0 propagates only on the cone $ct = \pm|x - x_0|$. This is called *Huygens' principle*. Poisson's formula in Theorem 2.8 show that Huygens' principle does not hold in 2-dimensions.

2.4.5 *Duhamel's principle*

Duhamel's principle is a general idea for solving the equation

$$\partial_t^2 w = Aw + h(t), \quad w\big|_{t=0} = 0, \quad \partial_t u_{t=0} = 0, \qquad (2.36)$$

where A is a linear operator defined in some vector spece. Assume that the initial value problem

$$\partial_t^2 u = Au, \quad u\big|_{t=s} = 0, \quad \partial_t u\big|_{t=s} = f, \qquad (2.37)$$

is uniquely solvable, and let

$$E(t, s) : f \to u(t, s)$$

be the operator assigning the solution $u(t, s)$ to the data at time $t = s$. It satisfies

$$(\partial_t^2 - A)E(t, s)f = 0, \quad E(s, s)f = 0, \quad \partial_t E(t, s)f\big|_{t=s} = f.$$

Let

$$w(t) = \int_0^t E(t, s)h(s)ds. \tag{2.38}$$

Then, obviously, $w\big|_{t=0}$, and

$$\partial_t w\big|_{t=0} = \left(E(t, t)h(t) + \int_0^t \partial_t E(t, s)h(s)ds \right)\bigg|_{t=0} = 0.$$

Moreover,

$$\partial_t^2 w = \partial_t E(t, s)h(s)\big|_{s=t} + \int_0^t \partial_t^2 E(t, s)h(s)ds.$$

Using $\partial_t^2 E(t, s)h(s) = AE(t, s)h(s)$, we have

$$\partial_t^2 w = h(t) + \int_0^t AE(t, s)h(s)ds = h(t) + Aw.$$

We have thus shown that $w(t)$ gives the solution to (2.36).

For the case of the wave equation, $A = c^2\Delta$, and

$$E(t, s)f = U(t - s)f = \frac{1}{4\pi c^2(t - s)} \int_{|x-y|=c(t-s)} f(s, y)dS_y.$$

Therefore,

$$w(t, x) = \int_0^t \frac{1}{4\pi c^2(t - s)} \int_{|x-y|=c(t-s)} h(s, y)dS_y ds$$

$$= \frac{1}{4\pi c^2} \int_{|x-y|<ct} \frac{h\left(t - \frac{|x-y|}{c}, y\right)}{|x - y|} dy.$$

We have thus obtained the following formula.

Theorem 2.9. *For $n = 3$, the solution of the equation $\partial_t^2 w = c^2\Delta w + h(t, x)$, $w(0, x) = 0$, $\partial_t w(0, x) = 0$ is given by*

$$w(t, x) = \frac{1}{4\pi c^2} \int_{|x-y|<ct} \frac{h\left(t - \frac{|x-y|}{c}, y\right)}{|x - y|} dy.$$

Note that by a direct computation, $\partial_t w(t, x)$ is written as

$$\partial_t w(t, x) = \frac{1}{4\pi c^2 t} \int_{|x-y|=ct} h(0, y)dS_y + \frac{1}{4\pi c^2} \int_{|x-y|<ct} \frac{\partial_t h\left(t - \frac{|x-y|}{c}, y\right)}{|x - y|} dy. \tag{2.39}$$

2.4.6 *Energy inequality*

The energy of the solution $u(t, x)$ of the wave equation $\partial_t^2 u = c^2 \Delta u$ in an open set $\Omega \subset \mathbf{R}^n$ is defined by

$$E(u(t); \Omega) = \frac{1}{2} \int_\Omega \left(\frac{1}{c^2} |\partial_t u(t, x)|^2 + |\nabla u(t, x)|^2 \right) dx. \qquad (2.40)$$

Let

$$B(x_0, R) = \{x \in \mathbf{R}^n \, ; \, |x - x_0| < R\}.$$

Theorem 2.10. *For any $x_0 \in \mathbf{R}^n$, and $R, t > 0$,*

$$E(u(t), B(x_0, R)) \leq E(u(0), B(x_0, R + t)).$$

Proof. By the change of variable $\tau = ct$, the theorem is reduced to the case $c = 1$. The proof is based on the following identity which holds for any C^∞-function $v(t, x)$:

$$\frac{1}{2} \partial_t \left(|\partial_t v|^2 + |\nabla v|^2 \right) - \nabla \cdot \left(\mathrm{Re}(\partial_t v \overline{\nabla v}) \right) = \mathrm{Re}((\partial_t^2 v - \Delta v) \overline{\partial_t v}). \qquad (2.41)$$

We then have, for the solution of the wave equation,

$$\frac{1}{2} \partial_t \left(|\partial_t u|^2 + |\nabla u|^2 \right) - \nabla \cdot \left(\mathrm{Re}(\partial_t u \overline{\nabla u}) \right) = 0. \qquad (2.42)$$

Let

$$D(T) = \{(x, t) \, ; \, 0 \leq t \leq T, \ x \in B(x_0, R + T - t)\},$$

whose boundaries are $B(x_0, R) \times \{T\}$, $B(x_0, R + T) \times \{0\}$ and

$$\Gamma = \{(x, t) \, ; \, 0 \leq t \leq T, \ |x - x_0| = R + T - t\}.$$

Let $\nu = (\nu_x, \nu_t) = (\nu_1, \cdots, \nu_n, \nu_{n+1})$ be the outer unit normal to Γ. Then, by the divergence theorem,

$$E(u(T), B(x_0, R)) - E(u(0), B(x_0, R + T))$$
$$+ \int_\Gamma \left(\frac{\nu_t}{2} (|\partial_t|^2 + |\nabla u|^2) - \nu_x \cdot (\mathrm{Re}\,(\partial_t u \overline{\nabla u})) \right) dS = 0.$$

Since $\nu_t = |\nu_x| = 1/\sqrt{2}$, we have

$$\frac{\nu_t}{2} (|\partial_t|^2 + |\nabla u|^2) - \nu_x \cdot (\mathrm{Re}\,(\partial_t u \overline{\nabla u}))$$
$$\geq \frac{1}{2\sqrt{2}} \left(|\partial_t|^2 + |\nabla u|^2 - 2|\partial_t u||\nabla u| \right) \geq 0,$$

from which follows the theorem. $\square$

Theorem 2.11. *The total energy of the solution to the wave equation is conserved, i.e.*

$$E(u(t), \mathbf{R}^n) = E(u(0), \mathbf{R}^n), \quad \forall t \in \mathbf{R}.$$

Proof. Theorem 2.10 implies $E(u(t), \mathbf{R}^n) \leq E(u(0), \mathbf{R}^n)$ for $t > 0$. We fix t, and put $v(s) = u(t - s)$. Then, regarding s as the time variable, $v(s)$ satisfies the wave equation. Hence

$$E(u(0), \mathbf{R}^n) = E(v(t), \mathbf{R}^n) \leq E(v(0), \mathbf{R}^n) = E(u(t), \mathbf{R}^n).$$

Therefore, the theorem is proved. □

Theorem 2.12. *The solution of the initial value problem (2.33), (2.34) is unique.*

Proof. Assume that the initial data is 0. Then, by Theorem 2.11, $\partial_t u(t, x) = 0$. Therefore, $u(t, x) = u(0, x) = 0$. □

Chapter 3

Potential theory

A charge distribution on a surface $S \subset \mathbf{R}^3$ generates an electric potential

$$S(x, \rho) = \int_S \rho(y) \frac{1}{|x - y|} dS_y, \tag{3.1}$$

called a *single layer potential*, where $\rho(y)$ is the charge density. A *double layer potential* is defined by

$$D(x, \rho) = \int_S \rho(y) \partial_{\nu(y)} \frac{1}{|x - y|} dS_y, \tag{3.2}$$

where $\partial_{\nu(y)} = \nu(y) \cdot \nabla_y$, and $\nu(y)$ is the outer unit normal to S at y. The functions $\partial_{\nu(x)} S(x, \rho)$ and $D(x, \rho)$ have jump discontinuities on S, which give rise to Dirac measures on S. This makes it possible to transform the interior or exterior boundary value problems for the Laplace operator to integral equations on the boundary. These jump properties of potentials are studied in Theorems 3.2, 3.3, 3.4 and 3.5 in §4. We then study their mapping properties in Hölder spaces in §6. Using Fredhlom's theorem, we solve the boundary value problems for the Laplace equation in §8. The boundary value problems for the Helmholtz equation are also investigated in §9.

Throughout this chapter, S is a compact $(n - 1)$-dimensional hypersurface in $\mathbf{R}^n$, $n \geq 2$. For the notational convenience, we denote functions on S as $f(x)$ where $x \in \mathbf{R}^n$. Therefore, if S is represented as $x = \varphi(u)$, $f(x) = f(\varphi(u))$ is actually a function of $u \in \mathbf{R}^{n-1}$.

3.1   Dirac measure and principal value

Let us start with the case where the kernel has weak singularities.

Lemma 3.1. *Let $g(x, y) \in C(\mathbf{R}^n \times S)$, and $0 < \gamma < n - 1$. Then the integral*

$$h(x) = \int_S \frac{g(x, y)}{|x - y|^\gamma} dS_y$$

is absolutely convergent and defines a continuous function on $\mathbf{R}^n$.

Proof. We have only to show the convergence of the integral and the continuity of $h(x)$ at $x_0 \in S$. Letting $x = (x', x_n)$, we can assume that near x_0, S is represented as $x_n = \varphi(x')$, and that as a function of y, $g(x, y)$ is supported in a small neighborhood of x_0. Then, we have

$$h(x) = \int_{\mathbf{R}^{n-1}} \frac{a(x, y')}{\left((x' - y')^2 + (x_n - \varphi(y'))^2\right)^{\gamma/2}} dy',$$

$$a(x, y') = g(x, y', \varphi(y')) J(y'), \quad J(y') = \left(1 + |\varphi'(y')|^2\right)^{1/2}.$$

Without loss of generality we assume that $x_0 = 0$ and $a(x, y') = 0$ if $|y'| > 1$. Letting $y' - x' = r\omega, \omega \in S^{n-2}$, we have if $|x| < 1/2$,

$$h(x) = \int_0^2 dr \int_{S^{n-2}} \frac{r^{n-2} a(x, x' + r\omega)}{\left(r^2 + (x_n - \varphi(x' + r\omega))^2\right)^{\gamma/2}} d\omega.$$

The integrand is dominated from above by $C r^{n-2-\gamma}$, where $-1 < n - 2 - \gamma$. Then, this intergal is absolutely convergent, and by Lebesgue's theorem $h(x)$ is continuous with respect to $x \in \mathbf{R}^n$. $\square$

If $\gamma = n - 1$, the above integral $h(x)$ diverges when x approaches S. To see what is going on in this case, let us recall the well-known formula

$$\int_{-\infty}^\infty \frac{f(x)}{x - a \mp i0} dx = \pm \pi i f(a) + \text{p.v.} \int_{-\infty}^\infty \frac{f(x)}{x - a} dx, \qquad (3.3)$$

which holds for any $a \in \mathbf{R}$ and $f(x) \in C_0^\infty(\mathbf{R})$. Here, p.v. denotes the principal value defined by

$$\text{p.v.} \int_{-\infty}^\infty \frac{f(x)}{x - a} dx = \lim_{\epsilon \to 0} \int_{|x-a| > \epsilon} \frac{f(x)}{x - a} dx.$$

Noting the formula

$$(\partial_x + i\partial_y) \frac{1}{2} \log \left((x - a)^2 + y^2\right) = \frac{1}{x - a - iy},$$

one can rewrite (3.3) as

$$\lim_{y \to \pm 0} \int_{-\infty}^{\infty} f(x) \left(\partial_x + i\partial_y\right) \frac{1}{2} \log\left((x-a)^2 + y^2\right) dx$$

$$= \pm \pi i \, f(a) + \text{p.v.} \int_{-\infty}^{\infty} \frac{f(x)}{x-a} dx.$$

From this formula, we obtain

$$\lim_{y \to \pm 0} \int_{-\infty}^{\infty} f(x) \partial_x \frac{1}{2} \log\left((x-a)^2 + y^2\right) dx = \text{p.v.} \int_{-\infty}^{\infty} \frac{f(x)}{x-a} dx, \qquad (3.4)$$

$$\lim_{y \to \pm 0} \int_{-\infty}^{\infty} f(x) \partial_y \frac{1}{2} \log\left((x-a)^2 + y^2\right) dx = \pm \pi \, f(a). \qquad (3.5)$$

We can rewrite (3.4) and (3.5) as

$$\lim_{y \to \pm 0} \partial_x \int_{-\infty}^{\infty} f(x') \frac{1}{2} \log\left((x-x')^2 + y^2\right) dx' = \text{p.v.} \int_{-\infty}^{\infty} \frac{f(x')}{x-x'} dx', \qquad (3.6)$$

$$\lim_{y \to \pm 0} \int_{-\infty}^{\infty} f(x') \frac{y}{(x-x')^2 + y^2} dx' = \pm \pi \, f(x). \qquad (3.7)$$

Note that (3.6) is the tangential derivative of the *logarithmic potential* (i.e. 2-dimensional single layer potential), and (3.7) is the 2-dimensional double layer potential.

Let us consider the general case. We put for $n \geq 2$

$$E(x) = -\int_{\mathbf{R}^{n-1}} \frac{(x'-y', x_n)}{\left((x'-y')^2 + x_n^2\right)^{n/2}} f(y') dy', \qquad (3.8)$$

where $x' = (x_1, \cdots, x_{n-1})$, and $f(y') \in C_0(\mathbf{R}^{n-1})$. Note that if $n \geq 3$

$$E(x) = \frac{1}{n-2} \nabla_x \int_{\mathbf{R}^{n-1}} \frac{f(y')}{\left((x'-y')^2 + x_n^2\right)^{(n-2)/2}} dy', \qquad (3.9)$$

and for $n = 2$, we take $-\frac{1}{2} \log((x'-y')^2 + x_2^2)$ instead of $(n-2)^{-1}((x'-y')^2 + x_n^2)^{-(n-2)/2}$ to get (3.9). Therefore, $E(x)$ is the gradient of the fundamental solution of the Laplacian integrated on $\mathbf{R}^{n-1}$, up to a constant.

It is obvious that $E(x) \in C^\infty(\mathbf{R}_+^n) \cup C^\infty(\mathbf{R}_-^n)$, where $\mathbf{R}_n^{\pm} = \{x \in \mathbf{R}^n; \pm x_n > 0\}$. We put

$$E_{tan}(x) = (E_1(x), \cdots, E_{n-1}(x)), \quad E_{nor}(x) = E_n(x).$$

For $F(x) \in C(\mathbf{R}_{\pm}^n)$, we put

$$(F(x', 0))_{\pm} = \lim_{x_n \to \pm 0} F(x', x_n). \qquad (3.10)$$

An appropriate property to deal with the integral like (3.8) is the *Hölder continuity*. A function $f(x)$ on $\mathbf{R}^n$ is said to be Hölder continuous if there exist constants $C > 0$ and $0 < \alpha \leq 1$ such that

$$|f(x) - f(y)| \leq C|x - y|^\alpha, \quad \forall x, y \in \mathbf{R}^n.$$

For an open set $U \subset \mathbf{R}^n$, and an integer $N \geq 0$, let $C^{N,\alpha}(U)$ be the set of $C^N(U)$-functions $f(x)$ satisfying

$$\|f\|_{N,\alpha} := \sup_{x \in U} \sum_{|\beta| \leq N} |\partial_x^\beta f(x)| + \sup_{x,y \in U} \sum_{|\beta| = N} \frac{|\partial_x^\beta f(x) - \partial_y^\beta f(y)|}{|x - y|^\alpha} < \infty. \quad (3.11)$$

The integral (3.8) converges absolutely only when $x_n \neq 0$. Therefore, we assume that $f(y')$ is Hölder continuous on $\mathbf{R}^{n-1}$, and modify its definition as the principal value:

$$\begin{aligned}
E_{tan}(x) &= -\text{p.v.} \int_{\mathbf{R}^{n-1}} \frac{x' - y'}{((x' - y')^2 + x_n^2)^{n/2}} f(y') dy \\
&= -\lim_{\epsilon \to 0} \int_{|x'-y'|>\epsilon} \frac{x' - y'}{((x' - y')^2 + x_n^2)^{n/2}} f(y') dy'.
\end{aligned} \quad (3.12)$$

As will be seen in the proof of the next lemma, this gives a consistent way to treat the integral (3.8) for all $x \in \mathbf{R}^n$.

Lemma 3.2. *Assume that $f(y') \in C^{0,\alpha}(\mathbf{R}^{n-1})$, $0 < \alpha \leq 1$, and compactly supported. Then defining $E_{tan}(x)$ by (3.12), we have:*
(1) $E_{tan}(x) \in C(\mathbf{R}^n)$.
(2) $(E_{nor}(x', 0))_\pm = \mp \dfrac{|S^{n-1}|}{2} f(x')$.

Proof. Suppose x varies over a bounded set $\{|x| < a\}$. Then, we can assume that the integral (3.12) is actually performed on a ball $\{|x'-y'| < A\}$ for some $A > 0$. Let

$$f(y') - f(x') = g(x', y'),$$

and split $E_{tan}(x)$ into two parts:

$$\begin{aligned}
E_{tan}(x) &= \lim_{\epsilon \to 0} \int_{\epsilon < |x'-y'| < A} \frac{y' - x'}{((x' - y')^2 + x_n^2)^{n/2}} f(x') dy' \\
&+ \lim_{\epsilon \to 0} \int_{\epsilon < |x'-y'| < A} \frac{y' - x'}{((x' - y')^2 + x_n^2)^{n/2}} g(x', y') dy'.
\end{aligned}$$

The 1st term vanishes, since the integrand is an odd function of $y' - x'$. The integrand of the 2nd term is dominated from above by $C|x' - y'|^{1+\alpha-n}$.

Therefore, by Lebesgue's theorem,

$$E_{tan}(x) = \int_{|x'-y'|<A} \frac{y'-x'}{(|x'-y'|^2 + x_n^2)^{n/2}} g(x',y')dy'$$

$$= \int_{|z|<A} \frac{z}{(|z|^2 + x_n^2)^{n/2}} g(x', x'+z)dz.$$

Since $|g(x', x'+z)| \leq C|z|^\alpha$, the right-hand side is continuous with respect to $x' \in \mathbf{R}^{n-1}$. This proves that $E_{tan}(x)$ is continuous for all $x \in \mathbf{R}^n$.

To compute $E_{nor}(x)$, we pass to the polar coordinates $y' = x' + r\omega, \omega \in S^{n-2}$. Letting $x_n = t$, we then have

$$E_{nor}(x) = -\int_0^A \int_{S^{n-2}} \frac{t}{(r^2+t^2)^{n/2}} f(x') r^{n-2} dr d\omega$$

$$- \int_0^A \int_{S^{n-2}} \frac{t}{(r^2+t^2)^{n/2}} r^{n-2} h(r,\omega) dr d\omega,$$

where $h(r,\omega) = g(x', x'+r\omega)$ is continuous for $r \geq 0, \omega \in S^{n-2}$. Moreover $|h(t,r,\omega)| \leq Cr^\alpha$. Noting that

$$\left| t r^{n-2} h(r,\omega) (r^2+t^2)^{-n/2} \right| = \left| \frac{t}{\sqrt{r^2+t^2}} \frac{r^{n-2}}{\left(\sqrt{r^2+t^2}\right)^{n-1}} h(r,\omega) \right| \leq Cr^{\alpha-1},$$

we see that the 2nd term vanishes as $t \to 0$ by Lebesgue's theorem. To compute the 1st term, we let $r = |t|s$ to have

$$E_{nor}(x) = -f(x') \frac{t}{|t|} \int_0^{A/|t|} \int_{S^{n-2}} \frac{s^{n-2}}{(s^2+1)^{n/2}} ds d\omega + o(1)$$

$$\to \mp f(x') |S^{n-2}| \int_0^\infty \frac{s^{n-2}}{(s^2+1)^{n/2}} ds = \mp |S^{n-2}| \frac{\sqrt{\pi}\Gamma(\frac{n-1}{2})}{2\Gamma(\frac{n}{2})} f(x'),$$

which proves the lemma. $\qquad\qquad\square$

Therefore, the tangential component of $E(x)$ is continuous across $\{x_n = 0\}$, while the normal component is discontinuous and converges to a Dirac measure. We prove it in a more general situation in the following sections.

As a corollary to the above computation, we have the following theorem.

Theorem 3.1. *If $f(x') \in C^{0,\alpha}(\mathbf{R}^{n-1})$, $0 < \alpha \leq 1$, and $f(x')$ is compactly supported, the boundary value problem*

$$\begin{cases} \Delta u(x) = 0, & for \quad x_n > 0, \\ u(x',0) = f(x') \end{cases} \qquad (3.13)$$

has a solution given by

$$u(x) = \frac{2}{|S^{n-1}|} \int_{\mathbf{R}^{n-1}} \frac{x_n}{\left((x'-y')^2 + x_n^2\right)^{n/2}} f(y')dy', \quad x_n > 0. \quad (3.14)$$

This solution is unique in the class of $C^2(\mathbf{R}_+^n) \cap C(\overline{\mathbf{R}_+^n})$-functions satisfying $u(x) \to 0$ as $|x| \to \infty$.

Proof.   Using the fundamental solution $R_0(x)$ to $-\Delta$, we can rewrite $u(x)$ as

$$u(x) = -2\frac{\partial}{\partial x_n} \int_{\mathbf{R}^{n-1}} R_0(x'-y', x_n)f(y')dy'.$$

Since $\Delta R_0(x) = 0$ for $x \neq 0$, it is clear that $\Delta u(x) = 0$ for $x_n > 0$. By Lemma 3.2, $u = f(x')$ for $x_n = 0$. The uniqueness can be proved in the same way as in Theorem 2.5.   □

3.2   Function spaces on S

Let $S \subset \mathbf{R}^n$ be a compact $(n-1)$-dimensional surface in $\mathbf{R}^n$. In the following arguments, we use three spaces $C(S)$, $L^2(S)$ and $C^{m,\alpha}(S)$.

The space $C(S)$ is the set of all continuous functions on S equipped with the norm

$$\|f\|_{C(S)} = \max_{x \in S} |f(x)|.$$

The space $L^2(S)$ is the set of all square integrable functions on S equipped with the inner product

$$(f, g)_{L^2(S)} = \int_S f(x)\overline{g(x)}dS_x$$

and the norm

$$\|f\|_{L^2(S)} = \sqrt{(f, f)_{L^2(S)}}.$$

The Hölder continuity on S is defined in the same way as in the case of $\mathbf{R}^n$ by taking local coordinates on S. By definition, there exists a finite number of open sets S_j, $j = 1, \cdots, N$, such that

$$S = \bigcup_{j=1}^{N} S_j, \quad (3.15)$$

and on each S_j is represented as $x_i = \varphi_j(x')$, $x' \in U_j$, where x' is x with x_i removed, (i depends on j), and U_j is an open set in $\mathbf{R}^{n-1}$. There exists also a partition of unity $\{\chi_j\}_{j=1}^N$ satisfying $\chi_j \in C_0^\infty(S_j)$ and $\sum_{j=1}^N \chi_j = 1$

on S. Then a function f on S belongs to $C^{m,\alpha}(S)$ if $\chi_j f \in C^{m,\alpha}(S_j)$ for any j. The norm of $C^{m,\alpha}(S)$ is defined by

$$\|f\|_{C^{m,\alpha}} = \sum_{j=1}^{N} \Big[\sup_{x' \in U_j} \sum_{|\beta| \le m} |\partial_{x'}^{\beta}(\chi_j(x)f(x))|$$

$$+ \sup_{x',y' \in U_j} \sum_{|\beta|=m} \frac{|\partial_{x'}^{\beta}(\chi_j(x)f(x)) - \partial_{y'}^{\beta}(\chi_j(y)f(y))|}{|x'-y'|^{\alpha}} \Big]. \tag{3.16}$$

This norm depends on the choice of local coordinates. However, they are equivalent in the sense that if $\|\cdot\|$ and $\|\cdot\|'$ are such norms, there exists a constant $C > 0$ such that

$$C\|\cdot\|' \le \|\cdot\| \le C^{-1}\|\cdot\|'.$$

We sometimes represent the Hölder continuity by the variables $x, y \in \mathbf{R}^n$. For example, $f \in C^{0,\alpha}(S)$ is equivalent to

$$|f(x) - f(y)| \le C|x-y|^{\alpha}, \quad x, y \in S. \tag{3.17}$$

To see it, assume that S_j is represented as $x_i = \varphi_j(x')$. We put

$$C_0 = \max_j \max_{x'} |\nabla_{x'} \varphi_j(x')|. \tag{3.18}$$

Then, by the mean value theorem, for $x, y \in S_j$,

$$|x'-y'| \le |x-y| \le \sqrt{1+C_0^2}\,|x'-y'|. \tag{3.19}$$

3.3 Continuity of tangential derivatives

Now, let us consider the integral

$$\widetilde{E}(x) = \text{p.v.} \int_S \frac{x-y}{|x-y|^n} f(x,y) dS_y$$

$$= \lim_{\epsilon \to 0} \int_{S \cap \{|x-y|>\epsilon\}} \frac{x-y}{|x-y|^n} f(x,y) dS_y, \tag{3.20}$$

where $f(x,y) \in C(\mathbf{R}^n \times S)$ and satisfies

$$|f(x,y) - f(x,\overline{y})| \le C|y-\overline{y}|^{\alpha}, \quad 0 < \alpha \le 1, \tag{3.21}$$

C being a constant independent of $(x,y), (x,\overline{y}) \in \mathbf{R}^n \times S$.

Lemma 3.3. *The integral* (3.20) *exists for all* $x \in \mathbf{R}^n$. *Moreover,* $\widetilde{E}(x) \in C(\mathbf{R}^n \setminus S) \cap C(S)$.

Remark that S splits $\mathbf{R}^n$ into two open sets, the interior domain D_{int} and the exterior domain D_{ext}. The lemma asserts that $\widetilde{E}(x)$ is continuous on each D_{int} and D_{ext}, and also on S. However, $\widetilde{E}(x)$ is discontinuous across S.

Proof. The existence and continuity of (3.20) is obvious on $\mathbf{R}^n \setminus S$. Let us consider (3.20) for $x \in S$. By using the partition of unity $\{\chi_j\}$ on S, we split $\widetilde{E}(x)$ as

$$\widetilde{E}(x) = \sum_j \text{p.v.} \int_S \frac{x-y}{|x-y|^n} f(x,y) \chi_j(y) dS_y.$$

Therefore, we have only to consider the case where, with respect to y, f is supported in a small compact set $K \subset S$, and x varies over a small neighborhood of K. We can assume that on K, S is reprensented as $y_n = \varphi(y')$, hence

$$\widetilde{E}(x) = \lim_{\epsilon \to 0} \int_{|x-y|>\epsilon} \frac{x-y}{|x-y|^n} g(x,y') dy',$$

$$g(x,y') = f(x,y',\varphi(y')) J(y'), \quad J(y') = \left(1 + \varphi'(y')^2\right)^{1/2}.$$

For $x, y \in K$, we pass to the polar coordinates $y' - x' = r\omega$, $\omega \in S^{n-2}$. By the Taylor expansion,

$$y - x = r(\omega, \varphi'(x') \cdot \omega) + O(r^2),$$

hence

$$\frac{y-x}{|y-x|^n} = \frac{1}{r^{n-1}} \frac{(\omega, \varphi'(x') \cdot \omega)}{(1 + \varphi'(x')^2)^{1/2}} + O\left(\frac{1}{r^{n-2}}\right).$$

Since the singularity of order $r^{-(n-2)}$ is integrable on S, we have only to consider

$$\widetilde{E}_1(x) = \lim_{\epsilon \to 0} \int_{\{|x-y|>\epsilon\} \cap \{r<1\}} \frac{1}{r^{n-1}} \frac{(\omega', \varphi'(x') \cdot \omega')}{(1 + \varphi'(x')^2)^{1/2}} g(x,y') dy'. \quad (3.22)$$

We split $g(x,y')$ as

$$g(x,y') = g(x,x') + h(x,y').$$

Since $|h(x,y')| \leq C|x' - y'|^\alpha$, the limit (3.22) with $g(x,y')$ replaced by $h(x,y')$ exists. It leads us to consider

$$\widetilde{E}_2(x) = \lim_{\epsilon \to 0} \int_{\{|x-y|>\epsilon\} \cap \{r<1\}} \frac{1}{r^{n-1}} \frac{(\omega', \varphi'(x') \cdot \omega')}{(1 + \varphi'(x')^2)^{1/2}} dy'. \quad (3.23)$$

Letting
$$\sigma = r(1 + (\varphi'(x') \cdot \omega)^2)^{1/2},$$
we have
$$|y - x| = \sigma + \sigma_1, \quad \sigma_1 = O(r^2).$$
We put
$$A = \{|y - x| > \epsilon\} \cap \{r < 1\}, \quad B = \{\sigma > \epsilon\} \cap \{r < 1\},$$
and shift the domain of integration from A to B,
$$\int_A = \int_B + \int_{A \setminus B} - \int_{B \setminus A}.$$
Since $\sigma_1 = O(\epsilon^2)$ on $(A \setminus B) \cup (B \setminus A)$, we have
$$\epsilon - C\epsilon^2 \le \sigma \le \epsilon + C\epsilon^2, \quad \text{on} \quad (A \setminus B) \cup (B \setminus A)$$
for a constant $C > 0$. Therefore, by the change of variable $r \to \sigma$,
$$\int_{(A \setminus B) \cup (B \setminus A)} \frac{1}{r^{n-1}} dy' \le C \int_{\epsilon - C\epsilon^2}^{\epsilon + C\epsilon^2} \frac{d\sigma}{\sigma} = C \log \frac{\epsilon + C\epsilon^2}{\epsilon - C\epsilon^2},$$
which converges to 0 as $\epsilon \to 0$. It remains to consider (3.23) on B, which vanishes since the integrand is an odd function of ω. $\qquad \square$

Take $x_0 \in S$, and for small $\delta > 0$ let
$$L(x_0) = \{x_0 + t\nu(x_0) \, ; \, |t| < \delta\}, \tag{3.24}$$
where $\nu(x_0)$ is the outer unit normal to S at x_0. Our concern is the behavior of the integral
$$E(x) = \text{p.v.} \int_S \frac{x - y}{|x - y|^n} f(y) dS_y$$
along $L(x_0)$. We first observe the contribution from the tangential part of the vector field $x - y$. We need the following lemma.

Lemma 3.4. *Let U be a small open set in S on which S is represented as $x_n = \varphi(x')$, $x = (x', x_n)$. For $x_0, y \in U$ and $x = x_0 + t\nu(x_0) \in L(x_0)$, we put $r = |y' - x_0'|$, where $y = (y', y_n)$, $x_0 = (x_0', x_{0n})$. Then, we have*
$$C(r + |t|) \le |x - y| \le C^{-1}(r + |t|), \tag{3.25}$$
$$\langle x - y, \nu(x_0) \rangle = t + O(r^2), \tag{3.26}$$
$$\langle x - y, \nu(y) \rangle = t + O(r^2(1 + |t|)) \tag{3.27}$$
for a constant $C > 0$.

Proof.   Using the polar coordinates $y' - x_0' = r\omega$, $r = |y' - x_0'|$, $\omega \in S^{n-2}$, and the Taylor expansion

$$\varphi(y') - \varphi(x_0') = \varphi'(x_0') \cdot r\omega + O(r^2),$$

we have

$$\begin{aligned}
x - y &= x_0 - y + t\nu(x_0) \\
&= -\big(r\omega, \varphi'(x_0') \cdot r\omega\big) + O(r^2) + t\nu(x_0).
\end{aligned} \tag{3.28}$$

Therefore,

$$|x - y|^2 = r^2\big(1 + (\varphi'(x_0') \cdot \omega)^2\big) + t^2 + O\big(r^2(r + |t|)\big),$$

which implies

$$|x - y| = \sqrt{r^2(1 + (\varphi'(x_0') \cdot \omega)^2) + t^2}\Big(1 + O(r + |t|)\Big). \tag{3.29}$$

This proves (3.25). By (3.28), we have

$$\langle x - y, \nu(x_0)\rangle = t + O(r^2),$$

which proves (3.26). To prove (3), note

$$\langle x - y, \nu(y)\rangle = \langle x - y, \nu(x_0)\rangle + \langle x - y, \nu(y) - \nu(x_0)\rangle. \tag{3.30}$$

Taylor expansion yields

$$\nu(y) - \nu(x_0) = W(x_0)(y - x_0) + O(|y - x_0|^2), \tag{3.31}$$

where $W(x_0)$ is the Weingarten map (or shape operator), which maps $T_{x_0}(S)$ into itself. To see it, let $N(y') = \nu(y)$. Then, we have by the Taylor expansion

$$N(y') = N(x_0') + \sum_{j=1}^{n-1} \frac{\partial N}{\partial y_j}(x_0')(y_j - x_{0j}') + O(r^2).$$

Differentiating $|N(y')|^2 = 1$, we see that $\frac{\partial N}{\partial y_j}(y') \in T_y(S)$. Let e_i be the standard basis of $\mathbf{R}^{n-1}$: $e_1 = (1, 0, \cdots, 0), \cdots, e_{n-1} = (0, \cdots, 0, 1)$, and put $v_i = (e_i, \varphi'(x_0') \cdot e_i)$, which forms a basis of $T_{x_0}(S)$. Then, we have

$$\frac{\partial N}{\partial y_j}(x_0') = \sum_{i=1}^{n-1} a_{ij} v_i$$

for suitable constants a_{ij}, which implies

$$N(y') = N(x_0') + \sum_{i,j=1}^{n-1} a_{ij}(y_j - x_{0j}')v_i + O(r^2).$$

We define the linear map $W(x_0)$ by

$$W(x_0) : T_{x_0}(S) \ni \sum_{j=1}^{n-1} c_j v_j \rightarrow \sum_{i,j=1}^{n-1} a_{ij} c_j v_i \in T_{x_0}(S),$$

and extend it to be 0 for $\nu(x_0)$. Then, we obtain (3.31), since

$$y - x_0 = \sum_{i=1}^{n-1} (y_i - x'_{0i}) v_i + O(r^2).$$

We have, therefore,

$$\begin{aligned}\langle y - x, \nu(y) - \nu(x_0) \rangle &= \langle y - x_0, W(x_0)(y - x_0) \rangle + O(r^2) \\ &\quad - t \left(\langle \nu(x_0), W(x_0)(y - x_0) \rangle + O(r^2) \right) \\ &= O\left(r^2(1 + |t|)\right),\end{aligned}$$

which, together with (3.26) and (3.30), proves (3.27). $\qquad\square$

For small $\delta > 0$, let

$$S_\delta = \{ x \in \mathbf{R}^n \,;\, \mathrm{dist}(x, S) < \delta \}$$

be the δ-neighborhood of S in $\mathbf{R}^n$, where $\mathrm{dist}(x, S) = \min_{y \in S} |x - y|$. Then for any $x \in S_\delta$, there exists a unique $x_0 \in S$ such that

$$|x - x_0| = \mathrm{dist}(x, S). \tag{3.32}$$

For these $x \in S_\delta$ and $y \in S$, we define

$$X_{tan}(x, y) = x - y - \langle x - y, \nu(y) \rangle \nu(y), \tag{3.33}$$

and put for $x \in S_\delta$ and $f \in C^{0,\alpha}(S)$

$$E_{tan}(x) = \mathrm{p.v.} \int_S \frac{1}{|x - y|^n} X_{tan}(x, y) f(y) dS_y. \tag{3.34}$$

We also define for $F(x) \in C(\mathbf{R}^n \setminus S)$

$$(F(x_0))_\pm = \lim_{t \to \pm 0} F(x_0 + t\nu(x_0)), \quad x_0 \in S. \tag{3.35}$$

Here, by $\lim_{t \to +0}$ ($\lim_{t \to -0}$), we mean the limit $\lim_{t>0, t \to 0}$ ($\lim_{t<0, t \to 0}$).

Lemma 3.5. *Let* $f(x) \in C^{0,\alpha}(S)$ *for some* $0 < \alpha \leq 1$. *Then for any* $x_0 \in S$, $\left(E_{tan}(x_0) \right)_\pm$ *exist and coincide:*

$$(E_{tan}(x_0))_+ = (E_{tan}(x_0))_- .$$

Furthermore, $E_{tan}(x) \in C(S_\delta)$.

Proof. We can assume without loss of generality that S is represented as $x_n = \varphi(x')$ and supp f is sufficiently small. Then for $x \in S_\delta$, $E_{tan}(x)$ is written as

$$E_{tan}(x) = \lim_{\epsilon \to 0} \int_{|x-y|>\epsilon} \frac{1}{|x-y|^n} g(x, y') dy', \qquad (3.36)$$

$$g(x, y') = X_{tan}(x, y', \varphi(y')) f(y', \varphi(y')) J(y'), \qquad (3.37)$$

$$J(y') = \left(1 + |\varphi'(y')|^2\right)^{1/2}. \qquad (3.38)$$

We take $x_0 = (x_0', \varphi(x_0')) \in S$ and let $y' - x_0' = r\omega$, $\omega \in S^{n-1}$. Then, in view of (3.28) and (3.27), we have for $x \in L(x_0)$

$$X_{tan}(x, y) = x_0 - y + t\big(\nu(x_0) - \nu(y)\big) + O\big(r^2(1 + |t|)\big)$$
$$= -(1 + tW(x_0))(y - x_0) + O\big(r^2(1 + |t|)\big),$$

where $W(x_0)$ is the Weingarten map at x_0. Letting

$$v(\omega) = (\omega, \varphi'(x_0) \cdot \omega), \quad V(\omega, t) = -(1 + tW(x_0))v(\omega),$$

we have $y - x_0 = rv(\omega) + O(r^2)$, hence

$$X_{tan}(x, y) = rV(\omega, t) + O\big(r^2(1 + |t|)\big).$$

We split the integral (3.36) as follows:

$$\int_{|x-y|>\epsilon} \frac{g(x, y')}{|x-y|^n} dy' = f(x_0) J(x_0') \int_{|x-y|>\epsilon} \frac{X_{tan}(x, y)}{|x-y|^n} dy'$$
$$+ \int_{|x-y|>\epsilon} \frac{X_{tan}(x, y)}{|x-y|^n} \big(f(y') J(y') - f(x_0) J(x_0')\big) dy'.$$

$$(3.39)$$

By the Hölder continuity of f, we have

$$|f(y) J(y') - f(x_0) J(x_0')| \le C|y' - x_0'|^\alpha.$$

This and (3.25) imply that the integrand of the 2nd term of the right-hand side of (3.39) is dominated from above by $Cr^{\alpha-(n-1)}$. Lebesgue's theorem then shows the existence of the limt as $\epsilon \to 0$. The 1st term is rewritten as

$$f(x_0) J(x_0') \int_{|x-y|>\epsilon} \frac{rV(\omega, t)}{|x-y|^n} dy' + f(x_0) J(x_0') \int_{|x-y|>\epsilon} \frac{O\big(r^2(1 + |t|)\big)}{|x-y|^n} dy.$$

$$(3.40)$$

The 2nd term of (3.40) converges as $\epsilon \to 0$ uniformly with respect to t, hence defines a continuous function. Letting

$$\sigma = \sqrt{r^2(1 + (\varphi'(x_0') \cdot \omega)^2) + t^2},$$

we have, by (3.29), $1/|x - y| = 1/\sigma + O(1)$. Therefore, to deal with the 1st term of (3.40), we have only to consider

$$\int_{1 > |x-y| > \epsilon} \frac{rV(\omega, t)}{\sigma^n} dy'.$$

For $t \neq 0$, we let $\epsilon \to 0$. Then, since the integrand is an odd function of ω, the limit vanishes. This completes the proof of the lemma. $\qquad\square$

In particular, we have proven that the tangential components of

$$\int_S f(y) \nabla_x \frac{1}{|x - y|^{n-2}} dS_y, \quad n \neq 2,$$

$$\int_S f(y) \nabla_x \log |x - y| dS_y, \quad n = 2$$

are continuous across S if $f(y) \in C^{0,\alpha}(S)$. The limits of the normal components are computed in the next section.

3.4   Single and double layer potentials

Let $R_0(x)$ be the fundamental solution to $-\Delta$ (see (2.9)). For a Hölder continuous function $f(x)$ on S, the *single layer potential* $S(x, f)$ and the *double layer potential* $D(x, f)$ are defined by

$$S(x, f) = \int_S f(y) R_0(x - y) \, dS_y, \tag{3.41}$$

$$D(x, f) = \int_S f(y) \partial_{\nu(y)} R_0(x - y) \, dS_y. \tag{3.42}$$

Note that $D(x, f)$ is rewritten as

$$D(x, f) = \frac{1}{|S^{n-1}|} \int_S f(y) \frac{\cos \theta(x, y)}{|x - y|^{n-1}} dS_y, \tag{3.43}$$

where

$$\cos \theta(x, y) = \left\langle \frac{x - y}{|x - y|}, \nu(y) \right\rangle. \tag{3.44}$$

By Lemma 3.1 for $n \geq 3$ and a direct computation for $n = 2$, we have

Lemma 3.6. *If $f \in C(S)$, then $S(x, f) \in C(\mathbf{R}^n)$.*

Let D_{ext} and D_{int} be the open domains exterior and interior to S, respectively. Lemma 3.4 implies

$$|\cos\theta(x,y)| \le C|x-y|, \quad \forall x,\ y \in S.$$

Hence by Lemma 3.1, the integral representing $D(x,f)$ is absolutely convergent for $x \in \mathbf{R}^n$, and for $f \in C(S)$

$$D(x,f) \in C(D_{ext}) \cap C(S) \cap C(D_{int}).$$

We are going to show that $(D(x,f))_\pm$ exist, however, their values do not coincide. For this purpose, the following lemma plays a key role.

Lemma 3.7.

$$D(x,1) = \frac{1}{|S^{n-1}|}\int_S \frac{\cos\theta(x,y)}{|x-y|^{n-1}}dS_y = \begin{cases} -1, & x \in D_{int}, \\ -1/2, & x \in S, \\ 0, & x \in D_{ext}. \end{cases}$$

Proof. First we consider the case $n \ge 3$. If $x \in D_{ext}$, we have by Gauss' theorem

$$0 = \int_{D_{int}} \Delta_y \frac{1}{|x-y|^{n-2}}dy$$
$$= \int_S \nu(y)\cdot\nabla_y \frac{1}{|x-y|^{n-2}}dS_y = (n-2)\int_S \frac{\cos\theta(x,y)}{|x-y|^{n-1}}dS_y.$$

If $x \in D_{int}$, we have

$$0 = \int_{D_{int}\cap\{|x-y|>\epsilon\}} \Delta_y \frac{1}{|x-y|^{n-2}}dy$$
$$= \int_S \nu(y)\cdot\nabla_y \frac{1}{|x-y|^{n-2}}dS_y + \int_{|y-x|=\epsilon} \frac{x-y}{|x-y|}\cdot\nabla_y \frac{1}{|x-y|^{n-2}}dS_y$$
$$= (n-2)\int_S \frac{\cos\theta(x,y)}{|x-y|^{n-1}}dS_y + (n-2)|S^{n-1}|.$$

If $x \in S$, we compute

$$0 = \int_{D_{int}\cap\{|x-y|>\epsilon\}} \Delta_y \frac{1}{|x-y|^{n-2}}dy$$
$$= \int_{S\cap\{|x-y|>\epsilon\}} \nu(y)\cdot\nabla_y \frac{1}{|x-y|^{n-2}}dS_y$$
$$+ \int_{D_{int}\cap\{|y-x|=\epsilon\}} \frac{x-y}{|x-y|}\cdot\nabla_y \frac{1}{|x-y|^{n-2}}dS_y.$$

As $\epsilon \to 0$, this converges to

$$(n-2) \int_S \frac{\cos \theta(x, y)}{|x-y|^{n-1}} dS_y + \frac{n-2}{2} |S^{n-1}|.$$

These formulas prove the lemma when $n \geq 3$.

If $n = 2$, we compute for $x \in D_{ext}$

$$0 = \int_S \Delta_y \log |x-y| \, dS_y = - \int_S \frac{\cos \theta(x, y)}{|x-y|} dS_y.$$

The other cases are dealt with in a similar way. □

Lemma 3.8. *For $f(x) \in C(S)$ and $x_0 \in S$, $D(x, f) - D(x, f(x_0))$ is continuous with respect to $x \in L(x_0)$.*

Proof. In view of Lemma 3.4, we have for $x = x_0 + t\nu(x_0)$,

$$\frac{|\cos \theta(x, y)|}{|x-y|^{n-1}} \leq C \frac{|t| + r^2(1+|t|)}{(r+|t|)^n},$$

where C is independent of r, t. For any $\epsilon > 0$, there exists $\delta > 0$ such that $|f(y) - f(x_0)| < \epsilon$ if $|y - x_0| < \delta$, $y \in S$. Therefore

$$\left| \int_{S \cap \{|y-x_0|<\delta\}} \frac{\cos \theta(x, y)}{|x-y|^{n-1}} (f(y) - f(x_0)) \, dS_y \right| \tag{3.45}$$
$$\leq \epsilon C \int_0^1 \frac{|t| + r^2(1+|t|)}{(r+|t|)^n} r^{n-2} dr \leq \epsilon C',$$

where the last inequality can be checked by the substitution $r = |t|s$. On the other hand, $\int_{S \cap \{|y-x_0|>\delta\}} \frac{\cos \theta(x, y)}{|x-y|^{n-1}} (f(y) - f(x_0)) \, dS_y$ is a continuous function of $x \in L(x_0)$. Letting $\epsilon \to 0$, $I(x; f) - I(x; f(x_0))$ is continuous on $L(x_0)$, since it is a uniform limit of continuous functions. □

Theorem 3.2. *For $f \in C(S)$, the boundary values $(D(x_0, f))_{\pm}$ exist for any $x_0 \in S$, and the convergence is locally uniform on $\overline{D_{ext}}$ (for the case $+$) and $\overline{D_{int}}$ for the case $-$. Moreover,*

$$(D(x_0, f))_{\pm} = D(x_0, f) \pm \frac{1}{2} f(x_0), \quad x_0 \in S. \tag{3.46}$$

In particular, we have

$$(D(x_0, f))_+ - (D(x_0, f))_- = f(x_0), \quad x_0 \in S. \tag{3.47}$$

Proof.   We split $D(x, f)$ as

$$D(x, f) = D(x, f(x_0)) + D(x, f - f(x_0)).$$

By Lemma 3.8, the 2nd term of the right-hand side is continuous. Therefore, in view of Lemma 3.7,

$$D(x_0, f) = -\frac{1}{2}f(x_0) + D(x_0, f - f(x_0)),$$

$$(D(x_0, f))_+ = D(x_0, f - f(x_0)),$$

$$(D(x_0, f))_- = -f(x_0) + D(x_0, f - f(x_0)).$$

These three formulas prove the theorem.   □

We next study the behavior of derivatives of $S(x, f)$ across S. To see it, we extend $\nu(x)$ near S in a C^∞-fashion. Noting the formula

$$\partial_{\nu(x)}R_0(x - y) = -\partial_{\nu(y)}R_0(x - y) - (\nu(x) - \nu(y)) \cdot \nabla_y R_0(x - y),$$

we have

$$\partial_{\nu(x)}S(x, f) = -D(x, f) - \int_S f(y)(\nu(x) - \nu(y)) \cdot \nabla_y R_0(x - y)dS_y.$$

The 2nd term of the right-hand side is continuous across S by the inequality $|(\nu(x) - \nu(y)) \cdot \nabla_y R_0(x - y)| \le C|x - y|^{2-n}$. Therefore, $\partial_{\nu(x)}S(x, f)$ has the same jump property as $D(x, f)$ up to sign. This proves the following theorem.

Theorem 3.3. *For $f \in C(S)$, the boundary values $(\partial_{\nu(x)}S(x, f))_\pm$ on S exist, and the convergence is locally uniform on $\overline{D_{ext}}$ for the case $+$ and $\overline{D_{int}}$ for the case $-$. Moreover,*

$$(\partial_{\nu(x)}S(x, f))_\pm = \partial_{\nu(x)}S(x, f) \mp \frac{1}{2}f(x). \tag{3.48}$$

In particular, we have

$$(\partial_{\nu(x)}S(x, f))_+ - (\partial_{\nu(x)}S(x, f))_- = -f(x). \tag{3.49}$$

Theorem 3.4. *Let $f \in C^{0,\alpha}(S)$ with $0 < \alpha \le 1$. For $x \in S$, we have*

$$\left(\nabla_x S(x, f)\right)_\pm = \mp\frac{1}{2}f(x)\nu(x) + \nabla_x S(x, f), \tag{3.50}$$

$$\left(\nabla_x S(x, f)\right)_+ - \left(\nabla_x S(x, f)\right)_- = -f(x)\nu(x). \tag{3.51}$$

Proof. The case for $n > 2$ are proven by Theorem 3.3, Lemma 3.5, and the formula

$$\frac{1}{2-n}\nabla_x \frac{1}{|x-y|^{n-2}} = \frac{\langle x-y, \nu(x)\rangle\nu(x)}{|x-y|^n} + \frac{x-y-\langle x-y, \nu(y)\rangle\nu(y)}{|x-y|^n}$$
$$+ O(|x-y|^{2-n}).$$

The case $n = 2$ is proven by a similar manner. $\qquad\square$

To deal with the normal derivative of $D(x, f)$, we need more regularity on f. We first consider the C^2 case.

Lemma 3.9. *For $f \in C^2(S)$, the boundary values $\left(\partial_{\nu(x)}D(x, f)\right)_{\pm}$ exist, and the convergence is locally uniform on $\overline{D_{ext}}$ for the case $+$ and $\overline{D_{int}}$ for the case $-$. We have, furthermore,*

$$\left(\partial_{\nu(x)}D(x, f)\right)_+ = \left(\partial_{\nu(x)}D(x, f)\right)_-, \quad \forall x \in S. \tag{3.52}$$

Proof. We consider the case $n \geq 3$. The 2-dimensional case can be proven similarly. We take $x_0 \in S$, and draw a ball $B(x_0, \epsilon_0)$ centered at x_0 with small radius $\epsilon_0 > 0$. Put $V = B(x_0, \epsilon_0) \cap D_{int}$. Then $\partial V = $ the boundary of V_0 is split into $S_0 \cup S'$, where $S_0 = B(x_0, \epsilon_0) \cap S$, $S' = \partial B(x_0, \epsilon_0) \cap D_{int}$. We extend f smoothly on V. Take $x_1 \in \overset{\circ}{V} = $ the interior of V, and draw a ball $B_1 := B(x_1, \epsilon_1) \subset \overset{\circ}{V}$. By Green's formula, we then have

$$\int_{V\setminus B_1} \frac{1}{|x_1-y|^{n-2}}\left(\Delta_y f(y)\right)dy$$
$$= \int_{S'\cup S_0} \frac{1}{|x_1-y|^{n-2}}\left(\partial_{\nu(y)}f(y)\right)dS_y + \int_{\partial B_1} \frac{1}{|x_1-y|^{n-2}}\left(\partial_{\nu(y)}f(y)\right)dS_y$$
$$- \int_{S'\cup S_0} \left(\partial_{\nu(y)}\frac{1}{|x_1-y|^{n-2}}\right)f(y)dS_y - \int_{\partial B_1} \left(\partial_{\nu(y)}\frac{1}{|x_1-y|^{n-2}}\right)f(y)dS_y.$$

Letting $\epsilon_1 \to 0$, we have

$$\int_{\partial B_1} \frac{1}{|x_1-y|^{n-2}}\left(\partial_{\nu(y)}f(y)\right)dS_y \to 0,$$

$$\int_{\partial B_1} \left(\partial_{\nu(y)}\frac{1}{|x_1-y|^{n-2}}\right)f(y)dS_y \to (n-2)|S^{n-1}|f(x_1),$$

which imply

$$\int_{S_0} \left(\partial_{\nu(y)} \frac{1}{|x_1 - y|^{n-2}} \right) f(y) dS_y + \int_{S'} \left(\partial_{\nu(y)} \frac{1}{|x_1 - y|^{n-2}} \right) f(y) dS_y$$
$$= \int_{S' \cup S_0} \frac{1}{|x_1 - y|^{n-2}} \left(\partial_{\nu(y)} f(y) \right) dS_y - \int_V \frac{1}{|x_1 - y|^{n-2}} \left(\Delta_y f(y) \right) dy$$
$$- (n-2)|S^{n-1}| f(x_1).$$

(3.53)

When $x_1 \to S$, all terms of the right-hand side have continuous boundary values. Moreover, applying $\nu(x_1) \cdot \nabla_{x_1}$ and using Theorem 3.3, we see that their normal derivatives are shown to have boundary values on S along $L(x_0)$. Therefore, the left-hand side has the same property. In particular, we have proven the existence of $\left(\partial_{\nu(x)} D(x, f) \right)_-$. We can apply the same argument to the integral on the opposite subdomain to prove the existence of $\left(\partial_{\nu(x)} D(x, f) \right)_+$.

For the sake of simplicity, we rewrite (3.53) as

$$\psi_1 + \psi_2 = \psi_3 + \psi_4 - (n-2)|S^{n-1}| f(x_1). \quad (3.54)$$

We take the limit along $L(x_0)$ of the normal derivative, and have

$$\left(\partial_\nu \psi_1 \right)_- + \left(\partial_\nu \psi_2 \right)_- = \left(\partial_\nu \psi_3 \right)_- + \left(\partial_\nu \psi_4 \right)_- - (n-2)|S^{n-1}| \partial_\nu f(x_0). \quad (3.55)$$

Now, we consider the same subdomain as above, however, we take x_2 not in V, but in the opposite subdomain. Then by Green's formula, we have

$$\int_{S_0} \left(\partial_{\nu(y)} \frac{1}{|x_2 - y|^{n-2}} \right) f(y) dS_y + \int_{S'} \left(\partial_{\nu(y)} \frac{1}{|x_2 - y|^{n-2}} \right) f(y) dS_y$$
$$= \int_{S' + S_0} \frac{1}{|x_2 - y|^{n-2}} \left(\partial_{\nu(y)} f(y) \right) dS_y - \int_V \frac{1}{|x_2 - y|^{n-2}} \left(\Delta_y f(y) \right) dy.$$

The point is that the term $f(x_2)$ does not appear in the right-hand side. Taking the limit along $L(x_0)$ of the normal derivative from the opposite side, we then have

$$\left(\partial_\nu \psi_1 \right)_+ + \left(\partial_\nu \psi_2 \right)_+ = \left(\partial_\nu \psi_3 \right)_+ + \left(\partial_\nu \psi_4 \right)_+. \quad (3.56)$$

Since x_1 and x_2 do not touch S', we see that

$$\left(\partial_\nu \psi_2 \right)_+ = \left(\partial_\nu \psi_2 \right)_-, \quad \left(\partial_\nu \psi_4 \right)_+ = \left(\partial_\nu \psi_4 \right)_-. \quad (3.57)$$

By Theorem 3.3, we have

$$\left(\partial_\nu \psi_3 \right)_+ - \left(\partial_\nu \psi_3 \right)_- = (2-n)|S^{n-1}| \partial_\nu f(x_0). \quad (3.58)$$

Subtracting (3.56) from (3.55) and using (3.57), we have

$$\left(\partial_\nu \psi_1 \right)_+ - \left(\partial_\nu \psi_1 \right)_- = \left(\partial_\nu \psi_3 \right)_+ - \left(\partial_\nu \psi_3 \right)_- - (n-2)|S^{n-1}| \partial_\nu f(x_0).$$

Using (3.58), we have $\left(\partial_\nu \psi_1 \right)_+ = \left(\partial_\nu \psi_1 \right)_-$, which proves the theorem. $\square$

We have now arrived at the final theorem of this section

Theorem 3.5. *For $f \in C^{1,\alpha}(S)$ with some $0 < \alpha \leq 1$, the boudary values $(\partial_{\nu(x)}D(x,f))_{\pm}$ exist, and the convergence is locally uniform on $\overline{D_{ext}}$ for the case $+$ and $\overline{D_{int}}$ for the case $-$. We have, furthermore,*

$$\left(\partial_{\nu(x)}D(x,f)\right)_{+} = \left(\partial_{\nu(x)}D(x,f)\right)_{-}, \quad \forall x \in S. \tag{3.59}$$

Proof. Let $x_0 \in S$, and for $x = x_0 + t\nu(x_0)$ with t small enough, we define $\nu(x) = \nu(x_0)$. Then, by a direct computation, we have

$$\partial_{\nu(x)}\frac{\cos\theta(x,y)}{|x-y|^{n-1}} = \frac{\langle\nu(x),\nu(y)\rangle}{|x-y|^n} - n\frac{\langle x-y,\nu(x)\rangle\langle x-y,\nu(y)\rangle}{|x-y|^{n+2}}. \tag{3.60}$$

Without loss of generality, we can assume that f is supported near x_0 and S is represented as $x_n = \varphi(x')$. Then, $\partial_{\nu(x)}D(x,f)$ is rewritten as

$$\partial_{\nu(x)}D(x,f) = \int g(y')\partial_{\nu(x)}\frac{\cos\theta(x,y)}{|x-y|^{n-1}}dy', \tag{3.61}$$

$$g(y') = f(y',\varphi(y'))\sqrt{1+\varphi'(y')^2}.$$

By the Hölder continuity of g', we have

$$g(y') - g(x_0') = \int_0^1 g'(sy' + (1-s)x_0')ds \cdot (y' - x_0')$$
$$= g'(x_0') \cdot (y' - x_0') + O(|y' - x_0'|^{1+\alpha}).$$

By Lemma 3.4 and (3.60), the integral

$$\int O(|y' - x_0'|^{1+\alpha})\partial_{\nu(x)}\frac{\cos\theta(x,y)}{|x-y|^{n-1}}dy'$$

is continuous across S. Then we have only to prove the theorem for the case where $g(y')$ in (3.61) is replaced by $g(x_0') + g'(x_0') \cdot (y' - x_0')$, which has been done in Lemma 3.9. $\qquad\square$

3.5 Dual space of a Banach space

Before entering into the mapping properties of single layer and double layer potentials on function spaces, we remark here a notion of dual operator or conjugate operator. Given a complex Banach space $\mathcal{X}$, the space of continuous linear functionals $\mathcal{X}^* = \mathbf{B}(\mathcal{X};\mathbf{C})$ is called the *dual space* of $\mathcal{X}$. The *dual* operator or *conjugate* operator of $K \in \mathbf{B}(\mathcal{X})$ is defined as the operator $K' \in \mathbf{B}(\mathcal{X}^*)$ satisfying

$$(K'x^*)(y) = x^*(Ky), \quad x^* \in \mathcal{X}^*, \quad y \in \mathcal{X}. \tag{3.62}$$

In the case of a Hilbert space, well-known Riesz' theorem says that for any $A \in \mathcal{H}^*$, there exists a unique $x_A \in \mathcal{H}$ such that

$$A(y) = (y, x_A), \quad \forall y \in \mathcal{H}, \quad \|A\| = \|x_A\|. \tag{3.63}$$

Therefore, it is more convenient to define the conjugate operator (or *adjoint*) operator in terms of the inner product:

$$(Ky, x) = (y, K^*x), \quad \forall y, x \in \mathcal{H}. \tag{3.64}$$

Note that the mapping $K \to K'$ is linear by the definition (3.62), i.e.

$$(\alpha K + \beta L)' = \alpha K' + \beta L', \quad \alpha, \beta \in \mathbf{C},$$

however, the mapping $K \to K^*$ is anti-linear by the definition (3.64), i.e.

$$(\alpha K + \beta L)^* = \overline{\alpha} K^* + \overline{\beta} L^* \quad \alpha, \beta \in \mathbf{C}.$$

In our applications below, we need to use both of the Banach space $\mathcal{X} = C(S)$ and the Hilbert space $\mathcal{H} = L^2(S)$. In this case, any $\varphi \in \mathcal{X}$ can be regarded as an element of $\mathcal{X}^*$ by the formula

$$\varphi(\psi) = (\psi, \overline{\varphi}), \quad \forall \psi \in \mathcal{X},$$

where the right-hand side is the inner product of $L^2(S)$. Then, we have

$$(K'\varphi)(\psi) = \varphi(K\psi) = (K\psi, \overline{\varphi}) = (\psi, \overline{K'\varphi}).$$

Under appropriate assumptions, if K is an integral operator, so are K' and K^*. Therefore, for an integral operator K on $C(S)$ with kernel $K(x, y)$, K' is also an integral operator with kernel $K(y, x)$, however the kernel of K^* on $L^2(S)$ is $\overline{K(y, x)}$.

Let us also recall some more notions in functional analysis. Let $\mathcal{X}$ and $\mathcal{Y}$ be Banach spaces. A bounded operator $K \in \mathbf{B}(\mathcal{X}; \mathcal{Y})$ is said to be *compact* if for any bounded sequence $\{f_n\}_{n=1}^{\infty}$, there exists a subsequence $\{f_{n(j)}\}_{j=1}^{\infty}$ such that $\{Kf_{n(j)}\}_{j=1}^{\infty}$ is convergent in $\mathcal{Y}$. For example, when $\mathcal{X} = \mathcal{Y} = L^2(\Omega)$, Ω being an open set in $\mathbf{R}^n$, an integral operator A with kernel $A(x, y)$ such that $A(x, y) \in L^2(\Omega \times \Omega)$ (called *Hilbert-Schmidt* class) is compact. It is also known that if $\{K_n\}$ is a sequence of compact operators from $\mathcal{X}$ to $\mathcal{Y}$ satisfying $\|K_n - K\| \to 0$ in the operator norm, K is also compact.

3.6 Mapping properties

Let $R_0(x)$ be the Green function of $-\Delta$ and put

$$K_0(x, y) = \partial_{\nu(y)} R_0(x - y). \tag{3.65}$$

Let K_0 be the integral operator on S with kernel $K_0(x, y)$. As has been seen in the previous section, for an integral operator K on S, its dual operator K' in the Banach space S and the adjoint operator K^* in the Hilbert space $L^2(S)$ are different. However, in our case, $K_0(x, y)$ is real-valued. Therefore, K_0' and K_0^* coincide. Thus, we put

$$K_0'(x, y) = K_0(y, x) = K_0^*(x, y) = \partial_{\nu(x)} R_0(x - y), \tag{3.66}$$

and let $K_0' = K_0^*$ be an integral operator with kernel $K_0^*(x, y)$. By this definition

$$(K_0 f)(x) = D(x, f), \quad (K_0' f)(x) = (K_0^* f)(x) = \partial_{\nu(x)} S(x, f). \tag{3.67}$$

Theorem 3.6. *On $L^2(S)$, K_0 and K_0^* are compact, moreover*

$$(K_0 f, g)_{L^2(S)} = (f, K_0^* g)_{L^2(S)}, \quad \forall f, g \in L^2(S). \tag{3.68}$$

Proof. The compactness of K_0 and K_0^* follows from the following Lemma 3.10. It also shows that K_0 and K_0^* are bounded on $L^2(S)$. Then (3.68) is proven by a direct computation. $\qquad\square$

Lemma 3.10. *Let U be a bounded open set in $\mathbf{R}^m$ and K an integral operator on U with kernel $K(x, y)$ satsifying $|K(x, y)| \leq C|x - y|^{-\gamma}$ with $\gamma < m$. Then K is a compact operator on $L^2(U)$.*

Proof. For $\epsilon > 0$, let $\chi_\epsilon(t) = 0$ for $t > \epsilon$ and $\chi_\epsilon(t) = 1$ for $t < \epsilon$. Let K_ϵ be the operator with kernal $(1 - \chi_\epsilon(|x - y|))K(x, y)$. Then K_ϵ is a Hilbert-Schmidt operator. Integrating the inequality

$$|Kf(x) - K_\epsilon f(x)|^2 \leq \int |\chi_\epsilon(x, y) K(x, y)| dy \int |\chi_\epsilon(x, y) K(x, y)| |f(y)|^2 dy,$$

we see that $K_\epsilon \to K$ in the operator norm. This proves the lemma. $\qquad\square$

As is seen above, the compactness of single and double layer potentials is rather easy to show in $L^2(S)$. However, from the view point of the boundary value problem, the compactness in $C(S)$ and $C^{0,\alpha}(S)$ is more important. To show it, we prepare the following inequality.

Lemma 3.11. *There exists a constant $C > 0$ such that*

$$|K_0(x, y) - K_0(z, y)| \leq C \frac{|x - z|}{|x - y|^{n-1}}, \tag{3.69}$$

$$|K_0^*(x,y) - K_0^*(z,y)| \leq C \frac{|x - z|}{|x - y|^{n-1}}, \tag{3.70}$$

if $x, y, z \in S$ and

$$2\sqrt{1 + C_0^2} \, |x - z| \leq |x - y|, \tag{3.71}$$

where C_0 is defined by (3.18).

Proof.   We only have to consider the case in which x, y, z lie in one of S_j's in (3.15). Assuming that S is represented as $x_n = \varphi(x')$, we put

$$a(x', y') = \langle x - y, \nu(y) \rangle, \quad L(x', y') = |x - y|^2,$$

where $\nu(y) = (-\varphi'(y'), 1)/\sqrt{1 + \varphi'(y')^2}$. Then

$$K_0(x,y) = \frac{a(x', y')}{L(x', y')^{n/2}} =: \widetilde{K}_0(x', y').$$

Since $a(x', y') = (\varphi(x') - \varphi(y') - \varphi'(y') \cdot (x' - y'))(1 + \varphi'(y')^2)^{-1/2}$, we have by the Taylor expansion

$$a(x', y') = O(|x' - y'|^2).$$

This implies for $0 \leq t \leq 1$

$$\left| \frac{d}{dt} a(tz' + (1-t)x', y') \right| \leq C|tz' + (1-t)x' - y'||z' - x'|.$$

This and the inequality

$$\left| \frac{d}{dt} L(tz' + (1-t)x', y') \right| \leq CL(tz' + (1-t)x', y')^{1/2}|z' - x'|$$

yields

$$\left| \frac{d}{dt} \widetilde{K}_0(tz' + (1-t)x', y') \right| \leq C \frac{|x' - z'|}{|tz' + (1-t)x' - y'|^{n-1}},$$

where we have also used $L(x', y')^{1/2} \geq |x' - y'|$. In view of (3.19) and (3.71), we have

$$|x' - y'| \geq 2|x' - z'|,$$

which implies for $0 \leq t \leq 1$

$$|x' - y' + t(z' - x')| \geq |x' - y'| - |z' - x'| \geq \frac{1}{2}|x' - y'|.$$

Therefore

$$\left| \frac{d}{dt} \widetilde{K}_0(tz' + (1-t)x', y') \right| \leq C \frac{|x' - z'|}{|x' - y'|^{n-1}}.$$

Integrating it with respect to t, we obtain

$$|K_0(x,y) - K_0(z,y)| \leq C \frac{|x' - z'|}{|x' - y'|^{n-1}}.$$

Again using (3.19), we have proven (3.69). Noting that

$$K_0^*(x,y) = \frac{\langle x - y, \nu(y) \rangle}{|x - y|^n} - \frac{\langle x - y, \nu(y) - \nu(x) \rangle}{|x - y|^n}$$
$$= K_0(x,y) + O(|x - y|^{2-n}),$$

we obtain (3.71) by a similar computation. $\qquad\square$

Theorem 3.7. *For any $0 < \alpha < 1$,*
$$K_0, \ K_0' \in \mathbf{B}(C(S); C^{0,\alpha}(S)).$$

Proof. We prove the theorem for K_0. The proof for K_0^* is the same as for K_0. Let $C = 2(1 + C_0^2)^{1/2}$, where C_0 is defined by (3.18), and put

$$S^{(1)}(x,z) = S \cap \{y \in S \,;\, |x - y| \geq C|x - z|\},$$
$$S^{(2)}(x,z) = S \cap \{y \in S \,;\, |x - y| \leq C|x - z|\}.$$

We split $K_0 f(x) - K_0 f(z) = A^{(1)} + A^{(2)}$, where

$$A^{(i)}(x,z) = \int_{S^{(i)}(x,z)} \big(K(x,y) - K(z,y)\big) f(y) dS_y.$$

On $S^{(1)}(x,z)$, we have

$$\frac{|x - z|}{|x - y|^{n-1}} \leq \frac{|x - z|^\alpha}{|x - y|^{n-2+\alpha}} \left(\frac{|x - z|}{|x - y|}\right)^{1-\alpha} \leq C^{\alpha-1} \frac{|x - z|^\alpha}{|x - y|^{n-2+\alpha}},$$

hence by using (3.69)

$$|A^{(1)}(x,z)| \leq C \int_{S^{(1)}(x,z)} \frac{|x - z|^\alpha}{|x - y|^{n-2+\alpha}} |f(y)| dS_y \tag{3.72}$$
$$\leq C|x - z|^\alpha \|f\|_{C(S)}.$$

We split $S^{(2)}(x,z)$ into 2 parts, $S^{(2)}(x,z) = S^{(2,1)}(x,z) \cup S^{(2,2)}(x,z)$, where

$$S^{(2,1)}(x,z) = \{y \in S \,;\, |x - y| \leq C|x - z|, \ |z - y| \geq C|x - z|\},$$
$$S^{(2,2)}(x,z) = \{y \in S \,;\, |x - y| \leq C|x - z|, \ |z - y| \leq C|x - z|\}.$$

On $S^{(2,1)}(x,z)$, we have $|z - y| \leq |z - x| + |x - y| \leq (1 + C)|x - z|$, hence

$$\frac{1}{|x - y|^{n-2}} + \frac{1}{|z - y|^{n-2}} \leq \frac{C|x - z|^\alpha}{|x - y|^{n-2+\alpha}} + \frac{C|x - z|^\alpha}{|z - y|^{n-2+\alpha}}.$$

This holds also on $S^{(2,2)}(x,z)$. We have thus derived

$$|A^{(2)}(x,z)| \leq C \int_{S^{(2)}(x,z)} \left(\frac{1}{|x - y|^{n-2}} + \frac{1}{|z - y|^{n-2}}\right) |f(y)| dS_y \tag{3.73}$$
$$\leq C|x - z|^\alpha \|f\|_{C(S)}.$$

The inequalities (3.72) and (3.73) prove the theorem. $\qquad\square$

The double layer potential maps Hölder continuous functions to Hölder continuously differentiable functions. Furthermore, it maps Hölder continuously differentiable functions on S to Hölder continuously differentiable functions on $\mathbf{R}^n \setminus S$. Namely, letting D_{int} be a bounded domain whose boundary is S and, $D_{ext} = \mathbf{R}^n \setminus \overline{D}_{int}$, we have the following theorem.

Theorem 3.8. *For any $0 < \alpha < 1$, K_0 and K_0' satisfy*

$$K_0, K_0' \in \mathbf{B}(C^{0,\alpha}(S); C^{1,\alpha}(S)),$$
$$K_0, K_0' \in \mathbf{B}(C^{1,\alpha}(S); C^{1,\alpha}(D_{int})),$$
$$K_0, K_0' \in \mathbf{B}(C^{1,\alpha}(S); C^{1,\alpha}(D_{ext})).$$

We omit the proof of Theorems 3.8. See, Colton-Kress [22], pp. 56, 57. As for the single layer potential, we have the following theorem. Let

$$(L_0 f)(x) = \int_S R_0(x - y) f(y) dS_y.$$

Theorem 3.9. *For any $0 < \alpha < 1$, L_0 satisfies*

$$L_0 \in \mathbf{B}(C^{0,\alpha}(S); C^{1,\alpha}(S)),$$
$$L_0 \in \mathbf{B}(C^{0,\alpha}(S); C^{1,\alpha}(D_{int})),$$
$$L_0 \in \mathbf{B}(C^{0,\alpha}(S); C^{1,\alpha}(D_{ext})).$$

See [22], p. 51.

Let us recall the well-known Ascoli-Arzelà theorem for compactness of a set of continuous functions.

Theorem 3.10. *Let D be a compact set in $\mathbf{R}^m$. Suppose that a sequence $\{f_n\}_{n=1}^\infty \subset C(D)$ is uniformly bounded, i.e.*

$$\sup_{n \geq 1} \|f_n\|_{C(D)} < \infty,$$

and equi-continuous, i.e. for any $\epsilon > 0$, there is a constant $\delta > 0$ such that

$$|f_n(x) - f_n(y)| < \epsilon, \quad \forall n \geq 1, \quad \text{if} \quad |x - y| < \epsilon, \quad x, y \in D.$$

Then, there exists a subsequence $\{f_{n(j)}\}_{j=1}^\infty$ convergent in $C(D)$, i.e.

$$\lim_{i,j \to \infty} \|f_{n(i)} - f_{n(j)}\|_{C(D)} \to 0.$$

For the proof, see e.g. [102], p. 85.

Theorem 3.11. *K_0 and K_0' are compact operators on $C(S)$.*

Proof. Let $\{f_n\}_{n=1}^\infty$ be a bounded sequence in $C(S)$. Since $K_0 \in$ $\mathbf{B}(C(S))$ by Theorem 3.7, $\{K_0 f_n\}_{n=1}^\infty$ is also bounded in $C(S)$. By the Hölder continuity, letting $u(x) = K_0 f(x)$, we have

$$|u(x) - u(y)| \le C\|f\|_{C^{0,\alpha}(S)} |x - y|^\alpha, \quad 0 < \alpha < 1,$$

which shows that $\{K_0 f_n\}_{n=1}^\infty$ is equi-continuous. Using Theorem 3.10, we obtain the theorem. $\qquad\square$

Theorem 3.12. *For $0 < \alpha < 1$, K_0 and K_0' are compact operators on $C^{0,\alpha}(S)$.*

Proof. We prove the compactness of K_0. Let $\{f_n\}_{n=1}^\infty$ be a bounded sequence in $C^{0,\alpha}(S)$. Since it is also bounded in $C(S)$, by Theorem 3.11, one can select a subsequence $\{f_{n'}\}$ such that $\{K_0 f_{n'}\}$ is convergent in $C(S)$. We denote $\{f_{n'}\}$ by $\{f_n\}$ again, and put $u_{n,m} = K_0 f_n - K_0 f_m$. Take $\alpha < \beta < 1$. By Theorem 3.7, we have $C = \sup_{n,m} \|u_{n,m}\|_{C^{0,\beta}(S)} < \infty$. Take a sequence $\epsilon_1 > \epsilon_2 > \cdots 0$, and let $A = \sup_{x,y \in S} |x - y|^{\beta - \alpha}$. We then have

$$\frac{|u_{n,m}(x) - u_{n,m}(y)|}{|x - y|^\alpha} = \frac{|u_{n,m}(x) - u_{n,m}(y)|}{|x - y|^\beta} |x - y|^{\beta - \alpha}. \qquad (3.74)$$

We split it into two parts: one for which $|x - y|^{\beta - \alpha} < \epsilon_j$, the other $|x - y|^{\beta - \alpha} > \epsilon_j$. Then, (3.74) is estimated from above by

$$C\epsilon_j + \|u_{n,m}\|_{C(S)} \epsilon_j^{\beta/(\alpha - \beta)}.$$

Take N_j so that $\|u_{n,m}\|_{C(S)} \epsilon_j^{\beta/(\alpha - \beta)} \le C\epsilon_j$ if $n, m > N_j$. We then have

$$\frac{|u_{n,m}(x) - u_{n,m}(y)|}{|x - y|^\alpha} < 2C\epsilon_j, \quad \text{if} \quad n, m > N_j.$$

This shows that $\{K_0 u_n\}$ is converent in $C^{0,\alpha}(S)$. $\qquad\square$

3.7 Fredholm's alternative theorem

3.7.1 *Theory of Riesz-Schauder*

In this section, we recall the well-known theory of Riesz [110] and Schauder [111] on the spectral theory of compact operators on a Banach space $\mathcal{X}$. For a bounded operator T, let

$$\mathcal{R}(T) = \{Tu \,;\, u \in \mathcal{X}\}$$

be the range of T, and
$$\mathcal{N}(T) = \{u \,;\, Tu = 0\}$$
be the nullspace of T. The essential feature of a compact operator is the following closed range property.

Lemma 3.12. *Let K be a bounded operator on a Banach space $\mathcal{X}$. Then:*
(1) K is compact if and only if K' is compact.
(2) If K is compact, $\mathcal{R}(1-K)$ is closed.
(3) If K is compact, $\mathcal{N}(1-K)$ has a finite dimension, and
$$\dim \mathcal{N}(1-K) = \dim \mathcal{N}(1-K').$$

For $K \in \mathbf{B}(\mathcal{X})$, the *resolvent set* of K, denoted by $\rho(K)$, is the set of all complex numberz $z \in \mathbf{C}$ such that $K - z$ is 1 to 1, i.e. $\mathcal{N}(K-z) = \{0\}$, onto, i.e. $\mathcal{R}(K-z) = \mathcal{X}$, and has a bounded inverse
$$R(z) = (K-z)^{-1}$$
called the *resolvent* of K. Namely, for any $f \in \mathcal{X}$, the equation $(K-z)u = f$ is uniquely solvable, and $u = (K-z)^{-1}f$ satisfies $\|u\| \leq C\|f\|$, where C is a constant independent of f (but depends on z). It is known that $\rho(K)$ is an open set, and $R(K,z)$ is a $\mathbf{B}(\mathcal{X})$-valued analytic function of $z \in \rho(K)$. The set $\sigma(K) = \mathbf{C} \setminus \rho(K)$ is called the *spectrum* of K. It satisfies
$$\sigma(K) \subset \{z \in \mathbf{C} \,;\, |z| \leq \|K\|\}.$$
An important fact is that on the Banach space, for instance $C(S)$,
$$\sigma(K') = \sigma(K), \quad (R(K,z))' = R(K',z), \tag{3.75}$$
while on the Hilbert space, e.g. $L^2(S)$,
$$\sigma(K^*) = \overline{\sigma(K)}, \quad R(K,z)^* = R(K^*,\overline{z}), \tag{3.76}$$
where $\overline{z}$ denotes the complex conjugate of $z \in \mathbf{C}$.

A complex number $\lambda \in \mathbf{C}$ is said to be an *eigenvalue* of K, if there exists $0 \neq u \in \mathcal{X}$ such that $Ku = \lambda u$. The dimension of $\mathcal{N}(K-\lambda)$ is said to be the *multiplicity* of λ. The set of all eigenvalues of K is called the *point spectrum* of K, and is denoted by $\sigma_p(K)$.

Lemma 3.12 entails the following properties of the spectrum of a compact operator K on a Banach space.

Lemma 3.13. *For a compact operator K, we have:*
(1) $\sigma(K) \setminus \{0\} \subset \sigma_p(K)$.
(2) $\sigma_p(K) \setminus \{0\}$ consists of an at most countable number of eigenvalues of finite multiplicities, with a possible accumulation point only at 0.
(3) $\sigma_p(K) \setminus \{0\} = \sigma_p(K') \setminus \{0\}$ with the same multiplicities.

For subsets $\mathcal{A} \subset \mathcal{X}$ and $\mathcal{B} \subset \mathcal{X}^*$, we put

$$\mathcal{A}^\perp = \{x^* \in \mathcal{X}^* \,;\, x^*(x) = 0,\ \forall x \in \mathcal{A}\}, \tag{3.77}$$

$$^\perp\mathcal{B} = \{x \in \mathcal{X} \,;\, x^*(x) = 0,\ \forall x^* \in \mathcal{B}\}. \tag{3.78}$$

Given a parameter $\lambda \in \mathbf{C}$, we consider a pair of linear equations

$$(\lambda - K)u = f, \tag{3.79}$$

$$(\lambda - K')v^* = g^*. \tag{3.80}$$

The following theorem, first established by Fredholm for integral equations [105], shows a complete analogue of linear equations in a Banach space to the finite dimensional ones.

Theorem 3.13. *Let K be a compact operator on a Banach space $\mathcal{X}$, and $\lambda \in \mathbf{C} \setminus \{0\}$ a constant.*
(1) If $\lambda \in \rho(K)$, for any $f \in \mathcal{X}$ and $g^* \in \mathcal{X}^*$, (3.79) and (3.80) have a unique solution $u \in \mathcal{X}$ and $v^* \in \mathcal{X}^*$.
(2) If $\lambda \in \sigma(K)$,

$$\dim \mathcal{N}(\lambda - K) = \dim \mathcal{N}(\lambda - K').$$

Moreover, (3.79) has a solution if and only if $f \in {}^\perp\mathcal{N}(\lambda - K')$, and (3.80) has a solution if and only if $g^* \in \mathcal{N}(\lambda - K)^\perp$. Let u_0 be a solution of (3.79). Then

$$\{u \in \mathcal{X} \,;\, (\lambda - K)u = f\} = \{u_0 + w \,;\, w \in \mathcal{N}(\lambda - K)\}.$$

Let v_0^* be a solution of (3.80). Then

$$\{v^* \in \mathcal{X}^* \,;\, (\lambda - K')v^* = g^*\} = \{v_0^* + h^* \,;\, h^* \in \mathcal{N}(\lambda - K')\}.$$

For the proof of Lemmas 3.12, 3.13 and Theorem 3.13, see e.g. [102], p. 283.

3.7.2 Dual system

In general, it is not easy to determine the dual space $\mathcal{X}^*$ of a Banach space $\mathcal{X}$, which makes the adjoint problem (3.80) hard to study. However, one more assumption allows us to deal with the equation (3.79) only in the Banach space $\mathcal{X}$. We follow the argument in [22], which is due to [97].

Let $\langle\,,\,\rangle : \mathcal{X} \times \mathcal{X} \to \mathbf{C}$ be a bounded symmetric bilinear form on $\mathcal{X}$, i.e.

$$\langle u, v \rangle = \langle v, u \rangle$$
$$\langle \alpha u + \beta v, w \rangle = \alpha \langle u, w \rangle + \beta \langle u, w \rangle, \tag{3.81}$$
$$|\langle u, v \rangle| \le C \|u\| \|v\|,$$

hold $\forall \alpha, \beta \in \mathbf{C}$, $\forall u, v \in \mathcal{X}$, where C is a constant independent of u, v. Moreover, it is assumed to be *non-degenerate*, i.e.

$$\langle u, v \rangle = 0, \quad \forall u \in \mathcal{X} \Longrightarrow v = 0. \tag{3.82}$$

Then, there exists a linear operator $J : \mathcal{X} \ni v \to v^* \in \mathcal{X}^*$ defined by

$$v^*(u) = \langle u, v \rangle.$$

Since J is injective, i.e. $v^* = w^* \Longrightarrow v = w$, $\mathcal{X}$ is identified with a subspace of $\mathcal{X}^*$. Now, suppose we are given compact operators K, L on $\mathcal{X}$ satisfying

$$\langle Ku, v \rangle = \langle u, Lv \rangle, \quad \forall u, v \in \mathcal{X}. \tag{3.83}$$

Equivalently, this means $(Lv)^* = K'v^*$. Therefore, K' restricted on $J(\mathcal{X})$ is a compact operator on $J(\mathcal{X})$.

Lemma 3.14. *Let K, L be as above. Then we have*

$$\dim \mathcal{N}(1 - K) = \dim \mathcal{N}(1 - L).$$

Proof.   Since $L = K'|_{J(\mathcal{X})}$, we have

$$\dim \mathcal{N}(1 - L) \leq \dim \mathcal{N}(1 - K') = \dim \mathcal{N}(1 - K).$$

Similary, we have

$$\dim \mathcal{N}(1 - K) \leq \dim \mathcal{N}(1 - L') = \dim \mathcal{N}(1 - L).$$

These two inequalities prove the lemma.   □

By virtue of Lemma 3.14 and $(Lv)^* = K'v^*$, we have $\mathcal{N}(1 - K') = \{v^* \in \mathcal{X}^* \,;\, (1 - L)v = 0\}$, hence $^\perp \mathcal{N}(1 - K') = \{u \in \mathcal{X} \,;\, \langle u, v \rangle = 0, \; \forall v \in \mathcal{N}(1 - L)\}$. We can then rewrite Theorem 3.13 as follows.

Theorem 3.14. *Let K, L be compact operators on a Banach space $\mathcal{X}$ satisfying (3.83), and $\lambda \in \mathbf{C} \setminus \{0\}$ a constant.*
(1) If $\lambda \in \rho(K)$, for any $f \in \mathcal{X}$, (3.79) has a unique solution $u \in \mathcal{X}$.
(2) If $\lambda \in \sigma(K)$, (3.79) has a solution if and only if $\langle f, v \rangle = 0$, $\forall v \in \mathcal{N}(\lambda - L)$. Let u_0 be a solution of (3.79). Then

$$\{u \in \mathcal{X} \,;\, (\lambda - K)u = f\} = \{u_0 + w \,;\, w \in \mathcal{N}(\lambda - K)\}.$$

Let us finally note that $\langle u, v \rangle$ need not be symmetric. We have only to assume that it is bilinear, bounded and non-degenerate with respect to both of the variables u, v.

3.8 Boundary value problems for the Laplace equation

3.8.1 *Boundary components*

Let D be a domain (a connected open set) in $\mathbf{R}^n$ with C^∞-boundary $S = \partial D$ having several components, $S = \cup_{i=1}^N S_i$. The problem we address is to seek a solution u of the Laplace equation

$$\Delta u = 0, \quad \text{in} \quad D, \tag{3.84}$$

satisfying the *Dirichlet condition* on some boundary components

$$u = f_i, \quad \text{on} \quad S_i, \quad 1 \le i \le m, \tag{3.85}$$

and the *Neumann condition* on the other boundary components

$$\frac{\partial u}{\partial \nu} = f_i, \quad \text{on} \quad S_i, \quad m+1 \le i \le N. \tag{3.86}$$

There are two cases to consider, interior domains and exterior domains.

The *interior domain* is a bounded domain $D_{int} \subset \mathbf{R}^n$ written as

$$D_{int} = D^{(0)} \setminus \bigcup_{i=1}^N D^{(i)}, \tag{3.87}$$

where $D^{(0)}, D^{(i)}$ are bounded domains with C^∞-boundaries

$$\partial D^{(0)} = S_{out}, \quad \partial D^{(i)} = S_{inn}^{(i)}, \quad 1 \le i \le N. \tag{3.88}$$

We assume that each S_{out} and $S_{inn}^{(i)}$ has only one connected component, and for $1 \le i, j \le N$,

$$\overline{D^{(i)}} \subset D^{(0)}, \quad \overline{D^{(i)}} \cap \overline{D^{(j)}} = \emptyset, \quad i \ne j. \tag{3.89}$$

We call S_{out} the outer boundary, and $S_{inn}^{(i)}$ the inner boundary. They are mutually disjoint.

For example, for $D_{int} = \{1 < |x| < 2\}$, we take $D^{(0)} = \{|x| < 2\}$, $D^{(1)} = \{|x| < 1\}$, $S_{out} = \{|x| = 2\}$, $S_{inn}^{(1)} = \{|x| = 1\}$.

The *exterior domain* D_{ext} is defined as above with $D^{(0)}$ replaced by $\mathbf{R}^n$:

$$D_{ext} = \mathbf{R}^n \setminus \bigcup_{i=1}^N D^{(i)}. \tag{3.90}$$

Therefore, it is an unbounded domain having only inner boundaries

$$\partial D_{ext} = S = \bigcup_{i=1}^N S_{inn}^{(i)}.$$

There are two choices for the direction of the unit normal to the boundary. To avoid the confusion, for both cases of $D = D_{int}$ and $D = D_{ext}$, we choose the direction of the unit normal ν to ∂D as follows:

The unit normal ν to ∂D is directed to the exterior of $D^{(j)}$ (i.e. toward infinty) for all $0 \le j \le N$.

For example, for $D_{int} = \{1 < |x| < 2\}$, we take $\nu = x/|x|$ on $\partial D_{int} = \{|x| = 1\} \cup \{|x| = 2\}$, and for $D_{ext} = \{|x| > 3\}$, $\nu = x/|x|$ on $\partial D_{ext} = \{|x| = 3\}$.

3.8.2   *Uniqueness*

By a solution to the Laplace equation (3.84), we mean $u \in C^2(D) \cap C^1(\overline{D})$ satisfying (3.84). The following lemma is proved by applying Theorem 1.5 to the equation

$$0 = \int_{D_{int}} (\Delta u)\overline{u}\, dx = \int_{D_{int}} \Delta u\, dx.$$

Lemma 3.15. *Let u be a solution to the Laplace equation (3.84) on an interior domain D_{int}. Then we have*

$$\int_{D_{int}} |\nabla u|^2\, dx = \int_{S_{out}} \frac{\partial u}{\partial \nu}\overline{u}\, dS - \sum_{i=1}^{N} \int_{S_{inn}^{(i)}} \frac{\partial u}{\partial \nu}\overline{u}\, dS, \qquad (3.91)$$

$$\int_{S_{out}} \frac{\partial u}{\partial \nu}\, dS - \sum_{i=1}^{N} \int_{S_{inn}^{(i)}} \frac{\partial u}{\partial \nu}\, dS = 0. \qquad (3.92)$$

We impose either Dirichlet or Neumann condition on each boundary component, i.e. we assume

$$u = f_i, \quad \text{or} \quad \frac{\partial u}{\partial \nu} = f_i, \quad \text{on} \quad \partial D^{(i)},$$

where $f_i \in C(\partial D^{(i)})$ is the boundary data.

Theorem 3.15. *(1) The solution of the interior problem is unique, if the Dirichlet condition is imposed on at least one component of the boundary. (2) If the Neumann condition is imposed on all components of the boundary, for the solution to exist, the boundary data must satify*

$$\int_{S_{out}} f_0\, dS - \sum_{i=1}^{N} \int_{S_{inn}^{(i)}} f_i\, dS = 0. \qquad (3.93)$$

Furthermore, the solution is unique up to an additive constant.

Proof.   If the boundary data is 0, the right-hand side of (3.91) vanishes, hence u is a constant. This, together with (3.92), proves the lemma.   □

The solution of the exterior problem is unique under the assumption that $u(x) \to 0$ as $|x| \to \infty$. To prove it, we prepare the following lemma.

Lemma 3.16. *Suppose $u(x)$ satisfies $\Delta u(x) = 0$ on $\{|x| > R\}$ and $u(x) \to 0$ as $|x| \to \infty$. Then, we have,*

$$u(x) = O(|x|^{m(n)}), \quad \nabla u(x) = O(|x|^{m(n)-1}), \quad \text{as} \quad |x| \to \infty, \qquad (3.94)$$

where $m(n) = 2 - n$ for $n \geq 3$ and $m(n) = -1$ for $n = 2$. Moreover, the limit

$$\lim_{|x| \to \infty} |x|^{-m(n)} u(x) = a_0(\omega), \quad \omega = x/|x| \tag{3.95}$$

exists for any $\omega \in S^{n-1}$.

Proof. Take $\psi(t) \in C^\infty(\mathbf{R})$ such that $\psi(t) = 0$ for $t \leq R+1$, $\psi(t) = 1$ for $t > R+2$, and put $v(x) = \psi(|x|)u(x)$. We put $w(x) = \int_{\mathbf{R}^n} R_0(x-y)g(y)dy$, where $R_0(x)$ is the Green function of $-\Delta$ and $g = -\Delta v$. We then have $\Delta(v - w) = 0$ and $v - w \to 0$ as $|x| \to \infty$. By virtue of Theorem 2.5, we have $v = w$ on $\mathbf{R}^n$, in particular,

$$u(x) = \int_{\mathbf{R}^n} R_0(x - y)g(y)dy, \quad |x| > R + 2.$$

For $n \geq 3$, $R_0(x) = C_n |x|^{2-n}$, hence (3.94) and (3.95) follow from a direct computation. For $n = 2$, $R_0(x) = C_2 \log |x|$. Using

$$\log |x - y| = \log |x| - \frac{\omega \cdot y}{|x|} + O(|x|^{-2}), \quad |x| \to \infty,$$

we have

$$u(x) \sim \log |x| \int_{\mathbf{R}^2} g(y)dy - \frac{1}{|x|} \int_{\mathbf{R}^2} \omega \cdot y g(y)dy.$$

Since $u(x) \to 0$ as $|x| \to \infty$, we have $\int_{\mathbf{R}^2} g(y)dy = 0$, and proven (3.94) and (3.95). $\qquad\square$

Theorem 3.16. *The solution of the exteior problem satisfying Dirichlet or Neumann boundary conditions on $S_{inn}^{(i)}$, $i = 1, \cdots, N$, and $u(x) \to 0$ as $|x| \to \infty$ is unique.*

Proof. Let $D_R = D_{ext} \cap \{|x| < R\}$. By integration by parts, we have for the solution satisfying 0 boundary conditions

$$0 = \int_{D_R} (\Delta u)\overline{u}\, dx = \int_{|x|=R} \frac{\partial u}{\partial r}\overline{u}\, dS - \int_{D_R} |\nabla u|^2 dx.$$

Letting $R \to \infty$ and using (3.94) and (3.95), we have $\int_D |\nabla u|^2 dx = 0$. Then, u is a constant. Since $u(x) \to 0$ as $|x| \to \infty$, $u = 0$. $\qquad\square$

3.8.3 *Boundary integral equations on a single boundary*

We fix $0 < \alpha < 1$ arbitrarily, and assume that $f \in C^{0,\alpha}(S)$. For $\psi, \varphi \in C^{0,\alpha}(S)$, we consider the single and double layer potentials

$$V(x) = S(x, \psi) = \int_S \psi(y) R_0(x - y) dS_y, \qquad (3.96)$$

$$W(x) = D(x, \varphi) = \int_S \varphi(y) \partial_{\nu(y)} R_0(x - y) dS_y. \qquad (3.97)$$

Then, letting $D = D_{int}$ or D_{ext}, by Theorem 3.8, we have

$$V, \ W \in C^{1,\alpha}(\overline{D}) \cap C^2(D), \quad \text{and} \quad \Delta V = \Delta W = 0, \quad \text{in} \quad D. \qquad (3.98)$$

The solvabilities of Dirichlet and Neumann problems are mutually related. Let us consider a simple case in which the boundary of D has only one connected component S. By virtue of our choice of ν, Theorems 3.2, 3.3 and (3.67), we have for $x \in S$,

$$\begin{cases} (W(x))_\pm = \lim_{\epsilon \to 0} W(x \pm \epsilon \nu(x)) = (K_0 \varphi)(x) \pm \dfrac{1}{2} \varphi(x), \\[2mm] (\partial_{\nu(x)} V(x))_\pm = \lim_{\epsilon \to 0} \partial_\nu V(x \pm \epsilon \nu(x)) = (K_0' \psi)(x) \mp \dfrac{1}{2} \psi(x). \end{cases} \qquad (3.99)$$

Then, $W(x)$ solves the interior or exterior Dirichlet problem, if the double layer potential satisfies

$$(K_0 \mp 1/2)\varphi = f, \quad \text{on} \quad S, \qquad (3.100)$$

(we take $-$ for the interior problem and $+$ for the exterior problem), and $V(x)$ solves the interior or exterior Neumann problem, if the single layer potential satisfies

$$(K_0' \pm 1/2)\psi = f, \quad \text{on} \quad S, \qquad (3.101)$$

(we take $+$ for the interior problem and $-$ for the exterior problem). By Theorem 3.12, K_0, K_0' are compact on $C^{0,\alpha}(S)$. By the Fredholm theory (Theorem 3.14), the equation (3.100) is uniquely solvable if and only if $\pm 1/2 \notin \sigma_p(K_0)$, which is equivalent to $\pm 1/2 \notin \sigma_p(K_0')$. Therefore, the solvability of the *boundary integral equation* for the interior (exterior) Dirichlet problem is equivalent to that of the exterior (interior) Neumann problem:

Interior Dirichlet problem $\Longleftrightarrow$ Exterior Neumann problem,

Exterior Dirichlet problem $\Longleftrightarrow$ Interior Neumann problem.

Lemma 3.17. *Let D_{ext} and D_{int} be the domains outside and inside of S, respectively. Then:*

(1) $1/2 \notin \sigma_p(K_0) = \sigma_p(K_0')$.

(2) $-1/2 \in \sigma_p(K_0) = \sigma_p(K_0')$.

(3) $\mathcal{N}(K_0 + 1/2) = \mathbf{C}$.

(4) $\mathcal{N}(K_0' + 1/2) = \{c\psi_{ext}; c \in \mathbf{C}\}$, where $\psi_{ext} = (\partial_\nu v_{ext})_+$ and v_{ext} is a unique solution to the following exterior Dirichlet problem

$$\begin{cases} \Delta v = 0, & in \quad D_{ext}, \\ v = 1, & on \quad S, \\ v(x) \to 0, & as \quad |x| \to \infty. \end{cases} \tag{3.102}$$

Proof. Suppose there exists $\psi \in C^{0,\alpha}(S)$ such that $(K_0' - 1/2)\psi = 0$. Put $v(x) = S(x, \psi)$. Then, $\Delta v = 0$ in D_{ext}, $\partial_\nu v = 0$ on S by (3.99), and $v(x) \to 0$ as $|x| \to \infty$. Theorem 3.16 then implies that $v(x) = 0$ in D_{ext}. Since $v(x) \in C(\mathbf{R}^n)$ by Lemma 3.1, we have $v(x) = 0$ on S, moreover $\Delta v = 0$ in D_{int}. By Theorem 3.15, $v(x) = 0$ in D_{int}. Theorem 3.3 then implies $\psi = 0$. This proves (1).

Let $\varphi = 1$. Applying Lemma 3.7 to the domain D_{int}, we have $K_0\varphi = -1/2 = -\varphi/2$. This proves (2).

Suppose $(K_0 + 1/2)\varphi = 0$, and put $w(x) = D(x, \varphi)$, which satisfies $\Delta w = 0$ in $\mathbf{R}^n \setminus S$. By (3.99), $(w(x))_+ = 0$ on S, hence $w(x) = 0$ in D_{ext}, where we used Theorem 3.16. By Theorem 3.5, we then have $(\partial_\nu w(x))_- = (\partial_\nu w(x))_+ = 0$ on S. Here, we note that $\varphi \in C^{1,\alpha}(S)$ by virtue of Theorem 3.8. Then, $w(x)$ is a constant in D_{int} by Theorem 3.15. Finally, we use (3.47) to see that φ is a constant, which proves (3).

Suppose $(K_0' + 1/2)\psi = 0$. Putting $v = S(x, \psi)$, we have $\Delta v = 0$ in $\mathbf{R}^n \setminus S$ and $(\partial_\nu v)_- = 0$, $(\partial_\nu v)_+ = -\psi$ on S. Then $v = C$ in D_{int}. Since $v \in C(\mathbf{R}^n)$, v is a unique solution the equation $\Delta v = 0$ in D_{ext}, $v = C$ on S and $v \to 0$ as $|x| \to \infty$. This proves $\mathcal{N}(K_0' + 1/2) \subset \{cv_{ext}; c \in \mathbf{C}\}$. Since $\dim \mathcal{N}(K_0' + 1/2) \geq 1$, we have $\mathcal{N}(K_0' + 1/2) = \{cv_{ext}; c \in \mathbf{C}\}$. This proves (4). $\square$

3.8.4 *Boundary integral equations on multiple boundaries*

Given an interior, or exterior domain D, we impose Dirichlet or Neumann conditions on its inner boundaries:

$$\begin{cases} u = f_i, & on \quad S_{inn}^{(i)}, \quad 1 \leq i \leq m, \\ \dfrac{\partial u}{\partial \nu} = f_i, & on \quad S_{inn}^{(i)}, \quad m+1 \leq i \leq N, \end{cases} \tag{3.103}$$

and also on the outer boundary for the case of interior domain:

$$u = f_0, \quad \text{or} \quad \frac{\partial u}{\partial \nu} = f_0, \quad \text{on} \quad S_{out}, \tag{3.104}$$

or the condition at infinity for the case of exterior domain:

$$u(x) \to 0, \quad |x| \to \infty. \tag{3.105}$$

We put

$$S_0 = S_{out}, \quad S_i = S_{inn}^{(i)}, \quad i = 1, 2, \cdots, N.$$

Let $\nu_j(x)$ be the unit normal of S_j at x, $0 \leq j \leq N$, and

$$\begin{aligned} L(x,y) &= R_0(x-y), \quad L_j(x,y) = \partial_{\nu_j(y)} L(x,y), \\ {}_k L(x,y) &= \partial_{\nu_k(x)} L(x,y), \quad {}_k L_j(x,y) = \partial_{\nu_k(x)} \partial_{\nu_j(y)} L(x,y). \end{aligned} \tag{3.106}$$

Let $M(x,y)$ be any of the above functions. We define $M_{|j}$ to be the integral operator on S_j with kernel $M(x,y)$, i.e.

$$\left(M_{|j}\varphi \right)(x) = \int_{S_j} M(x,y)\varphi(y) dS_y, \quad x \in \mathbf{R}^n.$$

If, furthermore, we restrict x to S_k, we denote it by ${}_{k|}M_{|j}$, i.e.

$$\left({}_{k|}M_{|j}\varphi \right)(x) = \int_{S_j} M(x,y)\varphi(y) dS_y, \quad x \in S_k.$$

Let us give a remark about the notation ${}_{k|k}L_{j|j}$. We often extend the unit normal field $\nu_k(x)$ smoothly near S_k. Therefore, ${}_k L_j(x,y)$ is a smooth function near $S_k \times S_j$. We write ${}_{k|k}L_{j|j} = {}_{k|}({}_k L_j)_{|j}$ to make sure that we restrict x, y to S_k, S_j. Similar remark applies to ${}_{k|k}L_j = {}_{k|}({}_k L_j)$, and ${}_{k|}L_{j|j} = {}_{k|}(L_j)_{|j}$.

We fix $0 < \alpha < 1$, and let

$$S = \bigcup_{j=0}^{N} S_j, \quad \mathbf{X}_j = C^{0,\alpha}(S_j),$$

$$\mathbf{X} = C^{0,\alpha}(S) = \mathbf{X}_0 \oplus \mathbf{X}_1 \oplus \cdots \oplus \mathbf{X}_N.$$

Then, as can be checked easily, ${}_{k|}L_{|j}$, ${}_{k|}L_{j|j}$, ${}_{k|k}L_{|j}$, ${}_{k|k}L_{j|j}$ are compact operators from $\mathbf{X}_j$ to $\mathbf{X}_k$. Their adjoints are

$$\left({}_{k|}L_{|j} \right)' = {}_{j|}L_{|k}, \quad \left({}_{k|}L_{j|j} \right)' = {}_{j|j}L_{|k},$$

$$\left({}_{k|k}L_{|j} \right)' = {}_{j|}L_{k|k}, \quad \left({}_{k|k}L_{j|j} \right)' = {}_{j|j}L_{k|k}.$$

3.8.4.1 *Interior problem with at least one Dirichlet condition*

Let us first consider the interior problem imposing Dirichlet condition on at least one of its boundary components. For the sake of simplicity, we assume $m = 1$, $N = 2$. Let D be a bounded domain with boundary $S = S_0 \cup S_1 \cup S_2$.

When we impose the Dirichlet condition on S_0, we look for a solution in the form

$$u(x) = L_{0|0}\varphi_0 - L_{1|1}\varphi_1 + L_{|2}\varphi_2. \tag{3.107}$$

Applying (3.99) to $(u)_-$ on S_0 , $(u)_+$ on S_1 and $(\partial_\nu u)_+$ on S_2, we have the following boundary integral equations

$$
\begin{cases}
{}_{0|}L_{0|0}\varphi_0 - {}_{0|}L_{1|1}\varphi_1 + {}_{0|}L_{|2}\varphi_2 - \dfrac{1}{2}\varphi_0 = f_0, & \text{on } S_0, \\[2mm]
{}_{1|}L_{0|0}\varphi_0 - {}_{1|}L_{1|1}\varphi_1 + {}_{1|}L_{|2}\varphi_2 - \dfrac{1}{2}\varphi_1 = f_1, & \text{on } S_1, \\[2mm]
{}_{2|2}L_{0|0}\varphi_0 - {}_{2|2}L_{1|1}\varphi_1 + {}_{2|2}L_{|2}\varphi_2 - \dfrac{1}{2}\varphi_2 = f_2, & \text{on } S_2.
\end{cases}
\tag{3.108}
$$

This is rewritten into a matrix form

$$(\mathbf{K} - 1/2)\varphi = f, \tag{3.109}$$

$$\mathbf{K} = \begin{pmatrix} {}_{0|}L_{0|0} & -{}_{0|}L_{1|1} & {}_{0|}L_{|2} \\ {}_{1|}L_{0|0} & -{}_{1|}L_{1|1} & {}_{1|}L_{|2} \\ {}_{2|2}L_{0|0} & -{}_{2|2}L_{1|1} & {}_{2|2}L_{|2} \end{pmatrix}. \tag{3.110}$$

We show that

$$1/2 \notin \sigma_p(\mathbf{K}). \tag{3.111}$$

Suppose $\varphi \in \mathbf{X}$ satisfies $(\mathbf{K} - 1/2)\varphi = 0$, and let

$$v = L_{0|0}\varphi_0 - L_{1|1}\varphi_1 + L_{2|2}\varphi_2. \tag{3.112}$$

Then, $\Delta v = 0$ in D, and by (3.108),

$$
\begin{cases}
(v)_- = 0, & \text{on } S_0, \\
(v)_+ = 0, & \text{on } S_1, \\
(\partial_\nu v)_+ = 0, & \text{on } S_2.
\end{cases}
\tag{3.113}
$$

By Green's formula, we then have

$$0 = -\int_D (\Delta v)\bar{v}\, dx = \int_D |\nabla v|^2 dx = 0.$$

Therefore, v is a constant which is equal to 0 by the boundary condition on S_0. Observing $(v)_+$ on S_0, $(v)_-$ on S_1 and $(\partial_\nu v)_-$ on S_2, we have $\varphi = 0$. This proves (3.111).

When we impose the Neumann condition on S_0, we look for a solution in the form

$$w = -L_{|0}\varphi_0 - L_{1|1}\varphi_1 + L_{|2}\varphi_2. \tag{3.114}$$

Then, the boundary condition is transformed to the integral equation (3.109), where

$$\mathbf{K} = \begin{pmatrix} -_{0|0}L_{|0} & -_{0|0}L_{1|1} & _{0|0}L_{|2} \\ -_{1|}L_{|0} & -_{1|}L_{1|1} & _{1|}L_{|2} \\ -_{2|2}L_{|0} & -_{2|2}L_{1|1} & _{2|2}L_{|2} \end{pmatrix}. \tag{3.115}$$

For $\varphi \in \mathbf{X}$ satisfying $(\mathbf{K} - 1/2)\varphi = 0$, define w by (3.114). Then, it satisfies $\Delta w = 0$ in D, and

$$\begin{cases} (\partial_\nu w)_- = 0, & \text{on} \quad S_0, \\ (w)_+ = 0, & \text{on} \quad S_1, \\ (\partial_\nu w)_+ = 0, & \text{on} \quad S_2. \end{cases}$$

Arguing as above, we see that $w = 0$ by the Dirichlet condition on S_0. We thus have

$$1/2 \notin \sigma_p(\mathbf{K}). \tag{3.116}$$

The above facts are easily generalized to the boundary value problem with boundary condition (3.103) and (3.104).

Theorem 3.17. *Let D be a bounded domain with boundary $S = S_{out} \cup \cup_{i=1}^N S_{inn}^{(i)}$. Impose the Dirichlet condition on at least one of the boundary components. Then there exists a unique solution to the interior problem, $\Delta u = 0$ in D, satisfying the boundary condition (3.103) and (3.104).*

Proof. Since $1/2 \notin \sigma_p(\mathbf{K})$, by the Fredholm theorem (Lemma 3.13), $1/2 \in \rho(\mathbf{K})$. Hence there exists a solution $\varphi = (\mathbf{K} - 1/2)^{-1}f$ to (3.109). Then one can construct a solution by using φ and single and double layer potentials. The uniqueness of the solution is proven in Theorem 3.15. □

3.8.4.2 *Interior Neumann problem*

We consider the interior Neumann problem, i.e. we impose $\partial_\nu u = f_i$ on $S_i = S_{inn}^{(i)}$, $1 \le i \le N$, and $\partial_\nu u = f_0$ on $S_0 = S_{out}$. Assume $N = 1$ for the sake of simplicity, and look for a solution in the form

$$u = -L_{|0}\varphi_0 + L_{|1}\varphi_1. \tag{3.117}$$

Then, the boundary condition is transformed to the integral equation

$$(\mathbf{K} - 1/2)\varphi = f, \tag{3.118}$$

$$\mathbf{K} = \begin{pmatrix} -\,_{0|0}L_{|0} & _{0|0}L_{|1} \\ -\,_{1|1}L_{|0} & _{1|1}L_{|1} \end{pmatrix}. \tag{3.119}$$

For $\varphi \in \mathcal{N}(\mathbf{K} - 1/2)$, define v by (3.117). Then, $\Delta v = 0$ in $\mathbf{R}^n \setminus S$, and

$$\begin{cases} (\partial_\nu v)_- = 0, & \text{on} \quad S_0, \\ (\partial_\nu v)_+ = 0, & \text{on} \quad S_1. \end{cases} \tag{3.120}$$

Therefore, $v = c$, a constant, in D. Since v is continuous across S_1, $v = c$ inside S_1. By Lemma 3.17 (4), $v = c v_{ext}$ outside S_0, where v_{ext} is defined by (3.102) with S replaced by S_0. Using (3.99), we have $\varphi_0 = c(\partial_\nu v_{ext})_+$, $\varphi_1 = 0$. Hence $\dim \mathcal{N}(\mathbf{K} - 1/2) = 1$, and

$$\mathcal{N}(\mathbf{K} - 1/2) = \{c((\partial_\nu v_{ext})_+, 0)\,;\, c \in \mathbf{C}\}. \tag{3.121}$$

To discuss the solvability of the equation (3.118), we consider the adjoint equation

$$(\mathbf{K}' - 1/2)\psi = 0, \tag{3.122}$$

$$\mathbf{K}' = \begin{pmatrix} -\,_{0|}L_{0|0} & -\,_{0|}L_{1|1} \\ _{1|}L_{0|0} & _{1|}L_{1|1} \end{pmatrix}. \tag{3.123}$$

Lemma 3.18. $\mathcal{N}(\mathbf{K}' - 1/2) = \{(c, -c)\,;\, c \in \mathbf{C}\}.$

Proof. Lemma 3.7 implies $(c, -c) \in \mathcal{N}(\mathbf{K}' - 1/2)$. For $\psi \in \mathcal{N}(\mathbf{K}' - 1/2)$, we put $v = -L_{0|0}\psi_0 - L_{1|1}\psi_1$. Then, $(v)_+ = 0$ on S_0, $(v)_- = \psi_0$ on S_0, and $(v)_+ = -\psi_1$ on S_1, $(v)_- = 0$ on S_1. Since $\Delta v = 0$ outside S_0 and inside S_1, we have $v = 0$ outside S_0 and inside S_1 by Theorems 3.15 and 3.16. Since $\partial_\nu v$ is continuous across S_0 and S_1 by Theorem 3.5, we have

$$\Delta v = 0, \quad \text{in} \quad D,$$

$$\begin{cases} (\partial_\nu v)_- = 0, & \text{on} \quad S_0, \\ (\partial_\nu v)_+ = 0, & \text{on} \quad S_1. \end{cases}$$

Then v is equal to a constant, which proves that $\psi = (c, -c)$, $c \in \mathbf{C}$. □

Theorem 3.18. *Let D be a bounded domain with boundary $S = S_{out} \cup \cup_{i=1}^N S_{inn}^{(i)}$. Then there exists a unique solution to the interior Neumann problem: $\Delta u = 0$ in D, satisfying Neumann boundary condition $\partial_\nu u = f$ on S, where $f = (f_0, \cdots, f_N)$ is assumed to satisfy*

$$\int_{S_{out}} f_0 dS - \sum_{i=1}^N \int_{S_{inn}^{(i)}} f_i dS = 0. \tag{3.124}$$

Proof.   If f satisfies (3.124), by Lemma 3.18,

$$\int_S f \cdot \psi \, dS = 0, \quad \forall \psi \in \mathcal{N}(\mathbf{K}' - 1/2).$$

Fredholm's theorem then implies that there exists a solution $\varphi \in \mathbf{X}$ to the equation $(\mathbf{K} - 1/2)\varphi = f$. Then, $u = -\sum_{i=0}^N L_{i|i}\varphi_i$ is a desired solution. The uniqueness is proven in Theorem 3.15. $\square$

3.8.4.3   *Exterior problem with at least one Dirichlet condition*

Let us turn to the exterior domain D with boundary $S = \cup_{i=1}^N S_{inn}^{(i)}$. Let $S_i = S_{inn}^{(i)}$, and impose the Dirichlet boundary condition on S_1. For the sake of simplicity, we assume $m = 1$, $N = 2$. We look for a solution in the form

$$u = -L_{1|1}\varphi_1 + L_{|2}\varphi_2.$$

Then, the boundary integral equation becomes $(\mathbf{K} - 1/2)\varphi = f$, where

$$\mathbf{K} = \begin{pmatrix} -_{1|}L_{1|1} & _{1|}L_{|2} \\ -_{2|2}L_{1|1} & _{2|2}L_{|2} \end{pmatrix}. \tag{3.125}$$

Suppose $\varphi \in \mathcal{N}(\mathbf{K} - 1/2)$. Then, $u = -L_{1|1}\varphi_1 + L_{|2}\varphi_2$ satisfies $\Delta u = 0$ in $\mathbf{R}^n \setminus S$, $(u)_+ = 0$ on S_1, $(u)_- = \varphi_1$ on S_1, $(\partial_\nu u)_+ = 0$ on S_2, $(\partial_\nu u)_- = \varphi_2$ on S_2. By Theorem 3.16, this implies $u = 0$ in D. Hence $u = 0$ inside S_2, since u is continuous across S_2. As $\partial_\nu u$ is continuous across S_1, $u = C = $ const. inside S_1. We then see that $\varphi_1 = C$, $\varphi_2 = 0$. On the other hand, by Lemma 3.7, $\varphi = (1, 0)$ satisfies $(\mathbf{K} - 1/2)\varphi = 0$. Therefore, we have proven $\dim \mathcal{N}(\mathbf{K} - 1/2) = 1$ and

$$\mathcal{N}(\mathbf{K} - 1/2) = \{(c, 0) \, ; \, c \in \mathbf{C}\}. \tag{3.126}$$

Let us observe the adjoint operator

$$\mathbf{K}' = \begin{pmatrix} -_{1|1}L_{|1} & -_{1|1}L_{2|2} \\ _{2|}L_{|1} & _{2|}L_{2|2} \end{pmatrix}. \tag{3.127}$$

Since $\dim \mathcal{N}(\mathbf{K} - 1/2) = 1$, we have

$$\dim \mathcal{N}(\mathbf{K}' - 1/2) = 1. \tag{3.128}$$

Lemma 3.19. *Let v_{ext} be a unique solution to the equation*

$$\begin{cases} \Delta v = 0, & in \quad D, \\ (v)_+ = 1, & on \quad S_1, \\ (\partial_\nu v)_+ = 0, & on \quad S_2, \\ v(x) \to 0, & as \quad |x| \to \infty, \end{cases} \tag{3.129}$$

and $\psi_1 = (\partial_\nu v_{ext})_+$ on S_1, $\psi_2 = -(v_{ext})_+$ on S_2. Then

$$\mathcal{N}(\mathbf{K}' - 1/2) = \{c(\psi_1, \psi_2) \, ; \, c \in \mathbf{C}\}. \tag{3.130}$$

Proof. The uniqueness of the solution to (3.129) is shown in Theorem 3.16. For $\psi \in \mathcal{N}(\mathbf{K}' - 1/2)$, put $v = -L_{|1}\psi_1 - L_{2|2}\psi_2$. We show that v satisfies

$$\begin{cases} \Delta v = 0, & \text{in } \mathbf{R}^n \setminus S, \\ v(x) \to 0, & \text{as } |x| \to \infty, \end{cases} \tag{3.131}$$

$$\begin{cases} (v)_+ = (v)_- = \text{const.}, & \text{on } S_1, \\ (\partial_\nu v)_+ = \psi_1, \quad (\partial_\nu v)_- = 0, & \text{on } S_1, \end{cases} \tag{3.132}$$

$$\begin{cases} (v)_+ = -\psi_2, \quad (v)_- = 0, & \text{on } S_2, \\ (\partial_\nu v)_+ = (\partial_\nu v)_- = 0, & \text{on } S_2. \end{cases} \tag{3.133}$$

In fact, $(\partial_\nu v)_+ = \psi_1$ on S_1, $(\partial_\nu v)_- = 0$ on S_1, $(v)_+ = -\psi_2$ on S_2, $(v)_- = 0$ on S_2. This implies $v = C = \text{const.}$ inside S_1, hence $(v)_+ = C$ on S_1. Since $(v)_- = 0$ on S_2, $v = 0$ inside S_2. Since $\partial_\nu v$ is continuous across S_2, $(\partial_\nu v)_+ = (\partial_\nu v)_- = 0$ on S_2. The existence of the solution to (3.129) and (3.130) then follow from $\dim \mathcal{N}(\mathbf{K}' - 1/2) = 1$. $\qquad\square$

Let us now solve the equation

$$\begin{cases} \Delta u = 0, & \text{in } D, \\ (u)_+ = f_1, & \text{on } S_1, \\ (\partial_\nu u)_+ = f_2, & \text{on } S_2, \\ u(x) \to 0, & \text{as } |x| \to \infty. \end{cases} \tag{3.134}$$

The idea is to look for the solution in the form

$$u = -L_{1|1}\varphi_1 + L_{|2}\varphi_2 + \alpha v_{ext},$$

where v_{ext} is defined in Lemma 3.19, and α is a constant to be determined later. The boundary integral equation then becomes

$$(\mathbf{K} - 1/2)\varphi = (f_1, f_2) - (\alpha, 0) =: g.$$

We choose α so that g is orthogonal to $\mathcal{N}(\mathbf{K}' - 1/2)$. By Lemma 3.19, we have only to prove that

$$\int_{S_1} (f_1 - \alpha)(\partial_\nu v_{ext})_+ dS - \int_{S^2} f_2(v_{ext})_+ dS = 0$$

for a suitable choice of α. This is proven if we show $\int_{S_1} (\partial_\nu v_{ext})_+ dS \neq 0$. Let $D_R = D \cap \{|x| < R\}$. Then, by integration by parts

$$\int_{S_1} (\partial_\nu v_{ext})_+ dS = \int_{|x|=R} (\partial_r v_{ext}) v_{ext} dS - \int_{D_R} |\nabla v_{ext}|^2 dx.$$

The 1st term of the right-hand side vanishes as $R \to \infty$. Therefore, if the left-hand side vanishes, v_{ext} is a constant, hence $v_{ext} = 0$, and we arrive at a contradiction.

Using the Fredholm theory (Theorem 3.13), we have thus proven the following theorem.

Theorem 3.19. *If we assume the Dirichlet condition at least one of the boundaries, there exists a unique solution to the exterior problem.*

3.8.4.4　*Exterior Neumann problem*

Theorem 3.20. *If we assume the Neumann condition on all of the boundaries, there exists a unique solution to the exterior problem.*

Proof.　We have only to prove the existence of solutions. As above, we consider the case $N = 2$, and look for the solution in the form

$$u = L_{|1}\varphi_1 + L_{|2}\varphi_2.$$

Then, we have the boundary integral equation $(\mathbf{K} - 1/2)\varphi = f$, where

$$\mathbf{K} = \begin{pmatrix} {}_{1|1}L_{|1} & {}_{1|1}L_{|2} \\ {}_{2|2}L_{|1} & {}_{2|2}L_{|2} \end{pmatrix}. \tag{3.135}$$

For $\varphi \in \mathcal{N}(\mathbf{K} - 1/2)$, $u = L_{|1}\varphi_1 + L_{|2}\varphi_2$ satisfies $(\partial_\nu u)_+ = 0$ on S_1, $(\partial_\nu u)_- = \varphi_1$ on S_1, $(\partial_\nu u)_+ = 0$ on S_2, $(\partial_\nu u)_- = \varphi_2$ on S_2. Therefore, $u = 0$ in D, and $u = $ const. inside S_1 and S_2. Since u is continuous across S_1 and S_2, $u = 0$ inside S_1 and S_2. Therefore, $\varphi = 0$, hence

$$1/2 \notin \sigma(\mathbf{K} - 1/2). \tag{3.136}$$

By solving the boundary integral equation, we get the solution u.　□

3.9　Potential theory for the Helmholtz equation

The potential theory developed for the Laplace equation can be also extended to the Helmholtz equation. In this section, we consider its exterior boundary value problem.

3.9.1　*Radiation condition*

First let us note that, for $k > 0$, the fundamental solution $R_0(x; k^2)$ for the Helmholtz equation and the $R_0(x)$ for the Laplace equation have the same singularities:

$$R_0(x; k^2) = R_0(x)\,(1 + O(|x|)), \quad |x| \to 0,$$

which follows from (2.5), (2.9) and (2.14). Then, Lemmas and Theorems in Sections 3.3 and 3.4 remain valid with Δ replaced by $\Delta + k^2$. There is no change in the mapping properties in Section 3.6. The difference appears in the uniqueness assertions in Subsection 3.8.2. Solutions of the Helmholtz equation $(-\Delta - k^2)u = 0$ in an exterior domain D satisfying Dirichlet or Neumann boundary condition and decaying at infinity do not vanish identically. For example, consider $u = r^{-1}\sin(kr)$, $r = |x|$, in $\mathbf{R}^3$. It satisfies $(-\Delta - k^2)u = 0$ in $\mathbf{R}^3 \setminus \{0\}$, and $u = 0$ on $r = \pi/k$.

Let us return to the wave propagation phenomenon, and consider two waves $u_\pm = e^{ik(\pm r - t)}/r$ in $\mathbf{R}^3$, which satisfy the wave equation:

$$(\partial_t^2 - \Delta)u_\pm = 0, \quad x \neq 0.$$

They have the phase factor $\phi_\pm = \pm kr - kt$. Let $S_\pm(t) = \{x \in \mathbf{R}^3\,;\, \phi_\pm = 0\}$ $= \{r = \pm t\}$, called the wave front, on which the wave has the same phase. When t moves from $-\infty$ to 0, the wave front $S_-(t)$ comes from infinty to the origin, and when t moves from 0 to ∞, the wave front $S_+(t)$ moves from the origin to infinity. By this reason, the former is called the *incoming wave*, and the latter the *outgoing wave*. Note that

$$\left(\frac{\partial}{\partial r} \mp ik\right)\frac{e^{\pm ikr}}{r} = O(r^{-2}), \quad \text{as} \quad r \to \infty.$$

With this in mind, we define as follows.

Definition 3.1. Let D be an exterior domain in $\mathbf{R}^n$. A solution of the Helmholtz equation $(-\Delta - k^2)u = f$ in D, where $k > 0$, is said to satify the *outgoing radiation condition* if it satisfies

$$u = O(r^{-(n-1)/2}), \quad \left(\frac{\partial}{\partial r} - ik\right)u = o(r^{-(n-1)/2}), \quad \text{as} \quad r \to \infty,$$

and the *incoming radiation condition* if it satisfies

$$u = O(r^{-(n-1)/2}), \quad \left(\frac{\partial}{\partial r} + ik\right)u = o(r^{-(n-1)/2}), \quad \text{as} \quad r \to \infty.$$

3.9.2 *Uniqueness*

The radiation condition guarentees the uniqueness of solutions to the Helmholtz equation. In the following, when we consider $-\Delta - k^2$, we always assume $k > 0$.

Theorem 3.21. *Let u be a solution to the Helmholtz equation $(-\Delta - k^2)u = 0$ in an exterior domain D satisfying Dirichlet or Neumann boundary condition. If u satisfies the outgoing or incoming radiation condition, then $u = 0$ in D.*

Proof. We consider the case of outgoing radiation condition. Let $D_R = D \cap \{|x| < R\}$. Applying Green's formula to the integral

$$\int_{D_R} \left((\Delta + k^2)u\bar{u} - u\overline{(\Delta + k^2)\bar{u}} \right) dx,$$

we have

$$\int_{|x|=R} \left((\frac{\partial}{\partial r} - ik)u\bar{u} - u\overline{(\frac{\partial}{\partial r} - ik)u} \right) dS + 2ik \int_{|x|=R} |u|^2 dS = 0.$$

Using the radiation condition, we then have

$$\frac{1}{R} \int_{R<|x|<2R} |u(x)|^2 dx \to 0, \quad R \to \infty.$$

By Rellich-Vekua theorem (see Theorem 2.4 and (2.32)), we have $u = 0$ for $|x| > R_1$, where R_1 is a large constant. By the unique continuation theorem for 2nd order elliptic equations, we have $u = 0$ in D. $\qquad\square$

The *unique continuation theorem* is the following assertion.

Theorem 3.22. *Let Ω be a domain in $\mathbf{R}^n$, and L a differential operator defined in Ω*

$$L = \sum_{i,j=1}^{n} a_{ij}(x) \frac{\partial^2}{\partial x_i \partial x_j} + \sum_{i=1}^{n} b_i(x) \frac{\partial}{\partial x_i} + c(x),$$

where the coefficients are in $C^\infty(\Omega)$ and $a_{ij}(x)$ is real-valued. Assume that L is elliptic, i.e. there exists a constant $C_0 > 0$ such that

$$\sum_{i,j=1}^{n} a_{ij}(x)\xi_i\xi_j \geq C_0|\xi|^2, \quad \forall x \in \Omega, \quad \forall \xi \in \mathbf{R}^n.$$

Suppose $u \in C^\infty(\Omega)$ satisfies $Lu = 0$ in Ω, and there exists an open set $\Omega_1 \subset \Omega$ such that $u = 0$ on Ω_1. Then $u = 0$ on Ω.

There are lots of variations of this theorem. In this book, we regard it as the uniqueness question for the Cauchy problem. The classical theorem is the following result due to Holmgren. Let

$$D_{\epsilon_0} = \{(t,x) \in \mathbf{R}^{n+1} ; t + |x|^2 < \epsilon_0, \ t \geq 0\}, \quad \epsilon_0 > 0.$$

Theorem 3.23. *Assume that all coefficients $a_{i\alpha}(t,x)$ are analytic in D_{ϵ_0}. Then, there exists $0 < \epsilon < \epsilon_0$ such that any solution $u(t,x)$ to the equation*

$$\begin{cases} \partial_t^m u + \sum_{i=0}^{m-1} \sum_{i+|\alpha| \leq m} a_{i\alpha}(t,x) \partial_t^i \partial_x^\alpha u, & \text{in} \quad D_\epsilon = 0, \\ \partial_t^i u(0,x) = 0, \quad i = 0, \cdots, m-1, \quad |x| < \epsilon \end{cases} \tag{3.137}$$

satisfies $u(t,x) = 0$ in D_ϵ.

The extension of this theorem to the C^∞-coefficient case was proved by Calderón. We use the same notation as in Theorem 3.23.

Theorem 3.24. *Assume that $a_{i\alpha}(t,x) \in C^\infty(D_{\epsilon_0})$ for $i + |\alpha| \leq m$. Moreover, assume that $a_{i\alpha}(t,x)$ is real-valued for $i + |\alpha| = m$, and that all roots of the characteristic polynomial*

$$\lambda^m + \sum_{i+|\alpha|=m} a_{i,\alpha}(0,0)\xi^\alpha \lambda^i = 0$$

are distinct for $0 \neq \xi \in \mathbf{R}^n$. Then, the same conclusion holds as in Theorem 3.23.

For the proof of the above theorems and related recent results, see e.g. [42], [72].

3.9.3 Boundary integral equations

We return to the exterior domain D with boundary $S = \cup_{i=1}^N S_i$, $S_i = S_{inn}^{(i)}$. To study the exterior boundary value problem, we convert it into the boundary integral equation. The single and double layer potentials are defined as above using the fundamental solution of $-\Delta - k^2$. To fix the idea, we consider only the outgoing radiation condition. Namely, we replace $R_0(x)$ by $R_0(x; k^2)$, where we choose the branch of $\sqrt{\cdot}$ so that $\sqrt{k^2} = k$. For $1 \leq i \leq m$, we impose the Dirichlet condition on S_i, and adopt the double layer potential $-L_{i|i}$. For $m + 1 \leq i \leq N$, we impose the Neumann condition, and adopt the single layer potential $L_{|i}$. We look for the solution in the form $u = \sum_i L^{(i)}\varphi_i$, where $L^{(i)}$ is $\pm L_{i|i}$ or $\pm L_{|i}$ defined as above. We then get the boundary integral equation

$$(\mathbf{K}(k) - 1/2)\varphi = f.$$

For the sake of simplicity, we assume $m = 1$, $N = 2$. Then $\mathbf{K}(k)$ and $\mathbf{K}(k)'$ are written as

$$\mathbf{K}(k) = \begin{pmatrix} -{}_1|L_{1|1} & {}_1|L_{|2} \\ -{}_2|_2 L_{1|1} & {}_2|_2 L_{|2} \end{pmatrix}, \tag{3.138}$$

$$\mathbf{K}(k)' = \begin{pmatrix} -_{1|1}L_{|1} & -_{1|1}L_{2|2} \\ _{2|}L_{|1} & _{2|}L_{2|2} \end{pmatrix}. \tag{3.139}$$

Let D_i be the interior domain with boundary S_i, and $A_1 = -\Delta$ in D_1 and $A_2 = -\Delta$ in D_2, both of which are equipped with Dirichlet boundary condition. We put also $B_1 = -\Delta$ in D_1 and $B_2 = -\Delta$ in D_2, both of which are equiped with Neumann boundary condition. We put

$$\mathcal{N}(A_i - k^2) = \{u \in C^2(D_i) \cap C^{1,\alpha}(\overline{D_i}) \,;\, -\Delta u = k^2 u \text{ in } D_i, \ u = 0 \text{ on } S_i\}, \tag{3.140}$$

$$\mathcal{N}(B_i - k^2) = \{u \in C^2(D_i) \cap C^{1,\alpha}(\overline{D_i}) \,;\, -\Delta u = k^2 u \text{ in } D_i, \ \partial_\nu u = 0 \text{ on } S_i\}. \tag{3.141}$$

First let us note that $\mathcal{N}(\mathbf{K}(k) - 1/2)$ is related with the interior problem as follows.

Lemma 3.20. *Let* $\varphi = (\varphi_1, \varphi_2) \in \mathbf{X}$. *Then* $\varphi \in \mathcal{N}(\mathbf{K}(k) - 1/2)$ *if and only if there exist* $u_1 \in \mathcal{N}(B_1 - k^2)$ *and* $u_2 \in \mathcal{N}(A_2 - k^2)$ *such that*

$$(u_1)_- = \varphi_1, \quad \text{on} \quad S_1, \tag{3.142}$$

$$(\partial_\nu u_2)_- = \varphi_2, \quad \text{on} \quad S_2. \tag{3.143}$$

Proof. For $\varphi \in \mathcal{N}(\mathbf{K}(k) - 1/2)$, put $u = -L_{1|1}\varphi_1 + L_{|2}\varphi_2$. Then $(-\Delta - k^2)u = 0$ in $\mathbf{R}^n \backslash S$, and $(u)_+ = 0$ on S_1, $(u)_- = \varphi_1$ on S_1, $(\partial_\nu u)_+ = 0$ on S_2, $(\partial_\nu u)_- = \varphi_2$ on S_2. Since u satisfies the outgoing radiation condition, $u = 0$ in D by Theorem 3.21. As $\partial_\nu u$ is continuous across S_1, and u is continuous across S_2, letting $u_i = u\big|_{D_i}$, we have $u_1 \in \mathcal{N}(B_1 - k^2)$, $u_2 \in \mathcal{N}(A_2 - k^2)$ and (3.142), (3.143).

Conversely, suppose there exist $u_1 \in \mathcal{N}(B_1 - k^2)$, $u_2 \in \mathcal{N}(A_2 - k^2)$ satisfying (3.142), (3.143). Letting $w(x - y) = R_0(x - y; k^2)$, and applying Green's formula for $(\Delta_y u_i)w - u_i(\Delta_y w)$, we have

$$-L_{1|1}\varphi_1 = \begin{cases} u_1(x), & x \in D_1, \\ 0, & x \notin \overline{D_1}, \end{cases}$$

$$L_{|2}\varphi_2 = \begin{cases} u_2(x), & x \in D_2, \\ 0, & x \notin \overline{D_2}. \end{cases}$$

Then, we have

$$-L_{1|1}\varphi_1 + L_{|2}\varphi_2 = 0, \quad x \in D.$$

Letting $x \to S_1$ and $x \to S_2$, we have

$$-_{1|}L_{1|1}\varphi_1 + _{1|}L_{|2}\varphi_2 - \frac{\varphi_1}{2} = 0, \quad \text{on} \quad S_1,$$

$$-_{2|2}L_{1|1}\varphi_1 + _{2|2}L_{|2}\varphi_2 - \frac{\varphi_2}{2} = 0, \quad \text{on} \quad S_2.$$

Hence $\varphi \in \mathcal{N}(\mathbf{K}(k) - 1/2)$. $\qquad\square$

Next we study the relation between $\mathcal{N}(\mathbf{K}(k)' - 1/2)$ and $\mathcal{N}(\mathbf{K}(k) - 1/2)$. Define the scalar product on $\mathbf{X} \times \mathbf{X}$ by

$$\langle f, g \rangle = \int_S f \cdot g \, dS = \int_{S_1} f_1 g_1 \, dS + \int_{S_2} f_2 g_2 \, dS.$$

Lemma 3.21. *Let* $\{\psi_j = (\psi_{j1}, \psi_{j2})\}_{j=1}^{\ell}$ *be a basis of* $\mathcal{N}(\mathbf{K}(k)' - 1/2)$, *and put* $v_j = -L_{|1}\psi_{j1} - L_{2|2}\psi_{j2}$.
(1) ψ *is computed from* v *as follows:*

$$\begin{cases} \psi_{j1} = (\partial_\nu v_j)_+, & \text{on} \quad S_1, \\ \psi_{j2} = -(v_j)_+, & \text{on} \quad S_2. \end{cases} \tag{3.144}$$

(2) Let $\varphi_j = (\varphi_{j1}, \varphi_{j2})$, $j = 1, \cdots, \ell$ *be defined by*

$$\begin{cases} \varphi_{j1} = \overline{v_j}\big|_{S_1}, \\ \varphi_{j2} = \partial_\nu \overline{v_j}\big|_{S_2}. \end{cases} \tag{3.145}$$

Then $\{\varphi_j\}_{j=1}^{\ell}$ *gives a basis of* $\mathcal{N}(\mathbf{K}(k) - 1/2)$.
(3) Let B *be the* $\ell \times \ell$ *matrix with entry* $B_{ij} = \langle \overline{\varphi_i}, \psi_j \rangle$. *Then* $\det B \neq 0$.

Proof. By the jump relation, v_j satisfies $(\partial_\nu v_j)_+ = \psi_{j1}$ on S_1, $(\partial_\nu v_j)_- = 0$ on S_1, $(v_j)_+ = -\psi_{j2}$ on S_2, $(v_j)_- = 0$ on S_2. Therefore $v_j \in \mathcal{N}(B_1 - k^2) \cap \mathcal{N}(A_2 - k^2)$. This proves (1).

To prove (2), first let us note that $v_j\big|_{S_1} = (v_j)_+ = (v_j)_-$ on S_1 and $\partial_\nu v_j\big|_{S_2} = (\partial_\nu v_j)_+ = (\partial_\nu v_j)_-$ on S_2. Since $\overline{v_j}\big|_{D_1} \in \mathcal{N}(B_1 - k^2)$ and $\overline{v_j}\big|_{D_2} \in \mathcal{N}(A_2 - k^2)$, by Lemma 3.20, $\varphi_j \in \mathcal{N}(\mathbf{K}(k) - 1/2)$. To show that they are linearly independent, suppose $\sum_{j=1}^{\ell} c_j \varphi_j = 0$. Then, letting $v = \sum_{j=1}^{\ell} \overline{c_j} v_j$, we have by Green's formula,

$$0 = \text{Im} \int_{D \cap \{|x| < R\}} (\Delta + k^2) v \overline{v} \, dx = \text{Im} \int_{|x|=R} \partial_r v \overline{v} \, dS - \text{Im} \int_S \partial_\nu v \overline{v} \, dS.$$

By the radiation condition, we have

$$\text{Im} \int_{|x|=R} \partial_r v \overline{v} \, dS = k \int_{|x|=R} |v|^2 \, dS + o(1).$$

On the other hand, $\overline{v} = \sum_{j=1}^{\ell} c_j \varphi_{j1} = 0$ on S_1, and $\overline{v} = \sum_{j=1}^{\ell} c_j \varphi_{j2} = 0$ on S_2. Therefore, we have $\int_S \partial_\nu v \overline{v} dS = 0$, and $\int_{|x|=R} |v|^2 dS \to 0$ as $R \to \infty$. Hence by Rellich-Vekua theorem, $v = 0$ in D. In view of (3.144), we then have $\sum_{i=1}^{\ell} c_j \psi_j = 0$, which yields $c_j = 0$, since ψ_j's are linearly independent. This proves (2).

Suppose $\sum_{i=1}^{\ell} c_i B_{ij} = 0$, $\ell = 1, \cdots, \ell$. Since the scalar product $\langle \, , \, \rangle$ is non-degenerate, we have $\sum_{i=1}^{\ell} c_i \overline{\varphi_i} = 0$. Since φ_i's are linearly independent, $c_i = 0$, which proves (3). $\qquad \square$

Theorem 3.25. *For any $k > 0$, there exists a unique solution to the exterior problem $(-\Delta - k^2)u = 0$ in D, with boundary condition $u = f_i$ on S_i, $i = 1, \ldots, m$, $\partial_\nu u = f_i$ on S_i, $i = m+1, \ldots, N$, satisfying the radiation condition.*

Proof. As above, we consider the case $m = 1$, $N = 2$. We convert the problem to the boundary integral equation $(\mathbf{K}(k) - 1/2)\varphi = f$. If $1/2 \notin \sigma(\mathbf{K}(k))$, this is uniquely solvable. If $1/2 \in \sigma(\mathbf{K}(k) - 1/2)$, we look for a solution in the form

$$u = -L_{1|1}\varphi_1 + L_{2|2}\varphi_2 + \sum_{i=1}^{\ell} c_i v_i,$$

where v_i is defined in Lemma 3.21. Letting $v^{(i)} = (v_i|_{S_1}, \partial_\nu v_i|_{S_2})$, we need to solve the equation

$$(\mathbf{K}(k) - 1/2)\varphi = f - \sum_{i=1}^{\ell} c_i v^{(i)}.$$

This has a solution if the right-hand side is orhogonal to $\mathcal{N}(\mathbf{K}(k)' - 1/2)$. Noting that $v_i = \overline{\varphi_{i1}}, \partial_\nu v_i = \overline{\varphi_{i2}}$, we need to solve the equation

$$\langle f, \psi_j \rangle - \sum_{i=1}^{n} c_i B_{ij} = 0.$$

Since, $\det(B_{ij}) \neq 0$ by Lemma 3.21 (3), we can solve this equation to determine c_i. $\qquad\square$

Chapter 4

Maxwell equation

4.1 Equations in the vacuum

The Maxwell equation in the vacuum is the following system of partial differential equations in $\mathbf{R}^3$:

$$\operatorname{div} E = \frac{\rho}{\epsilon_0}, \tag{4.1}$$

$$\operatorname{div} B = 0, \tag{4.2}$$

$$\partial_t E = c_0^2 \operatorname{curl} B - \frac{J}{\epsilon_0}, \tag{4.3}$$

$$\partial_t B = -\operatorname{curl} E, \tag{4.4}$$

$$\partial_t \rho = -\operatorname{div} J, \tag{4.5}$$

where $\partial_t = \partial/\partial t$. The physical quantities appearing here are:

$$
\begin{aligned}
E &= \text{electric field strength}, \\
B &= \text{magnetic flux density}, \\
J &= \text{electric current density}, \\
\rho &= \text{electric charge density},
\end{aligned}
\tag{4.6}
$$

where E, B, J are $\mathbf{R}^3$-valued functions, ρ is $\mathbf{R}^1$-valued, and c_0 is the speed of light in the vacuum. The following physical qunatities are also used:

$$
\begin{aligned}
H &= c_0^2 \epsilon_0 B = \text{magnetic field strength}, \\
D &= \epsilon_0 E = \text{electric displacement}.
\end{aligned}
\tag{4.7}
$$

The physical constants $\epsilon_0, \mu_0 = (c_0^2 \epsilon_0)^{-1}$ are called *dielectric permittivity*, and *magnetic permeability*. Their experimental values are given in the Appendix F. In materials, they are 3×3 symmetric matrices depending non-linearly on E and B.

In this section, we are intersted in the initial value problem, i.e. given $\rho(t, x)$ and $J(t, x)$, we seek $E(t, x)$ and $B(t, x)$ satisfying the above equation with the prescribed initial data

$$E(0, x) = E_0(x), \quad B(0, x) = B_0(x). \tag{4.8}$$

By the linearity of the equation, we can decompose the solution into two parts: $E^{(1)}(t, x) + E^{(2)}(t, x)$, $B^{(1)}(t, x) + B^{(2)}(t, x)$, where $E^{(2)}(t, x)$, $B^{(2)}(t, x)$ satisfy the inhomgeneous equation with ρ and J, and $E^{(1)}(t, x)$, $B^{(1)}(t, x)$ satisfy the homogeneous equation with $\rho = 0$, $J = 0$, as well as the initial data $E_0(x) - E^{(2)}(0, x)$, $B_0(x) - B^{(2)}(0, x)$. Note that, mathematically, it is not necessary to restrict E, B, ρ, J to be real-valued.

First, we consider the case $\rho = 0, J = 0$. Uisng (4.2)–(4.5) and the formula $\operatorname{curl} \operatorname{curl} = \operatorname{grad} \cdot \operatorname{div} - \Delta$, we obtain

$$\partial_t^2 E = c_0^2 \, \Delta E, \quad \partial_t^2 B = c_0^2 \, \Delta B.$$

Using $\operatorname{div} \cdot \operatorname{curl} = 0$, we also have

$$\partial_t \operatorname{div} E = \partial_t \operatorname{div} B = 0.$$

Therefore,

$$\operatorname{div} E(x, t) = \operatorname{div} B(x, t) = 0,$$

if they are satisfied for $t = 0$. Given the initial data $E(0, x) = E_0(x)$ and $B(0, x) = B_0(x)$, we have

$$\partial_t E(0, x) = c_0^2 \operatorname{curl} B_0(x), \quad \partial_t B(0, x) = -\operatorname{curl} E_0(x).$$

Thus, by Theorem 2.7, we have the following formulas.

Theorem 4.1. *Assume that $\rho(t, x) = 0$ and $J(t, x) = 0$, and the initial data satisfies $\operatorname{curl} E_0(x) = 0$, $\operatorname{div} B_0(x) = 0$. Then $E(t, x)$ and $B(t, x)$ are written as*

$$E(t, x) = \partial_t \left(\frac{1}{4\pi c_0^2 t} \int_{|y-x|=c_0 t} E_0(y) dS_y \right) + \frac{1}{4\pi t} \int_{|y-x|=c_0 t} \operatorname{curl} B_0(y) dS_y.$$

$$B(t, x) = \partial_t \left(\frac{1}{4\pi c_0^2 t} \int_{|y-x|=c_0 t} B_0(y) dS_y \right) - \frac{1}{4\pi c_0^2 t} \int_{|y-x|=c_0 t} \operatorname{curl} E_0(y) dS_y.$$

4.2　Gauge transformation

4.2.1　*Existence of the potential*

Next let us consider the non-zero currents. In the following, we often denote $\partial_j = \partial/\partial x_j$.

Lemma 4.1. *Let $B(x) \in C^\infty(\mathbf{R}^3; \mathbf{C}^3)$.*
(1) If $B(x)$ satisfies curl $B = 0$, there exists $f(x) \in C^\infty(\mathbf{R}^3; \mathbf{C}^1)$ such that $B = \nabla f$.
(2) If $B(x)$ satisfies div $B(x) = 0$, there exists $A(x) \in C^\infty(\mathbf{R}^3; \mathbf{C}^3)$ such that $B = \operatorname{curl} A$.

Proof.　(1) We put

$$f(x) = \int_0^1 \left(B_1(tx)x_1 + B_2(tx)x_2 + B_3(tx)x_3\right) dt.$$

Then we have, using curl $B = 0$,

$$\partial_1 f(x) = \int_0^1 \left(B_1(tx) + (\partial_1 B_1)(tx)tx_1 + (\partial_1 B_2)(tx)tx_2 + (\partial_1 B_3)(tx)tx_3\right) dt$$

$$= \int_0^1 \left(B_1(tx) + (\partial_1 B_1)(tx)tx_1 + (\partial_2 B_1)(tx)tx_2 + (\partial_3 B_1)(tx)tx_3\right) dt$$

$$= \int_0^1 \left(B_1(tx) + t\frac{d}{dt}B_1(tx)\right) dt$$

$$= \int_0^1 \frac{d}{dt}\left(tB_1(tx)\right) dt = B_1(x).$$

Similary, we can prove $\partial_i f(x) = B_i(x)$, $i = 2, 3$.
(2) Let $x' = (x_1, x_2)$ and $a_1(x') = -\int_0^{x_2} B_3(x_1, y_2, 0)dy_2$, $a_2(x') = 0$. Then we have

$$B_3(x', 0) = \partial_1 a_2 - \partial_2 a_1.$$

We put

$$A_1(x) = \int_0^{x_3} B_2(x', y_3)dy_3 + a_1(x'),$$

$$A_2(x) = -\int_0^{x_3} B_1(x', y_3)dy_3 + a_2(x'),$$

$$A_3(x) = 0.$$

Then,

$$\partial_2 A_3 - \partial_3 A_2 = B_1, \quad \partial_3 A_1 - \partial_1 A_3 = B_2,$$

and also

$$\partial_1 A_2 - \partial_2 A_1 = -\int_0^{x_3} (\partial_1 B_1 + \partial_2 B_2)\, dy_3 + (\partial_1 a_2 - \partial_2 a_1)$$

$$= \int_0^{x_3} \partial_3 B_3(x', y_3)\, dy_3 + B_3(x', 0) = B_3(x).$$

This proves the lemma.   □

Returning to the Maxwell equation, taking account of Lemma 4.1 (2), we look for $B(t, x)$ in the form

$$B(t, x) = \operatorname{curl} A(t, x). \tag{4.9}$$

Then, in view of (4.4),

$$\operatorname{curl}(E + \partial_t A) = 0.$$

Lemma 4.1 (1) then implies that there exists $A_0(t, x) \in C^\infty(\mathbf{R}^4; \mathbf{C}^1)$ such that

$$E(t, x) = -\nabla A_0(t, x) - \partial_t A(t, x). \tag{4.10}$$

We call (A_0, A) an *electromagnetic potential* or a *gauge*. Then, the Maxwell equation is transformed into the following one:

$$\partial_t (\nabla \cdot A) + \Delta A_0 = -\frac{\rho}{\epsilon_0}, \tag{4.11}$$

$$\partial_t^2 A - c_0^2\, \Delta A = \frac{J}{\epsilon_0} - c_0^2 \nabla (\nabla \cdot A) - \partial_t (\nabla A_0), \tag{4.12}$$

$$\partial_t \rho = -\nabla \cdot J. \tag{4.13}$$

This choice of gauge is unique up to derivatives of scalar functions.

Lemma 4.2. (A_0, A) *and* $(\widetilde{A_0}, \widetilde{A})$ *are electromagnetic potentials for E and B if and only if there exists* $\psi(t, x) \in C^\infty(\mathbf{R}^4; \mathbf{C}^1)$ *such that*

$$\widetilde{A_0} = A_0 - \partial_t \psi, \quad \widetilde{A} = A + \nabla\psi. \tag{4.14}$$

Proof. If (A_0, A) and $(\widetilde{A_0}, \widetilde{A})$ are electromagnetic potentials, by (4.9), there exists ϕ such that $\widetilde{A} - A = \nabla\phi$. By (4.10), we have $\nabla(\widetilde{A_0} - A_0 + \partial_t \phi) = 0$. Therefore, $\widetilde{A_0} - A_0 + \partial_t \phi = C(t)$. Letting $\psi = \phi - \int_0^t C(t)dt$, we obtain (4.14). The converse direction is checked by a direct computation.   □

4.2.2 Coulomb gauge

We assume that for each fixed $t \in \mathbf{R}$, $\rho(t,x)$ and $J(t,x)$ are compactly supported wth respect to x. The *Coulomb gauge* is the case $\nabla \cdot A = 0$. To obtain it, we define $A_0(t,x)$ to be a solution to the Poisson equation

$$\Delta A_0 = -\frac{\rho}{\epsilon_0}.$$

We seek $A(t,x)$ as the solution of

$$\begin{cases} (\partial_t^2 - c_0^2 \Delta)A = \dfrac{J}{\epsilon_0} - \partial_t \nabla A_0, \\ A\big|_{t=0} = \partial_t A\big|_{t=0} = 0 \end{cases}$$

by Duhamel's formula in Theorem 2.9. Using the equation $\partial_t \rho = -\nabla \cdot J$,

$$(\partial_t^2 - c_0^2 \Delta)\nabla \cdot A = 0, \quad \nabla \cdot A\big|_{t=0} = \partial_t(\nabla \cdot A)\big|_{t=0} = 0,$$

which yields $\nabla \cdot A = 0$.

4.2.3 Lorentz gauge

The *Lorentz gauge* is the case $c_0^2 \nabla \cdot A + \partial_t A_0 = 0$. We first solve the equation

$$\begin{cases} (\partial_t^2 - c_0^2 \Delta)A_0 = \dfrac{c_0^2 \rho}{\epsilon_0}, \\ (\partial_t^2 - c_0^2 \Delta)A = \dfrac{J}{\epsilon_0}. \end{cases}$$

We put

$$A_0(t,x) = A_0'(t,x) + A_0''(t,x),$$

$$A_0'(t,x) = \frac{1}{4\pi\epsilon_0} \int_{|x-y|<c_0 t} \frac{\rho(t - \frac{|x-y|}{c_0}, y)}{|x-y|} dy,$$

$$A_0''(t,x) = \partial_t \left(\frac{1}{4\pi c_0^2 t} \int_{|x-y|=c_0 t} f(y) dS_y \right), \qquad (4.15)$$

$$f(x) = \frac{1}{4\pi\epsilon_0} \int_{\mathbf{R}^3} \frac{\rho(0,y)}{|x-y|} dy,$$

$$A(t,x) = \frac{1}{4\pi\epsilon_0 c_0^2} \int_{|x-y|<c_0 t} \frac{J(t - \frac{|x-y|}{c_0}, y)}{|x-y|} dy,$$

and let $w := c_0^2 \nabla \cdot A + \partial_t A_0 = 0$. By Theorems 2.7 and 2.9, we have

$$(\partial_t^2 - c_0^2 \Delta)w = \frac{c_0^2}{\epsilon_0}(\nabla \cdot J + \partial_t \rho) = 0.$$

Let us check $w(0, x) = \partial_t w(0, x) = 0$, which are rewritten as

$$\begin{cases} c_0^2 \nabla \cdot A(0, x) + \partial_t A_0(0, x) = 0, \\ c_0^2 \nabla \cdot \partial_t A(0, x) + c_0^2 \Delta A_0(0, x) + \dfrac{c_0^2}{\epsilon_0} \rho(0, x) = 0. \end{cases}$$

Since $A_0'(0, x) = \partial_t A_0'(0, x) = 0$, $A(0, x) = \partial_t A(0, x) = 0$ and $A_0''(0, x) = f(x)$, $\partial_t A''(0, x) = 0$, they are satisfied. Therefore, we have $w(t, x) = 0$.

We show

$$A_0''(t, x) = \frac{1}{4\pi\epsilon_0} \int_{|x-y|>c_0 t} \frac{\rho(0, y)}{|x - y|} dy. \tag{4.16}$$

In fact, we have

$$A_0''(t, x) = \frac{1}{4\pi\epsilon_0} \int_{\mathbf{R}^3} \partial_t \left(\frac{1}{4\pi c_0^2 t} \left(\int_{|x-y|=c_0 t} \frac{dS_y}{|y - z|} \right) \right) \rho(0, z) dz.$$

Letting

$$\Phi(x) = \frac{1}{4\pi} \int_{S^2} \frac{d\omega}{|x - \omega|} = \begin{cases} \dfrac{1}{|x|}, & |x| > 1, \\ 1, & |x| \leq 1, \end{cases} \tag{4.17}$$

we have

$$\frac{1}{4\pi} \int_{|x-y|=c_0 t} \frac{dS_y}{|y - z|} = c_0 t \, \Phi\left(\frac{x - z}{c_0 t} \right).$$

Hence

$$A_0''(t, x) = \frac{1}{4\pi\epsilon_0 c_0} \int_{\mathbf{R}^3} \partial_t \Phi\left(\frac{x - z}{c_0 t} \right) \rho(0, z) dz.$$

Noting that

$$\partial_t \Phi\left(\frac{x - z}{c_0 t} \right) = \begin{cases} \dfrac{c_0}{|x - z|}, & \text{if} \quad |x - z| > c_0 t, \\ 0, & \text{if} \quad |x - z| < c_0 t, \end{cases}$$

we obtain (4.16). We have thus proven the following lemma.

Lemma 4.3. *The Lorentz gauge is given by* (A_0, A), *where*

$$A_0(t, x) = \frac{1}{4\pi\epsilon_0} \int_{|x-y|<c_0 t} \frac{\rho(t - \frac{|x-y|}{c_0}, y)}{|x - y|} dy + \frac{1}{4\pi\epsilon_0} \int_{|x-y|>c_0 t} \frac{\rho(0, y)}{|x - y|} dy,$$

$$A(t, x) = \frac{1}{4\pi\epsilon_0 c_0^2} \int_{|x-y|<c_0 t} \frac{J(t - \frac{|x-y|}{c_0}, y)}{|x - y|} dy.$$

4.3 Biot-Savart's law

In the previous section, we have seen that $B(t,x)$ is written as
$$B(t,x) = \operatorname{curl} A(t,x) + B'(t,x),$$
where $A(t,x)$ is the gauge in Lemma 4.3, and $B'(t,x)$ is the solution to the homogeneous equation given in Theorem 4.1. Since $B'(t,x)$ is a solution to the wave equation in 3-dimension, it propagates to infinity as $t \to \infty$. We assume that $J(t,x) = 0$ for $|x| > R$, where $R > 0$ is a constant independent of $t > 0$, and $J(t,x) \to J(x)$ together with its derivatives. Noting that
$$A(t,x) = \frac{\mu_0}{4\pi} \int_{|z| < c_0 t} \frac{J(t - \frac{|z|}{c_0}, x - z)}{|z|} dz,$$
we have
$$\operatorname{curl} A(t,x) \to \frac{\mu_0}{4\pi} \int_{\mathbf{R}^3} \frac{\operatorname{curl}_x J(x - z)}{|z|} dz$$
$$= -\frac{\mu_0}{4\pi} \int_{\mathbf{R}^3} \frac{\operatorname{curl}_z J(x - z)}{|z|} dz$$
$$= \frac{\mu_0}{4\pi} \int_{\mathbf{R}^3} J(x - z) \times \operatorname{grad}_z \frac{1}{|z|} dz.$$
We have thus seen that $B(t,x) \to B(x)$ on any compact set in $\mathbf{R}^3$, where
$$B(x) = \frac{\mu_0}{4\pi} \int_{\mathbf{R}^3} \frac{J(y) \times (x - y)}{|x - y|^3} dy. \tag{4.18}$$

As was discussed in (1.53), by shrinking the domain of integration to a curve, we can obtain a line integral. Let $C = \{c(t) \, ; \, t \in I\}$ be a non self-intersecting curve in $\mathbf{R}^3$. For $J(y) \in C_0^\infty(\mathbf{R}^3)$, we put
$$B(x) = \frac{\mu_0}{4\pi} \int_C \frac{J(y) \times (x - y)}{|x - y|^3} dC$$
$$= \frac{\mu_0}{4\pi} \int_I \frac{J(c(t)) \times (x - c(t))}{|x - c(t)|^3} |\dot{c}(t)| dt. \tag{4.19}$$
Physically, it represents the magnetic field strength generated by the current density $J(y)$ on C. The formula (4.19) is called *Biot-Savart's law*.

4.4 Linking number

Biot-Savart's law has an interesting property from the view point of topology. Let C be a closed curve in $\mathbf{R}^3$, and $\mu_0 J(y)$ the unit tangential field along C. Then (4.19) is rewritten as
$$B(x) = -\frac{1}{4\pi} \int_C \frac{(x - y) \times dy}{|x - y|^3}. \tag{4.20}$$

In fact, representing C as $\{c(t)\,;\, 0 \le t \le 1\}$, the right-hand side is equal to

$$-\frac{1}{4\pi}\int_0^1 \frac{(x - c(t)) \times \dot{c}(t)dt}{|x - c(t)|^3} = \frac{1}{4\pi}\int_0^1 \frac{\frac{\dot{c}(t)}{|\dot{c}(t)|} \times (x - c(t))}{|x - c(t)|^3}|\dot{c}(t)|dt.$$

Let us also note that

$$B(x) = \int_C \left(\nabla_x \frac{1}{4\pi|x - y|}\right) \times dy. \tag{4.21}$$

Lemma 4.4. *Let $B(x)$ be as in (4.20). Then we have for $x \notin C$*

$$\operatorname{curl} B(x) = 0, \quad \operatorname{div} B(x) = 0.$$

Proof.   For any vector field $V(x)$ and a constant vector c, we have (see (A.21), (A.19))

$$\operatorname{curl}(V(x) \times c) = (c \cdot \nabla_x)V(x) - (\operatorname{div} V(x))c, \tag{4.22}$$

$$\operatorname{div}(V(x) \times c) = c \cdot \operatorname{curl} V(x). \tag{4.23}$$

Using (4.22) and $\operatorname{div} \operatorname{grad}|x - y|^{-1} = 0$, we have

$$\nabla_x \times \left(\nabla_x \frac{1}{4\pi|x - y|} \times dy\right) = -(dy \cdot \nabla_x)\frac{x - y}{4\pi|x - y|^3}.$$

Parametrizing C as $C = \{y(t)\,;\, 0 \le t \le 1\}$ with $y(0) = y(1)$, we then have

$$\begin{aligned}
\operatorname{curl} B(x) &= -\int_0^1 \dot{y}(t) \cdot \nabla_x \frac{x - y(t)}{4\pi|x - y(t)|^3}dt \\
&= \int_0^1 \frac{d}{dt}\frac{x - y(t)}{4\pi|x - y(t)|^3}dt = \frac{x - y(t)}{4\pi|x - y(t)|^3}\Big|_{t=0}^{t=1} = 0.
\end{aligned}$$

Using (4.23) and $\operatorname{curl} \operatorname{grad} = 0$, we have

$$\operatorname{div} B(x) = \int_C dy \cdot \left(\operatorname{curl} \operatorname{grad} \frac{1}{4\pi|x - y|}\right) = 0. \qquad \square$$

We denote $\int_C \cdots dy = \int_{y \in C} \cdots dy$ to stress that the line integral is performed with respect to the variable y. For a closed curve C in $\mathbf{R}^3$, we associate the following 1-form

$$\begin{aligned}
\omega(C) = \omega_B^{(1)} &= -\frac{1}{4\pi}\int_{y \in C} \frac{(x - y) \times dy}{|x - y|^3} \cdot dx \\
&= -\frac{1}{4\pi}\int_{y \in C} \frac{(x_2 - y_2)\,dy_3 - (x_3 - y_3)\,dy_2}{|x - y|^3}dx_1 \\
&\quad -\frac{1}{4\pi}\int_{y \in C} \frac{(x_3 - y_3)\,dy_1 - (x_1 - y_1)\,dy_3}{|x - y|^3}dx_2 \\
&\quad -\frac{1}{4\pi}\int_{y \in C} \frac{(x_1 - y_1)\,dy_2 - (x_2 - y_2)\,dy_1}{|x - y|^3}dx_3.
\end{aligned} \tag{4.24}$$

Lemma 4.5. *In the region* $\mathbf{R}^3 \setminus C$, $\omega(C)$ *is a harmonic 1-form, and for two closed curves* C, C' *such that* $C \cap C' = \emptyset$,

$$\int_{C'} \omega(C) = \int_C \omega(C'). \tag{4.25}$$

Proof. Lemma 4.4 and the formulas (1.79), (1.83) imply $d\omega(C) = 0$, $\delta\omega(C) = 0$. Therefore, $\omega(C)$ is a harmonic 1-form.

For 3-vectors a, b, c, it holds that

$$(a \times b) \cdot c = \det(a, b, c).$$

We then have

$$\int_{C'} \omega(C) = -\frac{1}{4\pi} \iint_{(x,y) \in C' \times C} \frac{\det(x - y, dy, dx)}{|x - y|^3},$$

which is symmetric with respect to x and y. This proves (4.25). $\square$

Let C be a closed oriented curve in $\mathbf{R}^3$, and S a surface such that $\partial S = C$. We equip S with the orientation which is compatible with C (see Subsection 1.7.4). A closed curve C' is said to be *not linked* with C when C, C' and S can be smoothly deformed so that C' does not pass through S. A closed curve C' is said to be *linked positively once* with C when C, C' and S can be deformed smoothly so that C' passes through S once and, at the intersection point, C' has the same direction as S. Namely, when we parametrize $C' = \{x(t)\,;\, 0 \leq t \leq 1\}$, there exists a unique t_0 such that $C' \cap S = \{x(t_0)\}$, and

$$\nu_{x(t_0)} \cdot \dot{x}(t_0) > 0,$$

where ν_x denotes the unit normal at $x \in S$ whose direction is compatible with C.

Theorem 4.2. *Let* C, C' *be oriented closed curves in* $\mathbf{R}^3$. *Then we have*

$$\int_{C'} \omega(C) = \begin{cases} 1 & \text{if } C' \text{ is linked positively once with } C, \\ 0 & \text{if } C' \text{ is not linked with } C. \end{cases}$$

Proof. Since $d\omega = 0$, by Stokes' theorem, we can deform C and C' arbitrarily. If C and C' are not linked, we can assume that C' lies in a simply connected domain U such that $U \cap C = \emptyset$. Then on U, $\omega = df$ for some f. Hence $\int_{C'} \omega = 0$.

Suppose C and C' are linked positively once. Then one can assume without loss of generality that C is a circle $\{y_1^2 + y_2^2 = 1,\ y_3 = 0\}$, and C' is the curve

$$\{(0, 0, t)\,;\, -R \leq t \leq R\} \cup \{x_2^2 + x_3^2 = R^2,\ x_2 \geq 0,\ x_1 = 0\} := I'(R) \cup C'(R).$$

Then we have

$$\int_{C'(R)} \omega(C) \to 0, \quad R \to \infty.$$

To compute the contribution from $I(R)$, we put $x = (0,0,t)$ and $y = (\cos\theta, \sin\theta, 0)$. Then we have

$$\left((x-y) \times dy\right) \cdot dx = -d\theta dt, \quad |x-y|^2 = 1 + t^2.$$

Hence we have

$$-\frac{1}{4\pi} \int_{x \in I'(R)} \int_{y \in C} \frac{(x-y) \times dy \cdot dx}{|x-y|^3}$$

$$= \frac{1}{4\pi} \int_0^{2\pi} d\theta \int_{-R}^R \frac{dt}{(1+t^2)^{3/2}} \to 1. \qquad \square$$

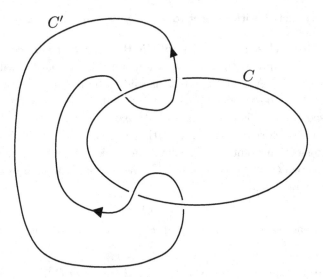

Fig. 4.1   Link $(C, C') = 2$

For two closed curves C and C' such that $C \cap C' = \emptyset$, their *linking number* Link (C, C') is defined as follows:

$$\text{Link}(C, C') = \int_{C'} \omega(C). \tag{4.26}$$

Let S be a surface with boundary $\partial S = C$. We can deform S, C, C' so that C' passes through S transversally, i.e. C' is not tangent to S. Letting $C' = \{x(t)\,;\, 0 \le t \le 1\}$ and $S \cap C' = \{x(t_i)\,;\, i = 1, \cdots, k\}$, we put

$$\sigma\left(x(t_i)\right) = \frac{\nu_{x(t_i)} \cdot \dot{x}(t_i)}{|\nu_{x(t_i)} \cdot \dot{x}(t_i)|}.$$

Then, one can show that

$$\text{Link}\,(C, C') = \sum_{i=1}^{k} \sigma\,(x(t_i)). \qquad (4.27)$$

Note that even if the linking number is 0, C and C' may be linked (see Figure 4.2).

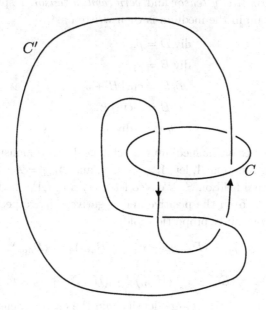

Fig. 4.2   Link $(C, C') = 0$

4.5   Maxwell equation in medium

The basic electromagnetic quantities in medium are

- D = electric displacement,
- E = electric field strength,
- B = magnetic flux density,
- H = magnetic field strength.

The effect of the medium appears in the *polarizations* P and M defined by

$$D = \epsilon_0 E + P, \quad B = \mu_0 H + M.$$

In general, D and B may depend on both of E and H (and also on the solution to the Maxwell equation itself). The standard assumption is that they have the form

$$D = \epsilon E, \quad B = \mu H, \tag{4.28}$$

where ϵ and μ are positive definite symmetric matrices depending only on x called *permittivity tensor* and *permeability tensor*, respectively. The Maxwell equation in the medium is then written as

$$\begin{aligned}
\operatorname{div} D &= \rho, \\
\operatorname{div} B &= 0, \\
\partial_t D &= \operatorname{curl} H + J, \\
\partial_t B &= -\operatorname{curl} E, \\
\partial_t \rho &= -\operatorname{div} J.
\end{aligned} \tag{4.29}$$

Physical properties of the medium may change discontinuously at a surface S. For a vector function A, let $A_{nor} = A \cdot \nu$, and $A_{tan} = A - A_{nor}\nu$, where ν is a unit normal field on S. We also let $(A)_+$ and $(A)_-$ be the boundary values of A at S from the positive and negative side, respectively. Then, the following continuity properties hold:

$$(D_{nor})_+ = (D_{nor})_- + \sigma_S, \quad (B_{nor})_+ = (B_{nor})_-, \tag{4.30}$$

$$(E_{tan})_+ = (E_{tan})_-, \quad (H_{tan})_+ = (H_{tan})_- + J_S \times \nu, \tag{4.31}$$

where σ_S and J_S are the charge density and the current density on S.

4.6 Conductor

4.6.1 *Electrostatic capacity*

We study electric potentials for the conductive body. Let D_{ext} be a domain exterior to bounded open sets $D_i, 1 \leq i \leq n$, i.e.

$$D_{ext} = \mathbf{R}^3 \setminus \overline{\bigcup_{i=1}^{n} D_i}, \tag{4.32}$$

where we assume that D_i has no holes inside and $\overline{D_i} \cap \overline{D_j} = \emptyset$ if $i \neq j$. Let $S = \partial D_{ext} = \cup_i^n S_i$, $S_i = \partial D_i$. Assume that each D_i is an electric conductor, and that ϵ, μ are piecewise constants, $\epsilon_i, \mu_i > 0$, inside D_i, and $\epsilon_0, \mu_0 > 0$ in D_{ext}. When charges are put inside the conductors at $t = 0$, as $t \to \infty$ (in fact, after a short time), the system tends to a static state, i.e.

$E(t,x) \rightarrow E(x)$, $H(t,x) \rightarrow H(x)$. Since the electrons move freely inside D_i, charges will accumulate on the boundary S_i. Hence we have $J(t,x) \rightarrow 0$ in $\mathbf{R}^3$, $\rho(t,x) \rightarrow 0$ pointwise in D_i. We put

$$\lim_{t\to\infty} \left(\int_{S_i} \sigma_S(t,x)dS + \int_{D_i} \rho(t,x)dx \right) = Q_i.$$

Since there is no current inside D_i, we have by Ohm's law

$$E(x) = 0, \quad \text{in} \quad D_i, \quad 1 \le i \le n. \tag{4.33}$$

By the equations (4.29), we have

$$\operatorname{div} E(x) = 0, \quad \operatorname{curl} E(x) = 0, \quad \text{in} \quad D_{ext}. \tag{4.34}$$

In view of (4.31),

$$(E(x)_{tan})_+ = 0, \quad \text{on} \quad S_i, \quad 1 \le i \le n. \tag{4.35}$$

Therefore, $E(x)$ is orthogonal to S_i. For a finite time, we have by the divergence theorem

$$\int_{S_i} (\nu \cdot E(t,x))_- \, dS = \int_{D_i} \nabla \cdot E(t,x)dx = \int_{D_i} \frac{\rho(t,x)}{\epsilon_i} dx.$$

Letting $t \rightarrow \infty$, and taking account of (4.30), we will have

$$\epsilon_0 \int_{S_i} (\nu \cdot E(x))_+ dS_x = Q_i, \quad 1 \le i \le n. \tag{4.36}$$

Recall that a domain D is said to be *simply connected* if any simple closed curve in D can be continuously shrunk to a point in D, passing through only inside D.

Lemma 4.6. *Let D be a simply connected domain in $\mathbf{R}^3$ and $V(x) \in C^\infty(D; \mathbf{R}^3)$, a vector field in D. If $\nabla \times V(x) = 0$ in D, there exists a $f(x) \in C^\infty(D; \mathbf{R})$ such that $\nabla f(x) = V(x)$ in D.*

Proof. Fix a point $x_0 \in D$, and for any $x \in D$, consider a curve L_x in D with end points x and x_0. Let $f(x)$ be defined by the line integral

$$f(x) = \int_{L_x} \langle V(y), v_{tan} \rangle dC_y,$$

where v_{tan} is the unit tangential field along L. For two such curves L_x and L'_x, we consider a surface S in D with boundary consisting of L'_x and $(L_x)^{-1}$, where $(L_x)^{-1}$ is the curve L_x with opposite direction. The assumption of simply connectedness makes it possible to construct such a surface. Then, by Stokes' Theorem (Theorem 1.6),

$$\int_{L_x'} \langle V(y), v_{tan} \rangle dC_y - \int_{L_x} \langle V(y), v_{tan} \rangle dC_y = \int_S \langle \operatorname{curl} V(y), \nu_S \rangle dS_y = 0.$$

Therefore, $f(x)$ is defined independently of the curve L_x. Then, we have

$$f(x_1 + s, x_2, x_3) - f(x_1, x_2, x_3) = \int_{x_1}^{x_1+s} V_1(y_1, x_2, x_3) dy_1,$$

hence, $\partial f / \partial x_1 = V_1(x)$. Similiary, we can prove $\partial f / \partial x_i = V_i(x)$, for $i = 2, 3$. □

By virtue of Lemma 4.6, if the domain is simply connected, the static electric field is written by a potential, which is constructed by solving the Dirichlet problem.

Theorem 4.3. *Given constants* $Q_i, 1 \leq i \leq n$, *arbitrarily, there exists a unique* $u(x) \in C(\mathbf{R}^3) \cap C^\infty(\mathbf{R}^3 \setminus S)$ *such that*

$$\Delta u(x) = 0, \quad x \in \mathbf{R}^3 \setminus S, \tag{4.37}$$

$$|u(x)| + |x||\nabla u(x)| \leq C(1 + |x|)^{-1}, \quad x \in D_{ext}, \tag{4.38}$$

$$\int_{S_i} (\partial_\nu u)_+ \, dS = Q_i, \quad 1 \leq i \leq n, \tag{4.39}$$

and

$$u = \text{constant}, \quad on \quad S_i, \quad 1 \leq i \leq n. \tag{4.40}$$

To prove this theorem, we first consider the exterior Dirichlet problem:

$$\begin{cases} \Delta u = 0, & \text{in} \quad D_{ext}, \\ u = V_i, & \text{on} \quad S_i, \quad 1 \leq i \leq n, \\ u(x) \to 0, & \text{as} \quad |x| \to \infty, \end{cases} \tag{4.41}$$

V_i being a constant, and the *voltage-to-charge* map:

$$V = (V_1, \cdots, V_n) \to Q = (Q_1, \cdots, Q_n).$$

By Theorem 3.19, there exists a unique solution $u(x)$ to (4.41). Note that $u(x)$ is real-valued if and only if V_i is real. In fact, if $V_i \in \mathbf{R}$, $u(x)$ and $\overline{u(x)}$ satisfy the same equation and the same boundary condition. Hence, $u(x) = \overline{u(x)}$ by the uniqueness of the solution.

For the solution $u(x)$ of (4.41), we put

$$\int_{S_i} (\partial_\nu u)_+ \, dS = Q_i, \tag{4.42}$$

and consider the linear map

$$C : (V_1, \cdots, V_n) \to Q = (Q_1, \cdots, Q_n). \tag{4.43}$$

Lemma 4.7. $C : \mathbf{R}^n \to \mathbf{R}^n$ *is a linear isomorphism, and*

$$(CV, V') = (V, CV'), \quad V, V' \in \mathbf{R}^n, \tag{4.44}$$

where $(\ ,\ )$ is the standard inner product of $\mathbf{R}^n$.

Proof. Suppose $Q = 0$. Letting $D_R = D_{ext} \cap \{|x| < R\}$, we have, by Green's formula,

$$0 = \int_{D_R} (\Delta u) u \, dx = \int_{|x|=R} (\partial_r u) u \, dS - \int_S (\partial_\nu u)_+ u \, dS - \int_{D_R} |\nabla u|^2 \, dx.$$

Since u is a constant on S_i and $Q_i = 0$, the 2nd term of the right-hand side vanishes. As $|u(x)| \leq C|x|^{-1}$ and $|\nabla u(x)| \leq C|x|^{-2}$, the 1st term of the right-hand side also vanishes as $R \to \infty$. Therefore, we have $\int_{D_{ext}} |\nabla u|^2 dx = 0$, hence $u(x) = $ a constant, which is equal to 0, since $u(x) \to 0$ as $|x| \to \infty$. We then have $V = 0$, which proves that C is an isomorphism.

Letting u' be the solution of (4.41) with V replaced by V', we have, by Green's formula,

$$\int_S \{(\partial_\nu u)_+ u' - u(\partial_\nu u')_+\} \, dS = \int_{|x|=R} \{(\partial_r u) u' - u(\partial_r u')\} \, dS.$$

Letting $R \to \infty$, we get (4.44). $\qquad \square$

Proof of Theorem 4.3. To prove the uniqueness, let $u(x)$ be the solution for $Q = 0$. In the proof of Lemma 4.7, we have seen that $u(x) = 0$ in D_{ext}. Since $\Delta u = 0$ in D_i and $u = 0$ on S_i, we have $u = 0$ in D_i. This proves the uniqueness.

To prove the existence, let $u(x)$ be the solution to (4.41), and extend $u(x)$ to be V_i in D_i. By Lemma 4.7, we can choose V so that the charges on S_i become the given Q_i. $\qquad \square$

Physically, Lemma 4.7 asserts that the total charge Q uniquely determines the voltage of the conductor. By using the standard basis of $\mathbf{R}^n$, C is represented by an $n \times n$ symmetric matrix (C_{ij}). The (i, i) entry C_{ii} is called the *electrostatic capacity* or the *capacitance* of the body D_i.

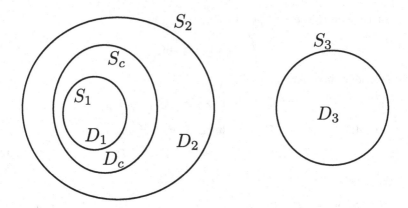

Fig. 4.3   Conductor and cavity

4.6.2   *Electrostatic shield*

Now, we consider the case in which there are *cavities* (vacuum holes) inside conductive bodies. We are given bounded domains $D_1, D_2, \cdots, D_n$ in $\mathbf{R}^3$ as follows. There are bounded domains $\mathcal{O}_i$, $i = 1, \frac{3}{2}, 2$, such that $\overline{\mathcal{O}_1} \subset \mathcal{O}_{\frac{3}{2}} \subset \overline{\mathcal{O}_{\frac{3}{2}}} \subset \mathcal{O}_2$, and $D_1 = \mathcal{O}_1$, $D_2 = \mathcal{O}_2 \setminus \overline{\mathcal{O}_{\frac{3}{2}}}$. We put $D_c = \mathcal{O}_{\frac{3}{2}} \setminus \overline{\mathcal{O}_1}$. Assume also that for $3 \le i \le n$, D_i is outside D_2, and $\overline{D_i} \cap \overline{D_j} = \emptyset$ for $i \ne j, 2 \le i, j \le n$. Let

$$D_{ext} = \mathbf{R}^3 \setminus \overline{\mathcal{O}_2 \cup D_3 \cup \cdots \cup D_n},$$

$$S_c = \partial \mathcal{O}_{\frac{3}{2}}, \quad S_2 = \partial \mathcal{O}_2, \quad S_i = \partial D_i, \quad i = 1, 3, \cdots, n,$$

$$S = S_c \cup \left(\cup_{i=1}^n S_i \right).$$

We regard $D_i, 1 \le i \le n$, as conductors, and D_{ext}, D_c as vacuum. Thus, D_c is a cavity between the conductors D_1 and D_2.

Theorem 4.4. *For any constants V_i, $1 \le i \le n$, there exists a unique $u(x) \in C(\mathbf{R}^3) \cap C^\infty(\mathbf{R}^3 \setminus S)$ satisfying*

$$\begin{cases} \Delta u(x) = 0, & in \quad \mathbf{R}^3 \setminus S, \\ u(x) = V_i, & in \quad D_i, \quad 1 \le i \le n, \\ u(x) \to 0, & as \quad |x| \to \infty. \end{cases} \tag{4.45}$$

Let

$$Q_i = \int_{S_i} (\partial_\nu u)_+ \, dS, \quad 1 \le i \le n. \tag{4.46}$$

Then the map

$$C : V = (V_1, \cdots, V_n) \to Q = (Q_1, \cdots, Q_n) \tag{4.47}$$

is a linear isomorphism.

Proof. We first solve the exterior Dirichlet problem:

$$\begin{cases} \Delta u(x) = 0, & \text{in} \quad D_{ext}, \\ u(x) = V_i, & \text{on} \quad S_i, \quad i = 2, \cdots, n, \\ u(x) \to 0, & \text{as} \quad |x| \to \infty. \end{cases}$$

We define $u(x) = V_i$ in D_i, $i = 2, \cdots, n$, and then solve the interior Dirichlet problem in D_c:

$$\begin{cases} \Delta u(x) = 0, & \text{in} \quad D_c, \\ u(x) = V_2, & \text{on} \quad S_c, \\ u(x) = V_1, & \text{on} \quad S_1. \end{cases}$$

We finally define $u(x) = V_1$ in D_1. This proves the existence of $u(x)$. The uniqueness follows from that of the Dirichlet problem.

To show the bijectivity of C, let u be the solution to (4.45) for which $Q = 0$. Arguing as in the proof of Lemma 4.7, we have $u = 0$ in D_{ext}. Then, $V_2 = \cdots = V_n = 0$. which implies $u = 0$ in D_i, $i = 2, \cdots, n$. Applying Green's formula in D_c, we have

$$\int_{S_c} (\partial_\nu u)_- dS = \int_{S_1} (\partial_\nu u)_+ dS = Q_1 = 0.$$

Then, we have

$$0 = \int_{D_c} (\Delta u) u \, dx = - \int_{D_c} |\nabla u|^2 dx.$$

Therefore, u is a constant, which is equal to 0, since $u = 0$ on S_c. We then have $u = 0 = V_1$ on S_1, which proves that C is bijective. $\qquad \square$

As a consequence, we have the following corollary.

Corollary 4.1. *Given constants $Q_i, 1 \le i \le n$, arbitrarily, there exists a unique $u(x) \in C(\mathbf{R}^3) \cap C^\infty(\mathbf{R}^3 \setminus S)$ such that*

$$\Delta u(x) = 0, \quad x \in \mathbf{R}^3 \setminus S, \tag{4.48}$$

$$|u(x)| + |x||\nabla u(x)| \le C(1 + |x|)^{-1}, \quad x \in D_{ext}, \tag{4.49}$$

$$\int_{S_i} (\partial_\nu u)_+ dS = Q_i, \quad 1 \le i \le n, \tag{4.50}$$

$$u = \text{constant} =: V_i, \quad \text{in} \quad D_i, \quad 1 \le i \le n. \tag{4.51}$$

We observe the structure of the matrix $C = (C_{ij})$. Let u' be the solution of (4.45) with V replaced by V'. Then, applying Green's formula to the equation $\int_{D_{ext}} ((\Delta u)u' - u\Delta u')\, dx = 0$, we have

$$\sum_{i=2}^{n} Q_i V_i' = \sum_{i=2}^{n} V_i Q_i'. \tag{4.52}$$

Here, we note the following equivalence:

$$Q_1 = 0 \iff V_1 = V_2. \tag{4.53}$$

In fact, as has been discussed in the proof of Theorem 4.4, $Q_1 = 0$ implies u is constant in D_c, hence $V_1 = V_2$. Conversely, if $V_1 = V_2$, by the uniqueness of solutions to the Dirichlet problem, $u = V_1$ in D_c, hence $Q_1 = 0$.

Recall that $Q_i = \sum_{j=1}^{n} C_{ij} V_j$. In view of (4.53), we have $0 = (C_{11} + C_{12})V_1 + \sum_{j=3}^{n} C_{ij} V_j$ for any V_j. We then have

$$C_{12} = -C_{11}, \quad C_{13} = \cdots = C_{1n} = 0. \tag{4.54}$$

The equation (4.52) implies that if we remove the first column and the first row from C, the resulting matrix is symmetric. We then have

$$C_{21} = -C_{11}, \quad C_{31} = \cdots = C_{n1} = 0. \tag{4.55}$$

By (4.54) and (4.55), we have

$$\begin{aligned}
Q_1 &= C_{11}V_1 \ -C_{11}V_2, \\
Q_2 &= -C_{11}V_1 \ +C_{22}V_2 + \cdots + C_{2n}V_n, \\
Q_3 &= \qquad\quad C_{32}V_2 \ + \cdots + C_{3n}V_n, \\
\cdots &= \quad\cdots \qquad\qquad \cdots \\
Q_n &= \qquad\quad C_{n2}V_2 \ + \cdots + C_{nn}V_n.
\end{aligned} \tag{4.56}$$

Let us put $V_2 = 0$, which means that the conductor D_2 is grounded. Then, (4.56) shows that the charge Q_1 of D_1 depends only on the voltage V_1, and is independent of the voltages of the conductors outside D_2. Also, the charges of $D_3, \cdots, D_n$ are independent of the voltage V_1. Therefore, the conductor D_1 is electrically shielded by D_2 from the conductors $D_3, \cdots, D_n$ outside D_2.

Chapter 5

Cohomology

The existence of scalar or vector potentials is closely related to the topology of the domain. We summarize here an elementary part of the decomposition theory of vector fields in a domain in $\mathbf{R}^3$.

5.1 Bounded planar domain

Let $\mathcal{O}$ be an open set in $\mathbf{R}^2$, and U its open subset whose boundary consists of 2 closed curves C and C', i.e. $\partial U = C \cup C'$. Then, if a 1-form ω defined on $\mathcal{O}$ satisfies $d\omega = 0$, we have

$$\int_C \omega = \int_{C'} \omega. \tag{5.1}$$

In fact, this follows from Stokes' formula,

$$0 = \int_U d\omega = \int_{\partial U} \omega = \int_C \omega - \int_{C'} \omega.$$

Let $\mathcal{O}$ be a bounded domain in $\mathbf{R}^2$ with h holes. More precisely, $\mathcal{O}$ is written as

$$\mathcal{O} = D \setminus \left(\bigcup_{i=1}^{h} \overline{\mathcal{O}_i} \right), \tag{5.2}$$

where D and $\mathcal{O}_i$ are bounded simply connected open sets such that $\overline{\mathcal{O}_i} \subset D$ and $\overline{\mathcal{O}_i} \cap \overline{\mathcal{O}_j} = \emptyset, i \neq j$. In the following, the coefficients of differential forms on $\mathcal{O}$ are assumed to belong to $C^\infty(\overline{\mathcal{O}})$. Take a point $x^{(i)} = (x_1^{(i)}, x_2^{(i)}) \in \mathcal{O}_i$, and define a 1-form

$$\omega_i = \frac{1}{2\pi} \frac{(x_1 - x_1^{(i)})dx_2 - (x_2 - x_2^{(i)})dx_1}{(x_1 - x_1^{(i)})^2 + (x_2 - x_2^{(i)})^2}. \tag{5.3}$$

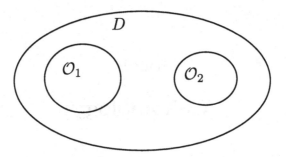

Fig. 5.1   Domain with holes

By the polar coordinates $x_1 - x_1^{(i)} = r\cos\theta, x_2 - x_2^{(i)} = r\sin\theta$, we have $\omega_i = d\theta/(2\pi)$, hence $d\omega_i = 0$ and

$$\int_{\partial\mathcal{O}_i} \omega_j = \delta_{ij}. \tag{5.4}$$

Note that the above $\omega_1, \cdots, \omega_h$ in (5.3) are harmonic, i.e.

$$d\omega_i = 0, \quad \delta\omega_i = 0. \tag{5.5}$$

Let ω be a closed 1-form, i.e. $d\omega = 0$. We put

$$c_i = \int_{\partial\mathcal{O}_i} \omega. \tag{5.6}$$

Then by Theorem 1.8,

$$\int_C \left(\omega - \sum_{i=1}^h c_i\omega_i\right) = 0 \tag{5.7}$$

for any closed curve C in $\mathcal{O}$. In fact, let I be the set of all i such that $\mathcal{O}_i$ is enclosed by C. Then by (5.1)

$$\int_C \omega = \sum_{i\in I}\int_{\partial\mathcal{O}_i}\omega, \quad \int_C\sum_{i=1}^h c_i\omega_i = \sum_{i\in I}\int_{\partial\mathcal{O}_i} c_i\omega_i,$$

which proves (5.7). The formula (5.7) and Theorem 1.11 then imply that for any curve C_x in $\mathcal{O}$ with fixed starting point p and end point x, the integral

$$f(x) = \int_{C_x}\left(\omega - \sum_{i=1}^h c_i\omega_i\right)$$

does not depend on the choice of the curve C_x. Then, $f(x)$ is a well-defined function on $\mathcal{O}$ and satisfies

$$df = \omega - \sum_{i=1}^{h} c_i \omega_i.$$

In fact, letting $\omega - \sum_{i=1}^{h} c_i \omega_i = a_1(x)dx_1 + a_2(x)dx_2$, we have

$$f(x_1 + \epsilon, x_2) - f(x_1, x_2) = \int_{x_1}^{x_1+\epsilon} a_1(t, x_2)dt,$$

which implies $f_{x_1} = a_1$. Similarly, $f_{x_2} = a_2$.

Let $Z^1(\mathcal{O})$ be the set of 1-forms ω such that $d\omega = 0$, and $B^1(\mathcal{O})$ be the set $\{df \, ; \, f \in C^\infty(\overline{\mathcal{O}})\}$. We put

$$H^1(\mathcal{O}) = Z^1(\mathcal{O})/B^1(\mathcal{O}), \tag{5.8}$$

and call it the *1st order cohomology group*. This means that $\omega, \omega' \in Z^1(\mathcal{O})$ are identified if there is $f \in C^\infty(\overline{\mathcal{O}})$ such that $\omega - \omega' = df$:

$$\omega = \omega' \quad \text{in} \quad H^1(\mathcal{O}) \Longleftrightarrow \omega - \omega' = df.$$

It is easy to see that $H^1(\mathcal{O})$ is a vector space. We show that $\omega_1, \cdots, \omega_h$ form a basis of $H^1(\mathcal{O})$. In fact, the above computation shows that any $\omega \in Z^1(\mathcal{O})$ is written as $\omega = \sum_{i=1}^{h} c_i \omega_i + df$, i.e. $\omega = \sum_{i=1}^{h} c_i \omega_i$ in $H^1(\mathcal{O})$. Suppose $\sum_{i=1}^{h} c_i' \omega_i = 0$ in $H^1(\mathcal{O})$, i.e. $\sum_{i=1}^{h} c_i' \omega_i = dg$. Using (5.4), we have

$$0 = \sum_i \int_{\partial \mathcal{O}_j} c_i' \omega_i = c_j',$$

which proves that $\omega_1, \cdots, \omega_h$ are linearly independent in $H^1(\mathcal{O})$.

In summary, we have proven the following theorem.

Theorem 5.1. *For a planar domain $\mathcal{O}$ with h-holes, the first Betti number of $\mathcal{O}$ is h, i.e.*

$$\beta_1(\mathcal{O}) = \dim H^1(\mathcal{O}) = h.$$

5.2 Closed 2-dimensional surface

Let us imagine a sphere S^2 and a tube $T = \{x = (x_1, x_2, x_3) \, ; \, x_1^2 + x_2^2 = \epsilon, \, 0 \le x_3 \le 1\}$, ϵ being a sufficiently small constant. One can then glue two ends of T on S^2 to get a surface with *handle*. One can also glue several handles to S^2. It is known that any compact 2-dimensional surface can be continuously deformed to a surface with handles. The number of handles of a surface S is called a *genus* and is denoted by $g(S)$. The genus $g(S)$ is equal to the number of holes of the surface S. By a *triangulation* of S, we mean to draw a net of curved triangles on S. There is a lot of ways of triangulation, however, the *Euler characteristics*

$$\chi(S) = V - E + F, \tag{5.9}$$

is independent of triangulation, where V is the number of vertices, E the number of edges and F the number of faces. The following formula is known:

$$\chi(S) = 2 - 2g(S). \tag{5.10}$$

If the surface S is parametrized by $\theta = (\theta_1, \theta_2)$, the 1-form $\omega = \sum_{i=1}^{3} a_i(x) dx_i$ on S is defined by

$$\sum_{i=1}^{3} a_i(x) dx_i = \sum_{i=1}^{3} a_i(x(\theta)) \left(\left(\frac{\partial x_i}{\partial \theta_1} \right) d\theta_1 + \left(\frac{\partial x_i}{\partial \theta_2} \right) d\theta_2 \right).$$

As in the case of the previous section, the 1st order cohomology group $H^1(S)$ is defined by

$$H^1(S) = Z^1(S)/B^1(S), \tag{5.11}$$

where $Z^1(S)$ is the set of closed 1-forms on S, and $B^1(S)$ is the set $\{df \, ; \, f \in C^\infty(S)\}$. Then, the 1st *Betti number* is defined by

$$\beta_1(S) = \dim H^1(S). \tag{5.12}$$

It is known that

$$\beta_1(S) = 2 - \chi(S) = 2g(S). \tag{5.13}$$

Let S be a compact 2-dimensional surface with h handles. It is regarded as a boundary of a compact 3-dimensional body D with h handles, which we call the *handle body*. This handle body is decomposed into $h + 1$ parts: $D = D_0 \cup D_1 \cup \cdots \cup D_h$, where D_0 is homeomorphic (i.e. continuously deformable) to the ball $\{|x| \leq 1\}$, and for $1 \leq i \leq h$, the handle D_i is homeomorphic to the pole $\{(x_1, x_2, x_3) \, ; \, x_1^2 + x_2^2 \leq \epsilon, \, 0 \leq x_3 \leq 1\}$. The handle D_i then makes a hole in D, which we call H_i. One can then consider closed curves $\gamma_i, i = 1, \cdots, h$, inside D such that γ_i and γ_j are not linked if $i \neq j$, and γ_i encircles the hole H_i. Let us also consider curves $\gamma_i', 1 \leq i \leq h$, outside D such that γ_i' and γ_j' are not linked if $i \neq j$, and γ_i' encircles the handle D_i. Let $\omega(\gamma_i), \omega(\gamma_i'), i = 1, \cdots, h$, be the 1-forms defined by (4.24).

Theorem 5.2. *Let S be a 2-dimensional compact surface in $\mathbf{R}^3$ with h handles. Then $\beta_1(S) = 2h$, and $H^1(S)$ has the following basis*

$$\omega(\gamma_i) = \left(\int_{y \in \gamma_i} \left(\nabla_x \frac{1}{4\pi|x-y|} \right) \times dy \right) \cdot dx, \quad i = 1, \cdots, h,$$

$$\omega(\gamma_i') = \left(\int_{y \in \gamma_i'} \left(\nabla_x \frac{1}{4\pi|x-y|} \right) \times dy \right) \cdot dx, \quad i = 1, \cdots, h.$$

This is proven in the same way as in Theorem 5.1 using Theorem 4.2. We omit the details.

5.3 Handle body

In the x_1-x_3 plane, we draw a disc $(x_1 - R)^2 + x_3^2 \le \rho^2$ with $0 < \rho < R$, and rotate it around the x_3-axis. Let Ω be the resulting *solid torus*, i.e. the 3-dimensional body. Take a circle $\gamma = \{(x_1 - R)^2 + x_3^2 = a^2,\ x_2 = 0\}$ with $\rho < a < R$ outside Ω, and also $\gamma' = \{x_1^2 + x_2^2 = \rho_1^2;\ x_3 = 0\}$, where $R - \rho < \rho_1 < R$, inside Ω. They are oriented so that γ and γ' are positively linked once. Then $\omega(\gamma)$ defined by (4.24) is a harmonic 1-form on Ω and

$$\int_{\gamma'} \omega(\gamma) = 1.$$

Take any closed 1-form ω on Ω, and put

$$c_0 = \int_{\gamma'} \omega.$$

For any closed curve C on Ω, there is an integer n such that $C \simeq n\gamma'$. Then

$$\int_C \omega = n \int_{\gamma'} \omega = nc_0 = c_0 \int_C \omega(\gamma).$$

Therefore $\omega - c_0\omega(\gamma) = df$ for some 0-form f, hence $\omega(\gamma)$ is a basis of the 1st order cohomology group $H^1(\Omega)$.

This can be generalized to any handle body. Let Ω be a bounded open set in $\mathbf{R}^3$ diffeomorphic to a ball glued with h handles. We take circles $\gamma_1, \cdots, \gamma_h$ encircling each handle outside Ω, and curves $\gamma_1', \cdots, \gamma_h'$ as above inside the handles. Then arguing as above, one can show that

Theorem 5.3. *Let Ω be a compact 3-dimensional body with h-handles. Then, the 1st Betti number of Ω is h and $\omega(\gamma_1), \cdots, \omega(\gamma_h)$ are the basis of $H^1(\Omega)$, where ω is defined by (4.24).*

5.4   Divergence free fields

Let Ω be a bounded domain in $\mathbf{R}^3$. It is well-known that any vector field A on Ω is decomposed as

$$A = \operatorname{grad} f + \operatorname{curl} B. \tag{5.14}$$

This fact, known as *Helmholtz decomposition*, is closely related with the topology of Ω and is a source of various mathematical problems. It plays important roles not only in mathematics, but also in many areas in physics such as electrodynamics, fluid dynamics and plasma physics. We explain below an elementary introduction to the associated mathematical framework, called the *Hodge theory*, in $\mathbf{R}^3$. We follow [18].

To fix the idea, we restrict ourselves to real vector fields, however the arguments below are valid also for complex vector fields. Let

$$\mathcal{V}(\Omega) = C^\infty(\overline{\Omega}\,;\mathbf{R}^3) \tag{5.15}$$

be the set of real C^∞-vector fields on $\overline{\Omega}$ equipped with the inner product

$$(A, B) = \int_\Omega A(x) \cdot B(x) dx. \tag{5.16}$$

We put

$$\mathcal{K}(\Omega) = \{A \in \mathcal{V}\,;\, \nabla \cdot A = 0 \text{ in } \Omega,\quad A \cdot \nu = 0 \text{ on } \partial\Omega\}, \tag{5.17}$$

$$\mathcal{G}(\Omega) = \{\nabla\varphi\,;\, \varphi \in C^\infty(\overline{\Omega};\mathbf{R})\}, \tag{5.18}$$

where ν is the outer unit normal field on $\partial\Omega$.

The first decomposition is as follows.

Lemma 5.1. *The orthogonal decomposition*

$$\mathcal{V}(\Omega) = \mathcal{K}(\Omega) \oplus \mathcal{G}(\Omega)$$

holds in the sense of the inner product (5.16).

Proof.   By intergration by parts, we have

$$\int_\Omega A(x) \cdot \nabla\varphi(x) dx = \int_{\partial\Omega} A(x) \cdot \nu\varphi(x) dS - \int_\Omega (\nabla \cdot A)\varphi dx,$$

which proves the orthogonality. If $V \in \mathcal{V}(\Omega)$ admits the decomposition $V = A + \nabla\varphi$ where $A \in \mathcal{K}(\Omega)$, we shoud have

$$\begin{cases} \Delta\varphi = \nabla \cdot V, & \text{in } \Omega, \\ \dfrac{\partial\varphi}{\partial\nu} = \nu \cdot V, & \text{on } \partial\Omega. \end{cases} \tag{5.19}$$

We look for φ in the form $\varphi = \varphi_0 + \psi$, where

$$\varphi_0 = -\int_\Omega R_0(x-y)\nabla \cdot V(y)dy,$$

$R_0(x)$ being the Green function for $-\Delta$. Then ψ satisfies

$$\begin{cases} \Delta\psi = 0, & \text{in } \Omega, \\ \partial_\nu\psi = \nu \cdot V + \int_\Omega \partial_{\nu_x} R_0(x-y)\nabla \cdot V(y)dy =: f, & \text{on } \partial\Omega. \end{cases} \tag{5.20}$$

By Lemma 3.7, we have

$$\int_{\partial\Omega} \partial_{\nu(x)} R_0(x-y)dS_x = -1, \quad y \in \Omega.$$

Therefore,

$$\int_{\partial\Omega} dS_x \int_\Omega \partial_{\nu_x} R_0(x-y)\nabla \cdot V(y)dy = -\int_\Omega \nabla \cdot V(y)dy$$
$$= -\int_{\partial\Omega} \nu \cdot V dS. \tag{5.21}$$

Then $\int_{\partial\Omega} f dS = 0$, and there exists a solution to the Neumann problem (5.20) by virtue of Theorem 3.18. Having constructed φ, we put $A = V - \nabla\varphi$. It then satisfies $\nabla \cdot A = 0$ in Ω and $\nu \cdot A = 0$ on $\partial\Omega$. $\qquad\square$

The following lemma proves (5.14).

Lemma 5.2. *For $A \in \mathcal{K}(\Omega)$, put*

$$B(x) = \frac{1}{4\pi} \int_\Omega \frac{A(y) \times (x-y)}{|x-y|^3} dy. \tag{5.22}$$

Then, $B(x) \in C^1(\overline{\Omega})$ and

$$\nabla \cdot B = 0, \quad \text{in } \Omega, \tag{5.23}$$

$$\nabla \times B = A, \quad \text{in } \Omega. \tag{5.24}$$

Proof. We put

$$G(x) = \frac{1}{4\pi} \int_\Omega \frac{A(y)}{|x-y|} dy,$$

and observe that $G(x) \in C^2(\overline{\Omega})$. In fact, it is easy to check $G(x) \in C^1(\overline{\Omega})$, and using $\nabla_x|x-y|^{-1} = -\nabla_y|x-y|^{-1}$,

$$\frac{\partial}{\partial x_i} G(x) = -\frac{1}{4\pi} \int_{\partial\Omega} A(y)\nu_i \frac{1}{|x-y|} dS_y + \frac{1}{4\pi} \int_\Omega \left(\frac{\partial}{\partial y_i} A(y)\right) \frac{1}{|x-y|} dy.$$

The 2nd term is in $C^1(\overline{\Omega})$. So is the 1st term by the property of single layer potential. This proves that $G(x) \in C^2(\overline{\Omega})$. We also see that $\text{div}\, G = 0$.

Take $A_\epsilon(x) \in C_0^\infty(\mathbf{R}^3)$ such that $A_\epsilon(x) \to A(x)$ in Ω, $A_\epsilon(x) \to 0$ outside Ω, and put

$$G_\epsilon(x) = \frac{1}{4\pi} \int_{\mathbf{R}^3} \frac{A_\epsilon(y)}{|x-y|}\, dy.$$

Then, $G_\epsilon \to G$ in Ω. Since $-\Delta G_\epsilon = A_\epsilon$ in $\mathbf{R}^3$, for any $\varphi \in C_0^\infty(\Omega)$,

$$(A_\epsilon, \varphi) = (-\Delta G_\epsilon, \varphi) = (G_\epsilon, -\Delta\varphi) \to (G, -\Delta\varphi).$$

Letting $\epsilon \to 0$, we have

$$(A, \varphi) = (-\Delta G, \varphi),$$

which proves $-\Delta G = A$ in Ω.

Noting that $B = \nabla \times G$, we have $\nabla \cdot B = \text{div}\,\text{curl}\, G = 0$, and also $\nabla \times B = \text{curl}\,\text{curl}\, G = \text{grad}\,\text{div}\, G - \Delta G = A$. $\qquad\square$

A 2-dimensional surface Σ in $\overline{\Omega}$ is said to be a *cross-sectional surface* if $\partial\Sigma \subset \partial\Omega$. Let $(\mathcal{CS})$ be the set of all cross-sectional surfaces in Ω. We define a subset $\mathcal{K}_0(\Omega) \subset \mathcal{K}(\Omega)$ by

$$\mathcal{K}_0(\Omega) = \left\{ A \in \mathcal{K}(\Omega);\; \int_\Sigma A \cdot \nu\, d\Sigma = 0, \quad \forall \Sigma \in (\mathcal{CS}) \right\}, \qquad (5.25)$$

where ν is the unit normal to Σ.

Here, let us introduce *surface gradient*. Suppose a surface S in $\mathbf{R}^3$ is parametrized by $\theta = (\theta^1, \theta^2)$, i.e. $S = \{x(\theta)\,;\, \theta \in U\}$, U being an open set in $\mathbf{R}^2$. Let

$$g_{ij} = \frac{\partial x}{\partial \theta^i} \cdot \frac{\partial x}{\partial \theta^j}, \quad (g^{ij}) = (g_{ij})^{-1}, \quad 1 \le i, j \le 2.$$

Then the surface gradient of a function $f = f(\theta)$ on S is defined by

$$\nabla_S f = g^{ij} \frac{\partial f}{\partial \theta^j} \frac{\partial x}{\partial \theta^i}. \qquad (5.26)$$

Recall that we are using Einstein's summation convention:

$$g^{ij} a_i b_j = \sum_{i,j=1}^{2} g^{ij} a_i b_j.$$

Lemma 5.3. *For $f \in C^\infty(\mathbf{R}^3)$, we have*

$$\nabla f = \frac{\partial f}{\partial \nu}\nu + \nabla_S f, \quad \text{on} \quad S,$$

where ν is a unit normal field on S.

Proof. We extend ν smoothly near S, and solve the differential equation

$$\frac{dy}{dt} = \nu, \quad y(0) = x(\theta) \in S. \tag{5.27}$$

Then the map $(x(\theta), t) \to y(t, \theta)$ is a diffeomorphism for $|t| < \epsilon$, where $\epsilon > 0$ is small enough. Letting $\theta_3 = t$, $g_S = (g_{ij})_{1 \le i,j \le 2}$, and $\overline{g}_{ij} = \frac{\partial y}{\partial \theta^i} \cdot \frac{\partial y}{\partial \theta^j}, 1 \le i, j \le 3$, we have

$$\left(\overline{g}_{ij}\right) = \begin{pmatrix} g_S & 0 \\ 0 & 1 \end{pmatrix} \quad \text{on} \quad S. \tag{5.28}$$

Let $\nabla_y f = a^i \frac{\partial y}{\partial \theta^i}$. Then we have $\frac{\partial y}{\partial \theta^j} \cdot \nabla_y f = a^i \overline{g}_{ij}$, hence

$$a^i = \overline{g}^{ij} \frac{\partial y}{\partial \theta^j} \cdot \nabla_y f = \overline{g}^{ij} \frac{\partial f}{\partial \theta^j}.$$

In view of (5.28), we have

$$a^i = g^{ij} \frac{\partial f}{\partial \theta^j} \quad (i = 1, 2), \quad a^3 = \frac{\partial f}{\partial \nu},$$

which comples the proof of the lemma. $\qquad\square$

Lemma 5.4. *For $A \in \mathcal{K}_0(\Omega)$, define B by (5.22) and put*

$$B_{tan} = \left(B - (\nu \cdot B)\nu\right)\big|_{\partial\Omega}, \tag{5.29}$$

ν being the outer unit normal to $\partial\Omega$. Then, there exists $f \in C^\infty(\partial\Omega)$ such that

$$B_{tan} = \nabla_{\partial\Omega} f = g^{ij} \frac{\partial f}{\partial \theta^j} \frac{\partial x}{\partial \theta^i}. \tag{5.30}$$

Proof. We show that for any closed curve C on $\partial\Omega$

$$\int_C B \cdot dx = \int_C B_{tan} \cdot dx = 0. \tag{5.31}$$

By deforming C continuously, we have only to consider the cases in which C is either a boundary of a surface Σ inside $\overline{\Omega}$ or outside Ω. In the former case, using Stokes' theorem (Theorem 1.6) and Lemma 5.2, we have

$$\int_C B \cdot dx = \int_\Sigma (\nabla \times B) \cdot \nu \, d\Sigma = \int_\Sigma A \cdot \nu \, d\Sigma = 0$$

by the definition of $\mathcal{K}_0(\Omega)$. Using $\nabla \times B = \text{grad div } G - \Delta G = 0$ outside Ω, we also have $\int_C B \cdot dx = 0$ if C is a boundary of a surface outside Ω.

Fix a point $x_0 \in \partial\Omega$ arbitrarily, and for any $x \in \partial\Omega$, draw a curve C_x in $\partial\Omega$ from x_0 to x. We define $f(x)$ by

$$f(x) = \int_{C_x} B \cdot dx.$$

By (5.31), $f(x)$ does not depend on the choice of the curve C_x. Take a small open set S in $\partial\Omega$ on which $\partial\Omega$ is parametrized by θ, i.e. $S = \{x(\theta) \, ; \, \theta \in U\}$. Then, taking $a = a(\theta_0) \in S$ arbitrarily, we have

$$f(x(\theta)) = f(a) + \int_{C_\theta} \left(B \cdot \frac{\partial x}{\partial \theta^1} d\theta^1 + B \cdot \frac{\partial x}{\partial \theta^2} d\theta^2 \right),$$

where C_θ is a curve in U form θ_0 to θ. Then, we have

$$\frac{\partial f}{\partial \theta^1} = B \cdot \frac{\partial x}{\partial \theta^1}, \quad \frac{\partial f}{\partial \theta^2} = B \cdot \frac{\partial x}{\partial \theta^2}.$$

Letting

$$B_{tan} = b^1 \frac{\partial x}{\partial \theta^1} + b^2 \frac{\partial x}{\partial \theta^2},$$

we have

$$\frac{\partial f}{\partial \theta^j} = B_{tan} \cdot \frac{\partial x}{\partial \theta^j} = b^i g_{ij},$$

hence we obtain (5.30). $\qquad\qquad\qquad\qquad\qquad\qquad\qquad\qquad\qquad\qquad\square$

We can now prove the following lemma.

Lemma 5.5.

$$\mathcal{K}_0(\Omega) = \left\{ \nabla \times U \, ; \, U \in \mathcal{V}(\Omega), \, \nabla \cdot U = 0 \text{ in } \Omega, \, U \times \nu = 0 \text{ on } \partial\Omega \right\}.$$

Proof. Let $A \in \mathcal{K}_0(\Omega)$ and B, f be as in (5.22) and (5.30). We solve the Dirichlet problem

$$\Delta\varphi = 0 \quad \text{in} \quad \Omega, \quad \varphi = f \quad \text{on} \quad \partial\Omega,$$

and let $U = B - \nabla\varphi$. Then we have

$$\nabla \cdot U = 0, \quad \nabla \times U = \nabla \times B = A.$$

Take a tangential vector field $T = T^\ell \frac{\partial x}{\partial \theta^\ell}$ on $\partial\Omega$. Then we have on $\partial\Omega$

$$B \cdot T = B_{tan} \cdot T = g^{ij} \frac{\partial f}{\partial \theta^j} \frac{\partial x}{\partial \theta^i} \cdot T = \frac{\partial f}{\partial \theta^\ell} T^\ell,$$

where we have used $g^{ij} g_{j\ell} = \delta^j_\ell$. Lemma 5.3 implies

$$\nabla\varphi = \frac{\partial\varphi}{\partial\nu}\nu + g^{ij} \frac{\partial f}{\partial \theta^j} \frac{\partial x}{\partial \theta^i},$$

hence

$$\nabla\varphi \cdot T = T^i \frac{\partial f}{\partial \theta^i}.$$

Therefore, we have $(B - \nabla\varphi) \cdot T = 0$ for any tangential vector field T, which proves $U \times \nu = 0$ on $\partial\Omega$.

To prove the opposite inclusion, put $A = \nabla \times U$, where U satisfies $\nabla \cdot U = 0$ and $U \times \nu = 0$. We then have $\nabla \cdot A = 0$ and by Stokes' theorem

$$\int_\Sigma A \cdot \nu \, d\Sigma = \int_\Sigma (\nabla \times U) \cdot \nu \, d\Sigma = \int_{\partial\Sigma} U \cdot dx = 0,$$

since U is orthogonal to $\partial\Omega$ and $\partial\Sigma$ is a curve on $\partial\Omega$. It remains to show $A \cdot \nu = 0$ on $\partial\Omega$. In view of Lemma 5.1, we have only to prove that A is orthogonal to $\mathcal{G}(\Omega)$. Take $\varphi \in C^\infty(\overline{\Omega})$. Using

$$\nabla \cdot (U \times \nabla\varphi) = (\nabla \times U) \cdot \nabla\varphi$$

(see (A.19)), we have

$$\int_\Omega A \cdot \nabla\varphi dx = \int_\Omega \nabla \cdot (U \times \nabla\varphi) dx = \int_{\partial\Omega} \nu \cdot (U \times \nabla\varphi) dS = 0,$$

since $(U \times \nabla\varphi) \cdot \nu = (\nu \times U) \cdot \nabla\varphi = 0$. □

5.5 Harmonic fields

We introduce another subset $\mathcal{K}_H(\Omega) \subset \mathcal{K}(\Omega)$ by

$$\mathcal{K}_H(\Omega) = \{A \in \mathcal{K}(\Omega) \, ; \, \nabla \times A = 0 \text{ in } \Omega\}.$$

In other words,

$$\mathcal{K}_H(\Omega) \ni A \Longleftrightarrow \begin{cases} \nabla \cdot A = 0, & \text{in} \quad \Omega, \\ \nabla \times A = 0, & \text{in} \quad \Omega, \\ A \cdot \nu = 0, & \text{on} \quad \partial\Omega. \end{cases} \tag{5.32}$$

Using $\operatorname{curl} \operatorname{curl} = \operatorname{grad} \operatorname{div} - \Delta$, we have $\Delta A = 0$ if $A \in \mathcal{K}_H(\Omega)$. We call $\mathcal{K}_H(\Omega)$ the set of *harmonic vector fields* on Ω.

Lemma 5.6. $\mathcal{K}_0(\Omega)$ *and* $\mathcal{K}_H(\Omega)$ *are orthogonal.*

Proof. For $A \in \mathcal{K}_H(\Omega)$ and $\nabla \times U \in \mathcal{K}_0(\Omega)$, we have by integration by parts

$$\int_\Omega (\nabla \times U) \cdot A dx = \int_{\partial\Omega} (\nu \times U) \cdot A dS + \int_\Omega U \cdot (\nabla \times A) dx,$$

which vanishes by the conditions on A and U. □

Suppose Ω has the handles $\mathcal{H}_j$, $j = 1, \cdots, h$, and let C_j be a closed curve in Ω enclosing the hole made by $\mathcal{H}_j$, and C'_j a closed curve in $\mathbf{R}^3 \setminus \overline{\Omega}$ encircling $\mathcal{H}_j$, so that the linking number of C_i with C'_j is δ_{ij}. Take a cross-sectional surface Σ_j in $\mathcal{H}_j$ so that C_j passes through Σ_j. We put

$$B_j(x) = \frac{1}{4\pi} \int_{C'_j} \frac{(y-x) \times dy}{|y-x|^3}. \tag{5.33}$$

Then by Lemma 4.4 and Theorem 4.2,

$$\nabla \cdot B_j = 0, \quad \nabla \times B_j = 0, \quad \int_{C_i} B_j \cdot dx = \delta_{ij}.$$

Consider the Neumann problem

$$\begin{cases} \Delta\varphi = 0, & \text{in} \quad \Omega, \\ \partial_\nu\varphi = \nu \cdot B_j, & \text{on} \quad \partial\Omega. \end{cases} \tag{5.34}$$

By Green's formula

$$\int_{\partial\Omega} \nu \cdot B_j dS = \int_\Omega \nabla \cdot B_j dx = 0.$$

Hence, (5.34) has a solution φ_j, which is unique up to a constant. We put

$$H_j = B_j - \nabla\varphi_j. \tag{5.35}$$

By the construction, $H_j \in \mathcal{K}_H(\Omega)$, moreover

$$\int_{C_i} H_j \cdot dx = \delta_{ij}. \tag{5.36}$$

This implies that $H_1, \cdots, H_h$ are linearly independent.

Lemma 5.7. *(1)* $H_1, \cdots, H_h$ *form a basis of* $\mathcal{K}_H(\Omega)$.
(2) The mappings

$$\mathcal{K}_H(\Omega) \ni A \to \left(\int_{C_1} A \cdot dx, \cdots, \int_{C_h} A \cdot dx \right) \in \mathbf{R}^h, \tag{5.37}$$

$$\mathcal{K}_H(\Omega) \ni A \to \left(\int_{\Sigma_1} A \cdot \nu \, d\Sigma_1, \cdots, \int_{\Sigma_h} A \cdot \nu \, d\Sigma_h \right) \in \mathbf{R}^h \tag{5.38}$$

are isomorphisms.

Proof. Take $A \in \mathcal{K}_H(\Omega)$ and let

$$\int_{C_j} A \cdot dx = c_j.$$

Letting $V = A - \sum_{j=1}^{h} c_j H_j$, we have $V \in \mathcal{K}_H(\Omega)$ and

$$\int_C V \cdot dx = 0$$

for any simple closed curve C in Ω. Then, V is written as $V = \nabla \varphi$. Since $\nabla \cdot V = 0$, this φ satisfies $\Delta \varphi = 0$ and $\partial_\nu \varphi = 0$. Hence, φ is a constant, which yields $V = 0$. This proves (1). By the formula (5.36), $\{C_i\}$ is a dual basis of $\{H_j\}$, which proves (5.37).

To prove (2), note that for any cross-sectional surface $\Sigma \subset \Omega$, there exists j such that

$$\int_\Sigma A \cdot \nu \, d\Sigma = \int_{\Sigma_j} A \cdot \nu \, d\Sigma_j. \tag{5.39}$$

In fact, for some j, Σ_j and Σ as well as a part of $\partial\Omega$ form the boundary of a 3-dimensional subset $\Omega' \subset \Omega$. Integrating $\nabla \cdot A = 0$ in Ω' and using $A \cdot \nu = 0$ on $\partial\Omega$, we obtain (5.39).

Now, we let

$$f_j = \int_{\Sigma_j} A \cdot \nu \, d\Sigma_j \tag{5.40}$$

and suppose $f_1 = \cdots = f_h = 0$. By (5.39), we have

$$\int_\Sigma A \cdot \nu \, d\Sigma = 0, \quad \forall \Sigma \in (\mathcal{CS}). \tag{5.41}$$

Hence $A \in \mathcal{K}_0(\Omega)$. By Lemma 5.6, we have $A = 0$. This proves that (5.38) is an isomorphism. $\square$

We put

$$\beta_{ij} = \int_{\Sigma_i} H_j \cdot \nu \, d\Sigma_i. \tag{5.42}$$

Since $H_1, \cdots, H_h$ are the basis of $\mathcal{K}_H(\Omega)$, Lemma 5.7 (2) implies that

$$\det\left(\beta_{ij}\right) \neq 0. \tag{5.43}$$

We can now prove the second decomposition.

Lemma 5.8.

$$\mathcal{K}(\Omega) = \mathcal{K}_0(\Omega) \oplus \mathcal{K}_H(\Omega).$$

Proof. Take $A \in \mathcal{K}(\Omega)$, and define f_j by (5.40). Using (5.43), we solve the equation

$$\sum_{j=1}^{h} \beta_{ij}\alpha_j = f_i, \quad i = 1, \cdots, h,$$

to obtain α_j and put $V = A - \sum_{j=1}^{h} \alpha_j H_j$. Then, $\int_{\Sigma_j} V \cdot \nu d\Sigma_j = 0$ holds. Therefore, $V \in \mathcal{K}_0(\Omega)$. This and Lemma 5.6 prove the lemma. $\square$

5.6   Gradient fields

We next decompose $\mathcal{G}(\Omega)$. We define

$$\mathcal{G}_g(\Omega) = \{\nabla\varphi \; ; \; \varphi \in C^\infty(\overline{\Omega}; \mathbf{R}), \; \varphi = 0 \text{ on } \partial\Omega\}, \qquad (5.44)$$

$$\mathcal{G}_h(\Omega) = \{\nabla\varphi \; ; \; \varphi \in C^\infty(\overline{\Omega}; \mathbf{R}), \; \Delta\varphi = 0 \text{ in } \Omega\}. \qquad (5.45)$$

Lemma 5.9.

$$\mathcal{G}(\Omega) = \mathcal{G}_g(\Omega) \oplus \mathcal{G}_h(\Omega).$$

Proof.   For $\nabla\varphi \in \mathcal{G}(\Omega)$, let ψ be the solution to the Dirichlet problem

$$\Delta\psi = 0 \quad \text{in} \quad \Omega, \quad \psi = \varphi \quad \text{on} \quad \partial\Omega.$$

Letting $\phi = \varphi - \psi$, we have $\nabla\phi \in \mathcal{G}_g(\Omega)$ and $\nabla\psi \in \mathcal{G}_h(\Omega)$.

Take $\nabla\phi \in \mathcal{G}_g(\Omega)$ and $\nabla\psi \in \mathcal{G}_h(\Omega)$. Then, by integration by parts

$$\begin{aligned}
(\nabla\phi, \nabla\psi) &= \int_\Omega \nabla \cdot (\phi\nabla\psi) dx \\
&= \int_{\partial\Omega} \nu \cdot (\phi\nabla\psi) dS = 0,
\end{aligned}$$

which proves the orthogonality. $\square$

We further decompose $\mathcal{G}_h(\Omega)$. Let $S^{(j)}$, $j = 1, \cdots, n$, be the connected components of $\partial\Omega$, and put

$$\mathcal{G}_{h0}(\Omega) = \{\nabla\varphi \in \mathcal{G}_h(\Omega) \; ; \; \int_{S^{(j)}} \partial_\nu\varphi dS = 0, \; j = 1, \cdots, n\}, \qquad (5.46)$$

$$\mathcal{G}_H(\Omega) = \{\nabla\varphi \in \mathcal{G}_h(\Omega) \; ; \; \nabla_{\partial\Omega}\varphi = 0 \text{ on } \partial\Omega\}. \qquad (5.47)$$

Note that $\nabla_{\partial\Omega}\varphi = 0$ means that φ is constant on each $S^{(j)}$.

Lemma 5.10. $\mathcal{G}_{h0}(\Omega)$ *and* $\mathcal{G}_H(\Omega)$ *are orthogonal.*

Proof.   Take $\nabla\varphi \in \mathcal{G}_{h0}(\Omega)$ and $\nabla\psi \in \mathcal{G}_H(\Omega)$. Then, by integration by parts

$$\int_\Omega \nabla\varphi \cdot \nabla\psi \, dx = \int_{\partial\Omega} (\partial_\nu\varphi)\psi dS - \int_\Omega (\Delta\varphi)\psi \, dx = 0.$$

This proves the lemma. $\square$

Suppose that Ω has s connected components

$$\Omega = \Omega_1 \cup \cdots \cup \Omega_s,$$

and the boundary $\partial\Omega_i$ of each component Ω_i has n_i connected components

$$\partial\Omega_i = S_{i1} \cup \cdots \cup S_{in_i}.$$

Since $\{S_{ij}\}$ is a rearrangement of $\{S^{(j)}\}$, we have

$$n = n_1 + \cdots + n_s.$$

Lemma 5.11.

$$\dim \mathcal{G}_H(\Omega) = n - s.$$

Proof. We have

$$\dim \mathcal{G}_H(\Omega) = \sum_{i=1}^{s} \dim \mathcal{G}_H(\Omega_i).$$

Let φ_{ij} be the solution to the Dirichlet problem

$$\begin{cases} \Delta\varphi_{ij} = 0, & \text{in} \quad \Omega_i, \\ \varphi_{ij} = 1, & \text{on} \quad S_{ij}, \\ \varphi_{ij} = 0, & \text{on} \quad S_{ik}, \quad k \neq j. \end{cases}$$

For any $\varphi_i \in \mathcal{G}_H(\Omega_i)$, we put

$$c_{ij} = \varphi_i\big|_{S_{ij}}. \tag{5.48}$$

Then, φ_i is written as $\varphi_i = \sum_{j=1}^{n_i} c_{ij}\varphi_{ij}$. Since $(c_{i1} - c, \cdots, c_{in_i} - c)$ determine the same $\nabla\varphi_i$ for all c, $(c_{i1}, \cdots, c_{in_i})$ determines $\nabla\varphi_i$ uniquely under the constraint $\sum_{j=1}^{n_i} c_{ij} = 0$. Hence $\dim \mathcal{G}_H(\Omega_i) = n_i - 1$. This proves the lemma. $\qquad\square$

The above proof shows that there is a linear bijection between $\mathcal{G}_H(\Omega_i)$ and the subspace $\mathcal{C}_i$ defined by

$$\mathcal{C}_i = \left\{ (c_{i1}, \cdots, c_{in_i}) \in \mathbf{R}^{n_i} \,;\, c_{i1} + \cdots + c_{in_i} = 0 \right\}. \tag{5.49}$$

For $\varphi_i \in \mathcal{G}_H(\Omega_i)$, we put

$$f_{ij} = \int_{S_{ij}} \partial_\nu \varphi_i dS_{ij}. \tag{5.50}$$

By the divergence theorem, we have the constraint $\sum_{j=1}^{n_i} f_{ij} = 0$. Define the subspace $\mathcal{F}_i$ by

$$\mathcal{F}_i = \left\{ (f_{i1}, \cdots, f_{in_i}) \in \mathbf{R}^{n_i} \,;\, f_{i1} + \cdots + f_{in_i} = 0 \right\}. \tag{5.51}$$

Lemma 5.12. *Define c_{ij} and f_{ij} by (5.48) and (5.50). Then, the mappings*

$$\mathcal{G}_H(\Omega_i) \ni \varphi_i \to (c_{i1}, \cdots, c_{in_i}) \in \mathcal{C}_i, \tag{5.52}$$

$$\mathcal{G}_H(\Omega_i) \ni \varphi_i \to (f_{i1}, \cdots, f_{in_i}) \in \mathcal{F}_i, \tag{5.53}$$

are linear isomorphisms.

Proof. We prove the assertion for (5.53). Assume that $(f_{i1}, \cdots, f_{in_i}) = 0$. Then, $\varphi_i \in \mathcal{G}_{h0}(\Omega_i) \cap \mathcal{G}_H(\Omega_i)$. Therefore, $\varphi_i = 0$ by Lemma 5.10. This proves the bijectivity. $\qquad\square$

Lemma 5.13.

$$\mathcal{G}_h(\Omega) = \mathcal{G}_{h0}(\Omega) \oplus \mathcal{G}_H(\Omega).$$

Proof. Take $\nabla\varphi_i \in \mathcal{G}_h(\Omega_i)$, and put $f_{ij} = \int_{S_{ij}} \partial_\nu \varphi_i dS$. By the divergence theorem, $\sum_j f_{ij} = 0$. We use Lemma 5.12 to construct $\nabla\psi_i \in \mathcal{G}_H(\Omega_i)$ such that $f_{ij} = \int_{S_{ij}} \partial_\nu \psi_i dS$. Then, $\nabla\varphi_i - \nabla\psi_i \in \mathcal{G}_{h0}(\Omega_i)$. This proves the decompostion. The orthogonality is proved in Lemma 5.10. $\qquad\square$

5.7 Hodge decomposition theorem

By virtue of Lemmas 5.1, 5.8, 5.9 and 5.13, we have proven the following theorem.

Theorem 5.4. *The following orthogonal decomposition holds:*

$$\mathcal{V}(\Omega) = \mathcal{K}_0(\Omega) \oplus \mathcal{K}_H(\Omega) \oplus \mathcal{G}_{h0}(\Omega) \oplus \mathcal{G}_H(\Omega) \oplus \mathcal{G}_g(\Omega).$$

The operators $div, curl, grad$ have the following relations to this decomposition.

Lemma 5.14.

$$
\begin{aligned}
\mathrm{curl}^{-1}\{0\} &= & \mathcal{K}_H(\Omega) \oplus \mathcal{G}_{h0}(\Omega) \oplus \mathcal{G}_H(\Omega) \oplus \mathcal{G}_g(\Omega), \\
\mathrm{grad}\, C^\infty(\overline{\Omega}) &= & \mathcal{G}_{h0}(\Omega) \oplus \mathcal{G}_H(\Omega) \oplus \mathcal{G}_g(\Omega), \\
\mathrm{div}^{-1}\{0\} &= \mathcal{K}_0(\Omega) \oplus \mathcal{K}_H(\Omega) \oplus \mathcal{G}_{h0}(\Omega) \oplus \mathcal{G}_H(\Omega), \\
\mathrm{curl}\, \mathcal{V}(\Omega) &= \mathcal{K}_0(\Omega) \oplus \mathcal{K}_H(\Omega) \oplus \mathcal{G}_{h0}(\Omega).
\end{aligned}
$$

Proof. The first three formulas can be checked directly by using definitions. We prove the last formula. Suppose $A = \nabla \times V$. Let φ be a solution to the Neumann problem

$$\Delta\varphi = 0 \quad \text{in} \quad \Omega, \qquad \partial_\nu\varphi = A \cdot \nu \quad \text{on} \quad \partial\Omega.$$

Since $\nabla \cdot A = 0$, $\int_S \nu \cdot A dS = 0$ for any closed surface S in Ω. Therefore, the solution φ exists and $\nabla\varphi \in \mathcal{G}_{h0}(\Omega)$. Let $U = A - \nabla\varphi$. Then, $\nabla \cdot U = 0$ and $\nu \cdot U = 0$ on $\partial\Omega$, hence $U \in \mathcal{K}(\Omega) = \mathcal{K}_0(\Omega) \oplus \mathcal{K}_H(\Omega)$. This proves $\mathrm{curl}\, \mathcal{V}(\Omega) \subset \mathcal{K}_0(\Omega) \oplus \mathcal{K}_H(\Omega) \oplus \mathcal{G}_{h0}(\Omega)$.

Suppose $A \in \mathcal{K}_0(\Omega) \oplus \mathcal{K}_H(\Omega) \oplus \mathcal{G}_{h0}(\Omega)$. Then,

$$\nabla \cdot A = 0 \quad \text{in} \quad \Omega, \qquad \int_{S^{(j)}} \nu \cdot A dS = 0$$

for any connected component $S^{(j)}$ of $\partial\Omega$. Put

$$B(x) = \frac{1}{4\pi} \int_\Omega \frac{A(y) \times (x - y)}{|x - y|^3} dy.$$

Then we have

$$\nabla_x \times B(x) + \frac{1}{4\pi} \nabla_x \int_{\partial\Omega} \frac{A(y) \cdot \nu}{|x - y|} dS_y = \begin{cases} A(x), & x \in \Omega, \\ 0, & x \in \Omega^c. \end{cases} \tag{5.54}$$

We show that the 2nd term of the left-hand side of (5.54) is written as $\nabla \times U$.

Take a big ball $\{|x| < R\}$ such that $\Omega \subset \{|x| < R/2\}$ and put $\widetilde\Omega = \{|x| < R\} \setminus \Omega$. Solve the Neumann problem

$$\Delta\varphi = 0 \quad \text{in} \quad \widetilde\Omega, \quad \partial_\nu\varphi = \begin{cases} -\nu \cdot A & \text{on} \quad \partial\Omega \cap \partial\widetilde\Omega, \\ 0 & \text{on} \quad \{|x| = R\}. \end{cases} \tag{5.55}$$

For the solution $\widetilde\varphi$, we put $\widetilde A = \nabla\widetilde\varphi$, and

$$\widetilde B(x) = \frac{1}{4\pi} \int_{\widetilde\Omega} \frac{\widetilde A(y) \times (x - y)}{|x - y|^3} dy.$$

Then we have

$$\nabla_x \times \widetilde B(x) + \frac{1}{4\pi} \nabla_x \int_{\partial\widetilde\Omega} \frac{\widetilde A(y) \cdot \nu}{|x - y|} dS_y = \begin{cases} \widetilde A(x), & x \in \widetilde\Omega, \\ 0, & x \in \widetilde\Omega^c. \end{cases} \tag{5.56}$$

On $\Omega \subset \widetilde\Omega^c$, we have

$$\nabla_x \times \widetilde B(x) = -\frac{1}{4\pi} \nabla_x \int_{\partial\widetilde\Omega} \frac{\widetilde A(y) \cdot \nu}{|x - y|} dS_y$$

$$= \frac{1}{4\pi} \nabla_x \int_{\partial\Omega} \frac{A(y) \cdot \nu}{|x - y|} dS_y.$$

We then have $A = \nabla \times (B + \widetilde B)$, which proves $\mathcal{K}_0(\Omega) \oplus \mathcal{K}_H(\Omega) \oplus \mathcal{G}_{h0}(\Omega) \subset \text{curl}\, \mathcal{V}(\Omega)$. □

Note that the Helmholtz decomposition (5.14) follows from the last formula.

The following lemma is easy to prove.

Lemma 5.15.

$$\mathcal{K}_H(\Omega) = \left(\text{curl}^{-1}\{0\}\right) \cap \left(\text{grad}\, C^\infty(\overline\Omega)\right)^\perp.$$

Let us rewrite the above results in terms of differential forms. For a vector-valued function $A = (A_1, A_2, A_3)$, we associate the 1-form $\omega_A^{(1)}$ and 2-form $\omega_A^{(2)}$ by

$$\omega_A^{(1)} = A_1 dx^1 + A_2 dx^2 + A3 dx^3,$$

$$\omega_A^{(2)} = A_1 dx^2 \wedge dx^3 + A_2 dx^3 \wedge dx^1 + A_3 dx^1 \wedge dx^2.$$

Lemma 5.14 shows that

$$d\omega_A^{(1)} = 0 \iff A \in \mathcal{K}_H(\Omega) \oplus \mathcal{G}(\Omega). \tag{5.57}$$

We have thus arrived at the main theorem of this Chapter. Let us repeat the definition of $\mathcal{K}_H(\Omega)$: the set of vector fields in Ω satisfying

$$\nabla \times A = 0, \quad \nabla \cdot A = 0 \quad \text{in} \quad \Omega,$$

$$A \cdot \nu = 0 \quad \text{on} \quad \partial\Omega,$$

which is often called the *Neumann harmonic vector field*.

Theorem 5.5. *The cohomology group $H^1(\Omega)$ is isomorphic to $\mathcal{K}_H(\Omega)$. In particular the first Betti number of Ω is h, and $H^1(\Omega)$ has a basis $\omega_{H_1}^{(1)}, \cdots, \omega_{H_h}^{(1)}$, where H_j is defined by (5.35).*

For example, consider disjoint soild tori $\Omega_1, \cdots, \Omega_h$ and a ball containing $\Omega_1, \cdots, \Omega_h$. Then the first Betti number of $B \setminus \cup_{i=1}^h \Omega_i$ is h.

For the compuation of the Betti number of an exterion domain in $\mathbf{R}^3$, see Werner [98]. Kress [60] computed the Neumann electric and magnetic fields for the non-zero frequency. For the Hodge theory on manifolds, see [87].

Chapter 6

Initial value problem for the Maxwell equation

In this chapter, we study the distribution theory based on the Fourier transformation, and solve the initial value problem for the free Maxwell equation in $\mathbf{R}^3$.

6.1 Fourier transform

6.1.1 *Rapidly decreasing functions*

The *Fourier transform* of $f \in L^1(\mathbf{R}^n)$ is defined by

$$\widehat{f}(\xi) = (2\pi)^{-n/2} \int_{\mathbf{R}^n} e^{-ix\cdot\xi} f(x)\, dx. \tag{6.1}$$

It is well-known that if $f, \widehat{f} \in L^1(\mathbf{R}^n)$, the following inversion formula holds (see [88], p. 11):

$$f(x) = (2\pi)^{-n/2} \int_{\mathbf{R}^n} e^{ix\cdot\xi} \widehat{f}(\xi)\, d\xi. \tag{6.2}$$

We define the operators F and $F^\dagger$ by

$$(Ff)(\xi) = (2\pi)^{-n/2} \int_{\mathbf{R}^n} e^{-ix\cdot\xi} f(x)\, dx, \tag{6.3}$$

$$(F^\dagger g)(x) = (2\pi)^{-n/2} \int_{\mathbf{R}^n} e^{ix\cdot\xi} g(\xi)\, d\xi. \tag{6.4}$$

Then (6.2) is rewritten as

$$f = F^\dagger F f. \tag{6.5}$$

For a non-negative integer m, define the norm p_m by

$$p_m(\varphi) = \sup_{x\in\mathbf{R}^n} (1+|x|)^m \sum_{|\alpha|\le m} |\partial_x^\alpha \varphi(x)|. \tag{6.6}$$

Schwartz' space of rapidly decreasing functions, denoted by $\mathcal{S}$, is defined to be the set of all $\varphi \in C^\infty(\mathbf{R}^n)$ such that

$$\mathcal{S} \ni \varphi \Longleftrightarrow p_m(\varphi) < \infty, \quad \forall m \geq 0. \tag{6.7}$$

By definition, $\mathcal{S}$ is the set of all C^∞-functions on $\mathbf{R}^n$ whose all derivatives decay faster than any polynomial of x at infinity. As can be checked easily,

$$F(\partial_x^\alpha \varphi) = (i\xi)^\alpha F\varphi, \quad \partial_\xi^\alpha F\varphi = F\big((-ix)^\alpha \varphi\big) \tag{6.8}$$

hold for any α and $\varphi \in \mathcal{S}$. This implies that

$$\mathcal{S} \ni \varphi \Longrightarrow F\varphi \in \mathcal{S}. \tag{6.9}$$

Therefore, the space $\mathcal{S}$ is invariant by F. In particular, the inversion formula (6.5) holds on $\mathcal{S}$.

6.1.2   *Fourier transforms of L^2-functions*

Let $(\ ,\ )$ and $\|\cdot\|$ be the inner product and the norm of $L^2(\mathbf{R}^n)$. Then, for $\varphi, \psi \in \mathcal{S}$, the following *Plancherel's formula* holds:

$$(\varphi, \psi) = (\widehat{\varphi}, \widehat{\psi}). \tag{6.10}$$

In fact, by using the inversion formula, we have

$$(\varphi, \psi) = \int_{\mathbf{R}^n} \varphi(x) \overline{\left((2\pi)^{-n/2} \int_{\mathbf{R}^n} e^{ix\cdot\xi} \widehat{\psi}(\xi) d\xi \right)} dx.$$

Changing the order of integration in the right-hand side, we obtain (6.10). Since $\mathcal{S}$ is dense in $L^2(\mathbf{R}^n)$, for any $f \in L^2(\mathbf{R}^n)$, there exists $\varphi_j \in \mathcal{S}$ such that $\|\varphi_j - f\| \to 0$. By virtue of (6.10), we then have $\|F\varphi_j - F\varphi_k\| \to 0$ as $j, k \to \infty$. Therefore, there exists $\widetilde{f} \in L^2(\mathbf{R}^n)$ such that $F\varphi_j \to \widetilde{f}$ in $L^2(\mathbf{R}^n)$. The Fourier transform of $f \in L^2(\mathbf{R}^n)$ is then defined by $Ff = \widehat{f} = \widetilde{f}$. This definition does not depend on the choice of sequence $\{\varphi_j\} \subset \mathcal{S}$ convergent to f. Plancherel's formula (6.10) also holds on $L^2(\mathbf{R}^n)$:

$$(f, g) = (\widehat{f}, \widehat{g}), \quad \forall f, g \in L^2(\mathbf{R}^n). \tag{6.11}$$

In particular, we have the following *Parseval's formula*

$$\|f\|_{L^2(\mathbf{R}^n)} = \|\widehat{f}\|_{L^2(\mathbf{R}^n)}. \tag{6.12}$$

The Fourier transformation F thus defined on $L^2(\mathbf{R}^n)$ is unitary on $L^2(\mathbf{R}^n)$. In fact, (6.12) shows that F is isometric on $L^2(\mathbf{R}^n)$. To show that it is onto, note that $F^\dagger$ has the same properties as F, in particular, $FF^\dagger = I$ on $\mathcal{S}$. For any $g(\xi) \in L^2(\mathbf{R}^n)$ take $\{\varphi_j(\xi)\} \subset \mathcal{S}$ such that $\|\varphi_j - g\| \to$ and put $\psi_j = F^\dagger \varphi_j$. Then $F\psi_j = \varphi_j \to g$ in $L^2(\mathbf{R}^n)$. Letting $h = \lim_{j\to\infty} F^\dagger \varphi_j$, we have $Fh = g$. By the general property of unitary operator, we have $F^{-1} = F^*$. Therefore, we obtain $F^\dagger = F^*$. Hence, in the following, we use F^* instead of $F^\dagger$.

6.1.3 *Tempered distributions*

If $f(x)$ grows up at most polynomially, i.e.

$$|f(x)| \le C(1 + |x|)^N, \quad x \in \mathbf{R}^n,$$

or more generally,

$$\int_{\mathbf{R}^n} (1 + |x|)^{-N} |f(x)|^2 dx < \infty,$$

for some $N > 0$, we can associate a linear map $T_f : \mathcal{S} \to \mathbf{C}$ by

$$\langle T_f, \varphi \rangle = \int_{\mathbf{R}^n} f(x)\varphi(x)dx, \quad \forall \varphi \in \mathcal{S}. \tag{6.13}$$

A linear map $T : \mathcal{S} \to \mathbf{C}$ is said to be a *tempered distribution* if it is continuous on $\mathcal{S}$, i.e.

$$p_m(\varphi_j) \to 0, \quad \forall m \quad \Longrightarrow \quad \langle T, \varphi_j \rangle \to 0, \tag{6.14}$$

where p_m is the norm in (6.6). In the following, we consider only tempered distributions, and often call them simply as distribution, although L. Schwatz's original definition is more general. The set of all tempered distributions on $\mathbf{R}^n$ is denoted by $\mathcal{S}'$. T_f in (6.13) is a tempered distribution. The single and double layer potentials as well as many measures on $\mathbf{R}^n$ define tempered distributions. In particular, for any $a \in \mathbf{R}^n$, one can associate a distribution δ_a, called *Dirac measure*, by

$$\langle \delta_a, \varphi \rangle = \varphi(a), \quad \varphi \in \mathcal{S}. \tag{6.15}$$

We often write this formula as

$$\langle \delta_a, \varphi \rangle = \int_{\mathbf{R}^n} \delta(x - a)\varphi(x)dx = \varphi(a),$$

where $\delta(x)$ is a symbolic notation. It is known that there is no integrable function $\delta(x)$ having this property. However, we often deal with $\delta(x)$ as if it were a usual function. It is customary to write δ_0 as δ, hence

$$\langle \delta, \varphi \rangle = \int_{\mathbf{R}^n} \delta(x)\varphi(x)dx = \varphi(0), \quad \forall \varphi \in \mathcal{S}.$$

If $f(x) \in L^1(\mathbf{R}^n)$, we have for $\varphi \in \mathcal{S}$

$$\int_{\mathbf{R}^n} \widehat{f}(\xi)\varphi(\xi)d\xi = \int_{\mathbf{R}^n} f(x)\widehat{\varphi}(x)dx.$$

With this formula in mind, we define the Fourier transform of $T \in \mathcal{S}'$ by

$$\langle \widehat{T}, \varphi \rangle = \langle T, \widehat{\varphi} \rangle, \quad \forall \varphi \in \mathcal{S}. \tag{6.16}$$

By this definition, $\widehat{T}$ is again a linear functional and not a usual functon. However, it often happens that $\widehat{T}$ can be written as T_f by some function f of polynomial growth. For example,

$$\langle \delta_a, \widehat{\varphi} \rangle = \langle T_f, \varphi \rangle, \quad f(x) = (2\pi)^{-n/2} e^{-ix \cdot a}.$$

The Fourier transform of $T \in \mathcal{S}'$ is often written symbolically as

$$\widehat{T}(\xi) = (2\pi)^{-n/2} \int_{\mathbf{R}^n} e^{-ix \cdot \xi} T(x) dx.$$

In the case of the Dirac measure, we then have

$$\widehat{\delta_a} = (2\pi)^{-n/2} \int_{\mathbf{R}^n} e^{-ix \cdot \xi} \delta(x - a) dx = (2\pi)^{-n/2} e^{-ia \cdot \xi}. \qquad (6.17)$$

In view of (6.16), it is natural to define the inverse Fourier transform of $T \in \mathcal{S}'$ by

$$\langle F^* T, \varphi \rangle = \langle T, F^* \varphi \rangle, \quad \forall \varphi \in \mathcal{S}.$$

Then, the inversion formula also holds on $\mathcal{S}'$,

$$T = F^* F T, \quad \forall T \in \mathcal{S}'. \qquad (6.18)$$

The following formal formula for the Dirac measure

$$\delta(x) = (2\pi)^{-n} \iint_{\mathbf{R}^{2n}} e^{i(x-y) \cdot \xi} dy d\xi. \qquad (6.19)$$

is thus justified by (6.17) and (6.18).

6.2   Distributions with support on surfaces

The support of a distribution $T \in \mathcal{S}'$ is defined as follows:

Definition 6.1. A point $x_0 \in \mathbf{R}^n$ *does not* belong to $\operatorname{supp} T$ if and only if there exists an open set $U \subset \mathbf{R}^n$ such that $x_0 \in U$ and

$$\langle T, \varphi \rangle = 0, \quad \forall \varphi \in C_0^\infty(U).$$

Thus, to check that $\operatorname{supp} T = K$, one has to show that for any $x_0 \in \mathbf{R}^n$ there exists an open neighborhood U_{x_0} of x_0 such that

$$\begin{cases} x_0 \notin K \implies \langle T, \varphi \rangle = 0, \quad \forall \varphi \in C_0^\infty(U_{x_0}), \\ x_0 \in K \implies \langle T, \varphi \rangle \neq 0, \quad \exists \varphi \in C_0^\infty(U_{x_0}). \end{cases}$$

One can easily check that

$$\operatorname{supp} \delta_a = \{a\},$$

and for single layer or double layer potentials on a surface $S \subset \mathbf{R}^n$, their supports are contained in S. We also consider a distribution with support in a curve $C \subset \mathbf{R}^n$ defined by the line integral

$$\langle T_C, \varphi \rangle = \int_C f(x)\varphi(x)ds, \quad \forall \varphi \in \mathcal{S}$$

under suitable assumptions on f.

A compactly supported distribution T can be extended to $C^\infty(\mathbf{R}^n)$. In fact, we have only to take $\chi \in C_0^\infty(\mathbf{R}^n)$ such that $\chi(x) = 1$ on $\operatorname{supp} T$ and define

$$\langle T, \psi \rangle = \langle T, \chi\psi \rangle, \quad \forall \psi \in C^\infty(\mathbf{R}^n).$$

In particular, the Fourier transform of a compactly supported distribution T can be computed as

$$\widehat{T}(\xi) = (2\pi)^{-n/2}\langle T, e^{-ix\cdot\xi} \rangle. \tag{6.20}$$

This can be seen from the following computation

$$\langle \widehat{T}, \varphi \rangle = \langle T, \widehat{\varphi} \rangle = \langle T, (2\pi)^{-n/2} \int e^{-ix\cdot\xi}\varphi(\xi)d\xi \rangle$$

$$= (2\pi)^{-n/2} \int \langle T, e^{-ix\cdot\xi} \rangle \varphi(\xi)d\xi.$$

The formula (6.20) shows that $\widehat{T}(\xi) \in C^\infty(\mathbf{R}^n)$. Furthermore, it is analytic with respect to $\xi \in \mathbf{C}^n$.

The Dirac measure, single and double layer potentials, and the measure supported on a line are not represented as T_f with some polynomially bounded function f on $\mathbf{R}^n$. However, their Fourier transforms are polynomially bounded functions. The behavior as $|\xi| \to \infty$ of these Fourier transforms depends on the dimension of their supports. The following theorem is due to Agmon-Hörmander [4], which plays a central role in the following arguments.

Theorem 6.1. *Let M be a compact surface of dimension m in $\mathbf{R}^n$, and u a distribution defined by*

$$\langle u, \varphi \rangle = \int_M u_0(x)\varphi(x)dM, \quad \forall \varphi \in \mathcal{S},$$

where $u_0 \in L^2(M)$. Then, $\widehat{u}$ is a polynomially bounded function, and there exists a constant $C > 0$ such that

$$\frac{1}{R^{n-m}} \int_{|\xi| < R} |\widehat{u}(\xi)|^2 d\xi \leq C\|u_0\|_{L^2(M)}^2, \quad \forall R > 1.$$

Proof.   By using a partition of unity we have only to consider the case in which the support of u_0 is sufficiently small. Then on supp u_0, M is represented as $x = \varphi(u)$, where $\operatorname{rank}\left(\frac{\partial \varphi}{\partial u_1}, \cdots, \frac{\partial \varphi}{\partial u_m}\right) = m$. Without loss of generality, we can assume that $\det\left(\frac{\partial \varphi_i}{\partial u_j}\right)_{1 \leq i,j \leq m} \neq 0$. Letting $x = (x', x'')$, where $x' = (x_1, \cdots, x_m)$, by the inverse function theorem we have $u = u(x')$. Therefore, M is represented as $x = (x', x''(x'))$. The Fourier transform is then computed as

$$
\begin{aligned}
\widehat{u}(\xi) &= (2\pi)^{-n/2} \int_M e^{-ix \cdot \xi} u_0 \, dM \\
&= (2\pi)^{-m/2} \int_{\mathbf{R}^m} e^{-ix' \cdot \xi'} e^{-ix''(x') \cdot \xi''} u_0(x') \sqrt{g(x')} \, dx',
\end{aligned}
$$

where $g(x')$ is the determinant of the induced metric on M. Parseval's formula (6.12) yields

$$
\int_{\mathbf{R}^m} |\widehat{u}(\xi', \xi'')|^2 d\xi' = \int_{\mathbf{R}^m} |u_0(x')|^2 g(x') dx' = C_m \int_M |u_0|^2 \sqrt{g} \, dM
$$

with a constant $C_m > 0$. Using this, we have

$$
\begin{aligned}
\frac{1}{R^{n-m}} \int_{|\xi| < R} |\widehat{u}(\xi', \xi'')|^2 d\xi &\leq \frac{1}{R^{n-m}} \int_{\xi' \in \mathbf{R}^m, |\xi''| < R} |\widehat{u}(\xi', \xi'')|^2 d\xi' d\xi'' \\
&= \frac{C_m}{R^{n-m}} \int_{|\xi''| < R} \left(\int_M |u_0|^2 \sqrt{g} \, dM \right) d\xi'' \\
&\leq C \|u_0\|_{L^2(M)}^2,
\end{aligned}
$$

which proves the therem.   □

The converse of Theorem 6.1 is also true.

Theorem 6.2. *Let M be as in Theorem 6.1. Suppose $\widehat{u} \in \mathcal{S}'$ satisfies* supp $u \subset M$ *and*

$$
\limsup_{R \to \infty} \frac{1}{R^{n-m}} \int_{|\xi| < R} |\widehat{u}(\xi)|^2 d\xi < \infty.
$$

Then, u is represented as

$$
\langle u, \varphi \rangle = \int_M u_0 \varphi \, dM, \quad \forall \varphi \in \mathcal{S},
$$

where $u_0 \in L^2(M)$ and

$$
\int_M |u_0|^2 dM \leq C \limsup_{R \to \infty} \frac{1}{R^{n-m}} \int_{|\xi| < R} |\widehat{u}(\xi)|^2 d\xi.
$$

Proof. Let $\chi \in C_0^\infty(\mathbf{R}^n)$ be such that $\chi(x) = 0$ for $|x| > 1$ and $\int_{\mathbf{R}^n} \chi(x)dx = 1$. We put $\chi_\epsilon(x) = \epsilon^{-n}\chi(x/\epsilon)$, and

$$u_\epsilon(x) = \int u(y)\chi_\epsilon(x-y)dy = \langle u, \widetilde{\chi}_{\epsilon,x}\rangle,$$

where $\widetilde{\chi}_{\epsilon,x}(y) = \chi_\epsilon(x-y)$. Then $u_\epsilon \in C_0^\infty(\mathbf{R}^n)$ and

$$\widehat{u}_\epsilon(\xi) = (2\pi)^{n/2}\widehat{u}(\xi)\widehat{\chi}_\epsilon(\xi) = (2\pi)^{n/2}\widehat{u}(\xi)\widehat{\chi}(\epsilon\xi).$$

By Parseval's formula,

$$(2\pi)^{-n}\|u_\epsilon\|^2 = \int_{\mathbf{R}^n} |\widehat{u}(\xi)|^2|\widehat{\chi}(\epsilon\xi)|^2 d\xi$$

$$= \int_{\epsilon|\xi|<1} |\widehat{u}(\xi)|^2|\widehat{\chi}(\epsilon\xi)|^2 d\xi + \sum_{j=1}^\infty \int_{2^{j-1}<\epsilon|\xi|<2^j} |\widehat{u}(\xi)|^2|\widehat{\chi}(\epsilon\xi)|^2 d\xi.$$

Assuming that $0 < \epsilon < 1$, and letting $\ell = n - m$, we estimate the 1st term of the right-hand side from above by

$$\sup_{\xi\in\mathbf{R}^n} |\widehat{\chi}(\xi)|^2 \, \epsilon^{-\ell} \sup_{1\le R\le 1/\epsilon} \frac{1}{R^\ell}\int_{|\xi|<R} |\widehat{u}(\xi)|^2 d\xi.$$

For $2^{j-1} < \epsilon|\xi| < 2^j$, we have

$$|\widehat{\chi}(\epsilon\xi)|^2 \le C\,(1 + |\epsilon\xi|)^{-2\ell} \le C\,2^{-j\ell}.$$

Hence

$$\int_{2^{j-1}<\epsilon|\xi|<2^j} |\widehat{u}(\xi)|^2|\widehat{\chi}(\epsilon\xi)|^2 d\xi \le C\,2^{-j\ell}\,\epsilon^{-\ell} \sup_{1\le R\le 1/\epsilon} \frac{1}{R^\ell}\int_{|\xi|<R} |\widehat{u}(\xi)|^2 d\xi.$$

We have thus obtained

$$\|u_\epsilon\|^2 \le C\epsilon^{-\ell} \sup_{1\le R\le 1/\epsilon} \frac{1}{R^\ell}\int_{|\xi|<R} |\widehat{u}(\xi)|^2 d\xi.$$

Let us note that $\operatorname{supp} u_\epsilon \subset M_\epsilon$, where M_ϵ is the ϵ-neighborhood of M, and

$$\langle u_\epsilon, \psi\rangle \to \langle u, \psi\rangle, \quad \forall\psi \in \mathcal{S}.$$

Moreover, we have

$$\epsilon^{-\ell} \int_{M_\epsilon} |\psi(x)|^2 dx \to C\int_M |\psi|^2 dM.$$

Therefore, letting $\epsilon \to 0$ in $|\langle u_\epsilon, \psi\rangle|^2$, we have

$$|\langle u, \psi\rangle|^2 \le C\int_M |\psi|^2 dM \limsup_{R\to\infty}\frac{1}{R^\ell}\int_{|\xi|<R} |\widehat{u}(\xi)|^2 d\xi,$$

which completes the proof of the theorem. $\qquad\square$

6.3   Sobolev spaces

For a domain $\Omega \subset \mathbf{R}^n$, let $L^1_{loc}(\Omega)$ be the set of functions $f(x)$ on Ω satisfying $\int_K |f(x)|dx < \infty$ for any compact set $K \subset \Omega$. Let $\alpha = (\alpha_1, \cdots, \alpha_n) \in \mathbf{Z}^n_+$ be a multi-index. For $f, g \in L^1_{loc}(\Omega)$, we say that g is the α-th *weak derivative* or *distribution derivative* of f, if

$$\int_\Omega f(x)\partial_x^\alpha \varphi(x)dx = (-1)^{|\alpha|} \int_\Omega g(x)\varphi(x)dx, \quad \forall \varphi \in C_0^\infty(\Omega) \qquad (6.21)$$

is satisfied. If $f \in C^{|\alpha|}(\Omega)$, integration by parts shows that $\partial_x^\alpha f(x) = g(x)$. Therefore, without loss of generality, the α-th weak derivative of f is denoted by $\partial_x^\alpha f(x)$.

For example, when $n = 1$, $f(x) = |x|$ is not in C^1, however, its distribution derivative exists and is given by

$$(|x|)' = \begin{cases} 1, & x > 0, \\ -1, & x < 0. \end{cases}$$

For an integer $m \geq 0$, let $W^{m,p}(\Omega)$ be the set of functions whose α-th weak derivative exists for all α such that $|\alpha| \leq m$, and satisfies $\partial_x^\alpha f \in L^p(\Omega)$, where $1 \leq p \leq \infty$. It becomes a Banach space equipped with the norm

$$\|f\|_{W^{m,p}(\Omega)} = \sum_{|\alpha| \leq m} \|\partial_x^\alpha f\|_{L^p(\Omega)}.$$

When $p = 2$, $W^{m,p}(\Omega)$ is often denoted by $H^m(\Omega)$. For $\Omega = \mathbf{R}^n$, we have an equivalent definition of $H^m(\mathbf{R}^n)$ by passing to the Fourier transform:

$$H^m(\mathbf{R}^n) \ni f \Longleftrightarrow \sum_{|\alpha| \leq m} \|\xi^\alpha \widehat{f}(\xi)\|_{L^2(\mathbf{R}^n)} < \infty.$$

Thus for $s \in \mathbf{R}^n$, we define the Sobolev space $H^s(\mathbf{R}^n)$ as a set of distributions $f \in \mathcal{S}$ whose Fourier transform $\widehat{f}$ satisfies $(1 + |\xi|)^s \widehat{f}(\xi) \in L^2(\mathbf{R}^n)$. It is the Hilbert space with the inner product

$$(f, g)_{H^s} = \int_{\mathbf{R}^n} (1 + |\xi|^2)^{s/2} \widehat{f}(\xi)\overline{\widehat{g}(\xi)}d\xi.$$

If $s > 0$ is not an integer, put $s = m + \sigma$, where m is a non-negative integer and $0 < \sigma < 1$. Then, $u \in H^s(\mathbf{R}^n)$ if and only if

$$\sum_{|\alpha| \leq m} \int_{\mathbf{R}^n} |\partial_x^\alpha u(x)|^2 dx + \sum_{|\alpha| = m} \int_{\mathbf{R}^{2n}} \frac{|\partial^\alpha u(x) - \partial^\alpha u(y)|^2}{|x - y|^{n+2\sigma}} dx dy < \infty. \qquad (6.22)$$

As is seen above, the elements in $W^{m,p}(\Omega)$ do not belong to $C^m(\Omega)$ in general. However, they have a certain regularity depending on m and p. This is made precise in the following Sobolev's imbedding theorem.

Theorem 6.3. *Let Ω be a domain in $\mathbf{R}^n$ with smooth boundary $\partial\Omega$, and $1 \leq p < \infty$.*

(1) If $\dfrac{1}{q} = \dfrac{1}{p} - \dfrac{m}{n} > 0$, we have $W^{m,p}(\Omega) \subset L^q(\Omega)$ and

$$\|u\|_{L^q(\Omega)} \leq C\|u\|_{W^{m,p}(\Omega)}.$$

(2) If $\dfrac{1}{p} < \dfrac{m-k}{n}$, we have $W^{m,p}(\Omega) \subset C^k(\Omega)$ and

$$\|u\|_{C^k(\Omega)} \leq C\|u\|_{W^{m,p}(\Omega)}.$$

The following Rellich's theorem is also very important.

Theorem 6.4. *Let Ω be a bounded domain with smooth boundary, $m \geq 1$ and $1 \leq p < \infty$. Then, the imbedding $W^{m,p}(\Omega) \subset W^{m-1,p}(\Omega)$ is compact.*

For the proof of Theorems 6.3, 6.4, see [1]. As an example, we prove Theorem 6.3 (2) for the case when $\Omega = \mathbf{R}^n$ and $p = 2$. Given $s > n/2$, take $k \in \mathbf{Z}_+$ such that $s > k + n/2$. Then, by Fourier's inversion formula

$$|\partial_x^\alpha \varphi(x)| \leq (2\pi)^{-n/2} \int_{\mathbf{R}^n} |\xi^\alpha \widehat{\varphi}(\xi)| d\xi \leq C_t \|\varphi\|_{H^{t+|\alpha|}}$$

holds if $t > n/2$. Therefore

$$\sum_{|\alpha| \leq k} |\partial_x^\alpha \varphi(x)| \leq C\|\varphi\|_{H^s},$$

which proves Theorem 6.3 (2) when $p = 2$ and $s > n/2$.

6.4 Fourier transformation and the Maxwell equation

As has been introduced in Section 4.1, the Maxwell equation in the vacuum is written as

$$\begin{cases} \epsilon_0 \partial_t E = \operatorname{curl} H, \\ \mu_0 \partial_t H = -\operatorname{curl} E, \\ \operatorname{div} E = 0, \\ \operatorname{div} H = 0, \end{cases} \tag{6.23}$$

where ϵ_0, μ_0 are positive constants such that

$$\epsilon_0 \mu_0 = c_0^{-2}, \tag{6.24}$$

c_0 being the speed of light. Then, passing to the Fourier transform, the equation (6.23) is rewritten as

$$\begin{cases} \epsilon_0 \partial_t \widehat{E} = i\xi \times \widehat{H}, \\ \mu_0 \partial_t \widehat{H} = -i\xi \times \widehat{E}, \\ \xi \cdot \widehat{E} = 0, \\ \xi \cdot \widehat{H} = 0. \end{cases} \tag{6.25}$$

In the following, it is convenient to identify the operator $\xi\times$ with the skew-symmetric matrix:

$$\xi\times = \begin{pmatrix} 0 & -\xi_3 & \xi_2 \\ \xi_3 & 0 & -\xi_1 \\ -\xi_2 & \xi_1 & 0 \end{pmatrix}. \tag{6.26}$$

We let $u = (E, H)$ (by which we mean the column vector), and define $\mathcal{L}(\xi)$ and $\mathcal{M}_0$ by

$$\mathcal{L}(\xi) = \begin{pmatrix} 0 & -\xi\times \\ \xi\times & 0 \end{pmatrix}, \tag{6.27}$$

$$\mathcal{M}_0 = \begin{pmatrix} \epsilon_0 I_3 & 0 \\ 0 & \mu_0 I_3 \end{pmatrix}, \tag{6.28}$$

where I_3 is the 3×3 identity matrix. Then, (6.25) takes the form

$$\mathcal{M}_0 \partial_t \widehat{u} = -i\mathcal{L}(\xi)\widehat{u}. \tag{6.29}$$

We introduce the inner product

$$\langle a, b \rangle_{\mathcal{M}_0} = \langle \mathcal{M}_0 a, b \rangle_{\mathbf{C}^6} = (\epsilon_0 a_E) \cdot \overline{b_E} + (\mu_0 a_H) \cdot \overline{b_H} \tag{6.30}$$

for $a = (a_E, a_H), b = (b_E, b_H) \in \mathbf{C}^6$. Then $\mathcal{M}_0^{-1}\mathcal{L}(\xi)$ is Hermitian with respect to this inner product. The eigenvalues are

$$\lambda_0(\xi) = 0, \quad \lambda_\pm(\xi) = \pm c_0|\xi|, \tag{6.31}$$

each of which has multiplicity 2. Letting $\omega_1, \omega_2, \omega_3 \in S^2$ be such that $\omega_i \cdot \omega_j = \delta_{ij}$ and

$$\omega_1 = \xi/|\xi|, \quad \omega_1 \times \omega_2 = \omega_3, \quad \omega_2 \times \omega_3 = \omega_1, \quad \omega_3 \times \omega_1 = \omega_2, \tag{6.32}$$

the associated eigenvectors are given by

$$(\omega_1, 0), \quad (0, \omega_1) \quad \text{for} \quad \lambda_0(\xi), \tag{6.33}$$

$$(-\sqrt{\mu_0}\omega_3, \sqrt{\epsilon_0}\omega_2), \quad (\sqrt{\mu_0}\omega_2, \sqrt{\epsilon_0}\omega_3) \quad \text{for} \quad \lambda_+(\xi), \tag{6.34}$$

$$(\sqrt{\mu_0}\omega_3, \sqrt{\epsilon_0}\omega_2), \quad (-\sqrt{\mu_0}\omega_2, \sqrt{\epsilon_0}\omega_3) \quad \text{for} \quad \lambda_-(\xi). \tag{6.35}$$

Therefore, $e^{itM_0^{-1}L(\xi)}$ is represented as

$$e^{itM_0^{-1}L(\xi)} = \mathcal{P}_0^{(0)}(\xi) + e^{it\lambda_+(\xi)}\mathcal{P}_+^{(0)}(\xi) + e^{-it\lambda_-(\xi)}\mathcal{P}_-^{(0)}(\xi), \qquad (6.36)$$

where $\mathcal{P}_0^{(0)}(\xi), \mathcal{P}_\pm^{(0)}(\xi)$ are the eigenprojections associated with $\lambda_0(\xi)$ and $\lambda_\pm(\xi)$. Then, the solution of the Maxwell equation (6.23) with initial data $u(0,x) = f(x)$ is given by

$$
\begin{aligned}
u(t,x) &= (2\pi)^{-3/2} \int_{\mathbf{R}^3} e^{ix\cdot\xi - itM_0^{-1}L(\xi)} \widehat{f}(\xi) d\xi \\
&= (2\pi)^{-3/2} \int_{\mathbf{R}^3} e^{ix\cdot\xi} \Big(\mathcal{P}_0^{(0)}(\xi) + e^{-it\lambda_+(\xi)}\mathcal{P}_+^{(0)}(\xi) \qquad (6.37) \\
&\qquad\qquad + e^{-it\lambda_-(\xi)}\mathcal{P}_-^{(0)}(\xi) \Big) \widehat{f}(\xi) d\xi.
\end{aligned}
$$

This means that the solution of the Maxwell equation is written as a superposition of waves of the form

$$e^{i(x\cdot\xi \mp tc_0|\xi|)}.$$

Therefore, we have the correspondence:

$$\frac{c_0|\xi|}{2\pi} = \text{frequency}, \qquad (6.38)$$

$$\frac{2\pi}{|\xi|} = \text{wave length}. \qquad (6.39)$$

However, we often omit 2π (even c_0) and call $|\xi|$ *frequency*, and $1/|\xi|$ *wave length*.

Let $\mathbf{H}_0$ be the Hilbert space $\left(L^2(\mathbf{R}^3)\right)^6$ equipped with the inner product

$$
\begin{aligned}
(f,g)_{\mathcal{M}_0} &= \int_{\mathbf{R}^3} \langle f(x), g(x) \rangle_{\mathcal{M}_0} dx \\
&= \int_{\mathbf{R}^3} \left(\epsilon_0 f_E(x) \cdot \overline{g_E}(x) + \mu_0 f_H(x) \cdot \overline{g_H}(x) \right) dx. \qquad (6.40)
\end{aligned}
$$

Then, the operator

$$U_0(t) : u(0,x) \to u(t,x) \qquad (6.41)$$

is unitary on $\mathbf{H}_0$. If the initial data $f = (f_E, f_H)$ satisfies $\xi \cdot \widehat{f}_E(\xi) = \xi \cdot \widehat{f}_H(\xi) = 0$, we have $\mathcal{P}_0^{(0)}(\xi)\widehat{f}(\xi) = 0$. Then

$$u(t,x) = (2\pi)^{-3/2} \int_{\mathbf{R}^3} e^{ix\cdot\xi} \left(e^{-it\lambda_+(\xi)}\mathcal{P}_+^{(0)}(\xi) + e^{it\lambda_-(\xi)}\mathcal{P}_-^{(0)}(\xi) \right) \widehat{f}(\xi) d\xi.$$

$$\qquad (6.42)$$

It is also clear that $\mathcal{P}_\pm^{(0)}(\xi)\widehat{f}(\xi)$ satisfies

$$\xi \cdot \left(\mathcal{P}_\pm^{(0)}(\xi)\widehat{f}(\xi)\right)_E = \xi \cdot \left(\mathcal{P}_\pm^{(0)}(\xi)\widehat{f}(\xi)\right)_H = 0.$$

Hence, $u(t,x) = (E(t,x), H(t,x))$ defined by (6.42) satisfies $\operatorname{div} E(t,x) = \operatorname{div} H(t,x) = 0$.

We also use the standard inner product:

$$(f,g) = \int_{\mathbf{R}^3} \left(f_E(x) \cdot \overline{g_E(x)} + f_H(x) \cdot \overline{g_H(x)}\right) dx. \tag{6.43}$$

For $\zeta, \eta \in \mathbf{R}^3$, define the projection

$$|\zeta\rangle\langle\eta| \, V = (\eta \cdot V)\zeta, \quad \forall V \in \mathbf{C}^3. \tag{6.44}$$

Using (6.32), we have

$$\xi \times \widehat{P}_\perp(\xi) = |\xi| \big(|\omega_3\rangle\langle\omega_2| - |\omega_2\rangle\langle\omega_3|\big). \tag{6.45}$$

Hence,

$$\left(\xi \times \widehat{P}_\perp(\xi)\right)^* = -\xi \times \widehat{P}_\perp(\xi), \tag{6.46}$$

where $*$ denotes the adjoint in $\mathbf{C}^3$ with standard inner product.

6.5   Hilbert space approach

6.5.1   *Self-adjoint operator and unitary group*

Let A be a linear operator defined in a Hilbert space $\mathbf{H}$. Assume that its domain $D(A)$ is dense in $\mathbf{H}$. It is said to be *symmetric* if

$$(Au,v) = (u,Av), \quad \forall u,v \in D(A). \tag{6.47}$$

A symmetric operator A is said to be *self-adjoint* if it has the following property: If $v,g \in \mathcal{H}$ satisfy $(Au,v) = (u,g), \forall u \in D(A)$, then $v \in D(A)$. It then follows that $Av = g$. Consider the following differential equation (evolution equation) for a $\mathbf{H}$-valued function $u(t)$

$$i\partial_t u(t) = Au(t), \quad u(0) = f \in \mathcal{H}. \tag{6.48}$$

It is known that if A is self-adjoint and $f \in D(A)$, this equation has a unique solution $u(t)$ having the following properties (see e.g. [85]):

- $u(t)$ is an $\mathbf{H}$-valued C^1-function and satisfies (6.48). Here, C^1 means the strong differentiability, i.e.

$$\lim_{\epsilon \to 0} \left\| \frac{u(t+\epsilon) - u(t)}{\epsilon} - \partial_t u(t) \right\| = 0$$

and that $\partial_t u(t)$ is continuous with respect to t.
- For any $t \in \mathbf{R}$, operator $U(t) : f \to u(t)$ is uniquely extended to a unitary operator on $\mathbf{H}$, and satisfies $U(t)U(s) = U(t+s)$ for any $t, s \in \mathbf{R}$. It is denoted as $U(t) = e^{-itA}$.

6.5.2 Maxwell equation in the vacuum

In $\mathbf{H}_0 = (L^2(\mathbf{R}^3))^6$ equipped with the inner product (6.40), we consider the differential operator $\mathcal{L}(D_x)$, where $\mathcal{L}(\xi)$ is defined by (6.27) and $D_x = -i\nabla_x = i(\partial_{x_1}, \partial_{x_2}, \partial_{x_3})$. Then, $\mathcal{M}_0^{-1}\mathcal{L}(D_x)$ with domain $(C_0^\infty(\mathbf{R}^3))^6$ is symmetric in $\mathbf{H}_0$. We make it self-adjoint by enlarging the domain.

We have already solved the eigenvalue problem for $\mathcal{M}_0^{-1}\mathcal{L}(\xi)$ in the previous section. Let us look at it in a different view point. Define projections $\widehat{P}_\parallel(\xi)$ and $\widehat{P}_\perp(\xi)$ on $\mathbf{C}^3$ by

$$\widehat{P}_\parallel(\xi)V = \left(\frac{\xi}{|\xi|} \cdot V\right)\frac{\xi}{|\xi|}, \quad \forall V \in \mathbf{C}^3, \tag{6.49}$$

$$\widehat{P}_\perp(\xi) = 1 - \widehat{P}_\parallel(\xi). \tag{6.50}$$

Then, letting

$$\mathcal{P}^{(0)}(\xi) = \mathcal{P}_+^{(0)}(\xi) + \mathcal{P}_-^{(0)}(\xi), \tag{6.51}$$

we have

$$\mathcal{P}_0^{(0)}(\xi) = \begin{pmatrix} \widehat{P}_\parallel(\xi) & 0 \\ 0 & \widehat{P}_\parallel(\xi) \end{pmatrix}, \quad \mathcal{P}^{(0)}(\xi) = \begin{pmatrix} \widehat{P}_\perp(\xi) & 0 \\ 0 & \widehat{P}_\perp(\xi) \end{pmatrix}. \tag{6.52}$$

The Fourier transformation is extended onto $\mathbf{H}_0$ by

$$\mathcal{F}_0 = (F, \cdots, F), \tag{6.53}$$

with F from (6.4). Then, we define the free Maxwell operator $\mathcal{H}_0$ by

$$\mathcal{H}_0 = \mathcal{F}_0^* \mathcal{M}_0^{-1} \mathcal{L}(\xi) \mathcal{F}_0, \tag{6.54}$$

with domain

$$D(\mathcal{H}_0) \ni u = (E, H) \iff |\xi|P_\perp \widehat{E}, |\xi|P_\perp \widehat{H} \in (L^2(\mathbf{R}^3))^3. \tag{6.55}$$

The operator $\mathcal{H}_0$ is easily seen to be symmetric. To show that it is self-adjoint, suppose

$$(\mathcal{H}_0 u, v)_{M_0} = (u, g)_{M_0}, \quad \forall u \in D(\mathcal{H}_0).$$

Letting $u = (u_E, u_H)$, $v = (v_E, v_H)$, $g = (g_E, g_H)$, we have in the standard inner product (6.43)

$$(\xi \times \widehat{u}_H, \widehat{v}_E) - (\xi \times \widehat{u}_E, \widehat{v}_H) = \epsilon_0(\widehat{u}_E, \widehat{g}_E) + \mu_0(\widehat{u}_H, \widehat{g}_H),$$

which implies, taking $\widehat{u}_H = 0$,

$$-(\xi \times \widehat{P}_\perp \widehat{u}_E, \widehat{v}_H) = \epsilon_0(\widehat{u}_E, \widehat{g}_E), \quad \forall u_E \in C_0^\infty(\mathbf{R}^3).$$

This implies, by (6.46), that $\xi \times \widehat{v}_H \in L^2(\mathbf{R}^3)$ and $\xi \times \widehat{v}_H = \epsilon_0\widehat{g}_E$. Similary, we obtain $-\xi \times \widehat{v}_E \in L^2(\mathbf{R}^3)$ and $\xi \times \widehat{v}_E = \mu_0\widehat{g}_H$. This proves that $v \in D(\mathcal{H}_0)$.

Since $\mathcal{H}_0$ is self-adjoint, the equation

$$i\partial_t u = \mathcal{H}_0 u, \quad u(0) = f \tag{6.56}$$

in $\mathbf{H}_0$ has a unique solution

$$u(t) = e^{-it\mathcal{H}_0}f = \mathcal{F}_0^*\left(e^{-it\mathcal{M}_0^{-1}\mathcal{L}(\xi)}\widehat{f}(\xi)\right). \tag{6.57}$$

6.5.3   *Maxwell equation in the medium*

The Maxwell equation in the medium was given by (4.29). We consider the case that $\epsilon(x)$ and $\mu(x)$ are positive definite symmetric matrix independent of time t and the solution itself, i.e. the linear equation with variable coefficients. To solve the initial value problem on $\mathbf{R}^3$, it is sufficient to show that the operator

$$\mathcal{H} = \mathcal{M}(x)^{-1}\mathcal{L}(D_x) \qquad (6.58)$$

with domain $D(\mathcal{H}) = D(\mathcal{H}_0)$ is self-adjoint, where

$$\mathcal{M}(x) = \begin{pmatrix} \epsilon(x) & 0 \\ 0 & \mu(x) \end{pmatrix}.$$

To guarantee it, we have only to assume that there exists a constant $C_0 > 0$ such that

$$C_0 I_3 \leq \epsilon(x) \leq C_0^{-1} I_3, \quad C_0 I_3 \leq \mu(x) \leq C_0^{-1} I_3. \qquad (6.59)$$

Let $\mathbf{H}$ be the Hilbert space $(L^2(\mathbf{R}^3))^6$ equipped with the inner product

$$(u, v)_{\mathcal{M}} = \int_{\mathbf{R}^3} \langle \epsilon(x)u_E(x), v_E(x)\rangle dx + \langle \mu(x)u_H(x), v_H(x)\rangle dx, \qquad (6.60)$$

where $u = (u_E, u_H)$, $v = (v_E, v_H)$, and $\langle\,,\,\rangle$ is the inner product of $\mathbf{C}^3$.

Assume $(\mathcal{H}u, v)_{\mathcal{M}} = (u, g)_{\mathcal{M}}$, $\forall u \in D(\mathcal{H})$. This is eqauivalent to

$$(\mathcal{H}_0 u, v)_{\mathcal{M}_0} = (u, \mathcal{M}_0^{-1}\mathcal{M}g)_{\mathcal{M}_0}, \quad \forall u \in D(\mathcal{H}_0).$$

By the self-adjointness of $\mathcal{H}_0$, $v \in D(\mathcal{H}_0)$ and $\mathcal{H}_0 v = \mathcal{M}_0^{-1}\mathcal{M}g$. This implies $v \in \mathcal{H}$ and $\mathcal{H}v = g$. Hence $\mathcal{H}$ is self-adjoint.

Therefore, the initial value problem for the equation

$$i\partial_t u = \mathcal{H}u, \quad u(0) = f \in D(\mathcal{H}) \qquad (6.61)$$

has a unique solution $u(t) = e^{-it\mathcal{H}}f$.

Chapter 7

Boundary value problem for the Maxwell equation

We give here the mathematical basis of the initial boundary value problem for the Maxwell equation. Setting suitable function spaces, we define a self-adjoint realization of the Maxwell operator and study its spectral properties together with the basic knowledge of spectral theory. We further study the asymptotic behavior in time of solutions to the Maxwell equation, and prove the asymptotic completeness of wave operators.

7.1 Curl and divergence spaces

We begin with introducing spaces $H(\text{div}\,;\Omega)$, $H(\text{curl}\,;\Omega)$ and the notion of trace operator which will play important roles in the boundary value problem for the Maxwell equation. All of them are well-known, and we omit the proof. The details are seen in e.g. [90], [35], [74]. Our arguments are restricted to the case of the Maxwell equation. For the general theory of boundary value problems for the system of partial differential equations, see [64].

7.1.1 *Trace operator*

Let S be a compact $(n-1)$-dimensional surface in $\mathbf{R}^n$. We first introduce positive order Sobolev spaces on S. Take a partition of unity on S, i.e. $\chi_j(x) \in C_0^\infty(\mathbf{R}^n)$, $1 \le j \le J$, such that $\sum_{j=1}^J \chi_j(x) = 1$ for $x \in S$, and on $S \cap \operatorname{supp}\chi_j$, S is represented by local coordinates $\theta \in U_j$, where U_j is an open set in $\mathbf{R}^{n-1}$. For $s \ge 0$, we define $H^s(S)$ to be the set of all functions $\varphi \in L^2(S)$ such that $\chi_j\varphi \in H^s(\mathbf{R}^{n-1})$ (with respect to the local coordinates θ) for $1 \leq j \leq J$. We put

$$\|\varphi\|_{H^s(S)} = \sum_{j=1}^{J} \|\chi_j \varphi\|_{H^s(\mathbf{R}^{n-1})}.$$

This norm depends on the choice of partition of unity and local coordinates, however, the set $H^s(S)$ is defined invariantly. The negative order Sobolev space $H^{-s}(S)$ is defined as the dual space of $H^s(S)$. Namely, it is the set of all bounded linear functionals $f : H^s(S) \to \mathbf{C}$ satisfying

$$|\langle f, \varphi \rangle| \leq C\|\varphi\|_{H^s(S)}, \quad \forall \varphi \in H^s(S).$$

The norm of $f \in H^{-s}(S)$ is defined by

$$\|f\|_{H^{-s}(S)} = \sup_{0 \neq \varphi \in H^s(S)} \frac{|\langle f, \varphi \rangle|}{\|\varphi\|_{H^s(S)}}.$$

Theorem 7.1. *Let $\Omega \subset \mathbf{R}^n$ be a bounded domain with C^∞-boundary $\partial\Omega$. Then for any integer $s > 1/2$, the mapping $C^\infty(\overline{\Omega}) \ni u \to u|_{\partial\Omega} \in C^\infty(\partial\Omega)$ is uniquely extended to an operator from $H^s(\Omega)$ to $H^{s-1/2}(\partial\Omega)$:*

$$\|u|_{\partial\Omega}\|_{H^{s-1/2}(\partial\Omega)} \leq C\|u\|_{H^s(\Omega)}.$$

The mapping: $H^s(\Omega) \ni u \to u|_{\partial\Omega} \in H^{s-1/2}(\partial\Omega)$ thus defined is called the *trace operator*. We can also define the Sobolev space $H^s(\Omega)$ for non-integer s (replace $\mathbf{R}^n$ in (6.22) by Ω), and Theorem 7.1 still holds for this case.

The trace operator: $H^s(\Omega) \ni u \to u|_{\partial\Omega} \in H^{s-1/2}(\partial\Omega)$ is onto.

Theorem 7.2. *If $s > 1/2$, there exists an operator of extension $\mathrm{Ext} \in \mathbf{B}(H^{s-1/2}(\partial\Omega); H^s(\Omega))$ such that for any $\varphi \in H^{s-1/2}(\partial\Omega)$*

$$\mathrm{Ext}\,\varphi = \varphi \quad on \quad \partial\Omega.$$

For the proof, see e.g. [99], Theorem 8.8.

Later, in Subsection 9.3, we give another definition of $H^s(S)$ for all $s \in \mathbf{R}$ in a unified manner by using the Laplace-Beltrami operator on S.

7.1.2 Spaces $H(\mathrm{div}\,; \Omega)$ and $H(\mathrm{curl}\,; \Omega)$

These spaces are defined on both bounded and unbounded sets. Let us begin with the case of bounded sets. Let $\Omega \subset \mathbf{R}^3$ be a bounded domain with C^∞-boundary. We define the space $H(\mathrm{div}\,; \Omega)$ by

$$H(\mathrm{div}\,; \Omega) = \{u \in L^2(\Omega)^3 \,; \nabla \cdot u \in L^2(\Omega)\}, \tag{7.1}$$

equipped with the norm

$$\|u\|_{H(\text{div}\,;\,\Omega)} = \left(\|u\|^2_{L^2(\Omega)} + \|\nabla \cdot u\|^2_{L^2(\Omega)}\right)^{1/2}. \qquad (7.2)$$

Here, the divergence is understood in the distribution sense. Namely, for $u \in L^2(\Omega)^3$, $\nabla \cdot u \in L^2(\Omega)$ means that there exists $f \in L^2(\Omega)$ such that

$$(f, \varphi)_{L^2(\Omega)} = -(u, \nabla \varphi)_{L^2(\Omega)^3}, \quad \forall \varphi \in C_0^\infty(\Omega).$$

This f is denoted by $\nabla \cdot u$.

Lemma 7.1. $C^\infty(\overline{\Omega})$ *is dense in* $H(\text{div}\,;\,\Omega)$.

Let ν be the outer unit normal to $\partial\Omega$.

Theorem 7.3. *The mapping* $C^\infty(\overline{\Omega}) \ni u \to \nu \cdot u|_{\partial\Omega} \in C^\infty(\partial\Omega)$ *is uniquely extended to a bounded operator from* $H(\text{div}\,;\,\Omega)$ *to* $H^{-1/2}(\partial\Omega)$:

$$\|\nu \cdot u\|_{H^{-1/2}(\partial\Omega)} \le C\|u\|_{H(\text{div}\,;\,\Omega)}.$$

Moreover, the following Green's formula holds:

$$(u, \nabla\phi) + (\nabla \cdot u, \phi) = \langle \nu \cdot u, \phi \rangle, \quad u \in H(\text{div}\,;\,\Omega), \quad \phi \in H^1(\Omega).$$

Here, $(\ ,\ )$ *is the inner product of* $L^2(\Omega)^3$ *and* $\langle\ ,\ \rangle$ *is the coupling of* $H^{-1/2}(\partial\Omega)$ *and* $H^{1/2}(\partial\Omega)$.

By Theorem 7.3, for $u \in H(\text{div}\,;\,\Omega)$,

$$\nu \cdot u = 0 \quad \text{on} \quad \partial\Omega \Longleftrightarrow (u, \nabla\phi) = -(\nabla \cdot u, \phi), \quad \forall \phi \in H^1(\Omega).$$

We define the space $H(\text{curl}\,;\,\Omega)$ by

$$H(\text{curl}\,;\,\Omega) = \{u \in L^2(\Omega)^3\,;\, \text{curl}\, u \in L^2(\Omega)^3\}, \qquad (7.3)$$

equipped with the norm

$$\|u\|_{H(\text{curl}\,;\,\Omega)} = \left(\|u\|^2_{L^2(\Omega)} + \|\text{curl}\, u\|^2_{L^2(\Omega)}\right)^{1/2}. \qquad (7.4)$$

Here, the curl is understood in the distribution sense. Namely, for $u \in L^2(\Omega)^3$, $\text{curl}\, u \in L^2(\Omega)^3$ means that there exists $w \in L^2(\Omega)^3$ such that

$$(u, \text{curl}\, \phi) = (w, \phi), \quad \forall \phi \in C_0^\infty(\Omega)^3.$$

This w is denoted by $\text{curl}\, u$.

Lemma 7.2. $C^\infty(\overline{\Omega})$ *is dense in* $H(\text{curl}\,;\,\Omega)$.

Theorem 7.4. *The mapping* $C^\infty(\overline{\Omega}) \ni u \to \nu \times u\big|_{\partial\Omega} \in C^\infty(\partial\Omega)$ *is uniquely extended to a bounded operator from* $H(\mathrm{curl}\,; \Omega)$ *to* $H^{-1/2}(\partial\Omega)$:

$$\|\nu \times u\|_{H^{-1/2}(\partial\Omega)} \leq C\|u\|_{H(\mathrm{curl}\,;\Omega)}.$$

Moreover, the following Green's formula holds:

$$(\nabla \times u, \phi) - (u, \nabla \times \phi) = \langle \nu \times u, \phi \rangle, \quad u \in H(\mathrm{curl}\,; \Omega), \quad \phi \in H^1(\Omega)^3.$$

Therefore, for $u \in H(\mathrm{curl}\,;\Omega)$,

$$\nu \times u = 0 \quad \text{on} \quad \partial\Omega \Longleftrightarrow (\nabla \times u, \phi) = (u, \nabla \times \phi), \quad \forall \phi \in H^1(\Omega)^3.$$

Let $H_0(\mathrm{div}\,; \Omega)$ be the closure of $C_0^\infty(\Omega)$ in $H(\mathrm{div}\,; \Omega)$.

Theorem 7.5. *(1)* $H_0(\mathrm{div}\,; \Omega) = \{u \in H(\mathrm{div}\,; \Omega)\,; \nu \cdot u\big|_{\partial\Omega} = 0\}.$
(2) $H(\mathrm{curl}\,; \Omega) \cap H_0(\mathrm{div}\,; \Omega) \subset H^1(\Omega)$, *and*

$$\|u\|_{H^1(\Omega)} \leq C(\|\mathrm{div}\,u\|_{L^2(\Omega)} + \|\mathrm{curl}\,u\|_{L^2(\Omega)} + \|u\|_{L^2(\Omega)})$$

holds for any $u \in H(\mathrm{curl}\,; \Omega) \cap H_0(\mathrm{div}\,; \Omega)$.

Let $H_0(\mathrm{curl}\,; \Omega)$ be the closure of $C_0^\infty(\Omega)$ in $H(\mathrm{curl}\,; \Omega)$.

Theorem 7.6. *(1)* $H_0(\mathrm{curl}\,; \Omega) = \{u \in H(\mathrm{curl}\,; \Omega)\,; \nu \times u\big|_{\partial\Omega} = 0\}.$
(2) $H_0(\mathrm{curl}\,; \Omega) \cap H(\mathrm{div}\,; \Omega) \subset H^1(\Omega)$, *and*

$$\|u\|_{H^1(\Omega)} \leq C(\|\mathrm{div}\,u\|_{L^2(\Omega)} + \|\mathrm{curl}\,u\|_{L^2(\Omega)} + \|u\|_{L^2(\Omega)})$$

holds for any $u \in H_0(\mathrm{curl}\,; \Omega) \cap H(\mathrm{div}\,; \Omega)$.

The above theorems are also true for the exterior domain. Let us note that for $\xi, v \in \mathbf{R}^3$,

$$|\xi|^2|v|^2 = |\xi \cdot v|^2 + |\xi \times v|^2,$$

which implies, by passing to the Fourier transform,

$$\|u\|_{H^1(\mathbf{R}^3)} \leq C(\|\mathrm{div}\,u\|_{L^2(\mathbf{R}^3)} + \|\mathrm{curl}\,u\|_{L^2(\mathbf{R}^3)} + \|u\|_{L^2(\mathbf{R}^3)}). \qquad (7.5)$$

Now, given an exterior domain $\Omega \subset \mathbf{R}^3$ such that $\partial\Omega \subset \{|x| < R - 1\}$, take $\chi_1(x) \in C^\infty(\mathbf{R}^3)$ such that $\chi_1(x) = 1$ for $|x| < R$, and $\chi_1(x) = 0$ for $|x| > R + 1$. Put $\chi_2(x) = 1 - \chi_1(x)$. Let $\Omega_1 = \Omega \cap \{|x| < R + 1\}$. Then, by multiplying $\chi_1(x)$ and $\chi_2(x)$, one can split $H(\mathrm{div}\,; \Omega)$ into $H(\mathrm{div}\,; \Omega_1)$ and $H(\mathrm{div}\,; \mathbf{R}^3)$. In view of (7.5), one can see that Theorems 7.3 $\sim$ 7.6 also hold for the exterior domain.

Let us review the well-known *Helmholtz decomposition*. In the following, the norm $\| \cdot \|_{L^2(\Omega)}$ is often denoted by $\| \cdot \|$. Assume that Ω is a bounded domain with C^∞ boundary. We define:

$$T_{har}(\Omega) = \{h \in C^\infty(\overline{\Omega}) \,;\, \mathrm{div}\, h = \mathrm{curl}\, h = 0 \text{ in } \Omega, \, \nu \cdot h = 0 \text{ on } \partial\Omega\},$$
$$N_{har}(\Omega) = \{h \in C^\infty(\overline{\Omega}) \,;\, \mathrm{div}\, h = \mathrm{curl}\, h = 0 \text{ in } \Omega, \, \nu \times h = 0 \text{ on } \partial\Omega\},$$
$$T(\Omega) = \{u \in L^2(\Omega) \,;\, \mathrm{div}\, u \in L^2(\Omega), \, \mathrm{curl}\, u \in L^2(\Omega), \, \nu \cdot u = 0 \text{ on } \partial\Omega\},$$
$$N(\Omega) = \{u \in L^2(\Omega) \,;\, \mathrm{div}\, u \in L^2(\Omega), \, \mathrm{curl}\, u \in L^2(\Omega), \, \nu \times u = 0 \text{ on } \partial\Omega\},$$
$$T_\sigma(\Omega) = \{u \in T(\Omega) \,;\, \mathrm{div}\, u = 0 \text{ in } \Omega\},$$
$$N_\sigma(\Omega) = \{u \in N(\Omega) \,;\, \mathrm{div}\, u = 0 \text{ in } \Omega\}.$$

Theorem 7.7. *For any $u \in L^2(\Omega)^3$, there exist $p \in H^1(\Omega), w \in N_\sigma(\Omega), h \in T_{har}(\Omega)$ satisfying*

$$u = \nabla p + \mathrm{curl}\, w + h, \tag{7.6}$$

$$\|\nabla p\| + \|w\| + \|h\| \leq C\|u\|. \tag{7.7}$$

This decomposition is unique: If u admits another decomposition $u = \nabla \widetilde{p} + \mathrm{curl}\, \widetilde{w} + \widetilde{h}$, it holds that

$$\nabla p = \nabla \widetilde{p}, \quad \mathrm{curl}\, w = \mathrm{curl}\, \widetilde{w}, \quad h = \widetilde{h}.$$

Theorem 7.8. *For any $u \in L^2(\Omega)^3$, there exist $p \in H_0^1(\Omega), w \in T_\sigma(\Omega), h \in N_{har}(\Omega)$ satisfying*

$$u = \nabla p + \mathrm{curl}\, w + h, \tag{7.8}$$

$$\|\nabla p\| + \|w\| + \|h\| \leq C\|u\|. \tag{7.9}$$

This decomposition is unique: If u admits another decomposition $u = \nabla \widetilde{p} + \mathrm{curl}\, \widetilde{w} + \widetilde{h}$, it holds that

$$\nabla p = \nabla \widetilde{p}, \quad \mathrm{curl}\, w = \mathrm{curl}\, \widetilde{w}, \quad h = \widetilde{h}.$$

For the proof, see [57], Theorem 2.1.

Theorem 7.9. *Suppose $\partial\Omega$ has $k+1$ connected components $\Gamma_0, \Gamma_1, \cdots, \Gamma_k$ such that $\Gamma_1, \cdots, \Gamma_k$ lie inside of Γ_0 and $\Gamma_i \cap \Gamma_j = \emptyset$ for $i \neq j$. Then*

$$\dim N_{har}(\Omega) = k. \tag{7.10}$$

Suppose furthermore there exist C^∞-surfaces $\Sigma_1, \cdots, \Sigma_\ell$ transversal to $\partial\Omega$ such that $\Sigma_i \cap \Sigma_j = \emptyset$ for $i \neq j$ and $\Omega \setminus \left(\cup_{i=1}^\ell \Sigma_j \right)$ is a simply connected domain, we have

$$\dim T_{har}(\Omega) = \ell. \tag{7.11}$$

In particular, if Ω is simply connected and has no holes,

$$\dim N_{har}(\Omega) = \dim T_{har}(\Omega) = 0. \tag{7.12}$$

See [57], p. 1862.

7.2   Local compactness

Let Ω be an interior or exterior domain in $\mathbf{R}^3$ with C^∞-boundary, and ν the outer unit normal to $\partial\Omega$. Suppose we are given two 3×3 matrix-valued functions $\epsilon(x)$, $\mu(x)$ on Ω, which are real-symmetric, positive definite, and satisfies

$$CI_3 \leq \epsilon(x) \leq C^{-1}I_3, \quad CI_3 \leq \mu(x) \leq C^{-1}I_3, \quad \text{on} \quad \Omega \tag{7.13}$$

for a constant $C > 0$, where I_3 is the 3×3 identity matrix. The starting point of the spectral analysis of the Maxwell operator is the following compactness theorems. We follow the arguments due to Weber [95].

Theorem 7.10. *Let $\{E_n\}$ be a sequence in $L^2(\Omega)^3$ satisfying*

$$\|E_n\| + \|\mathrm{curl}\, E_n\| + \|\mathrm{div}\,(\epsilon E_n)\| \leq C, \tag{7.14}$$

$$\nu \times E_n = 0 \quad on \quad \partial\Omega, \tag{7.15}$$

for any n, where C is a constant independent of n. Then, for any compact set $K \subset \overline{\Omega}$, there exists a subsequence $\{E_{n_i}\}$ which is convergent in $L^2(K)$.

Theorem 7.11. *Let $\{H_n\}$ be a sequence in $L^2(\Omega)^3$ satisfying*

$$\|H_n\| + \|\mathrm{curl}\, H_n\| + \|\mathrm{div}\,(\mu H_n)\| \leq C, \tag{7.16}$$

$$\nu \cdot (\mu H_n) = 0 \quad on \quad \partial\Omega, \tag{7.17}$$

for any n, where C is a constant independent of n. Then, for any compact set $K \subset \overline{\Omega}$, there exists a subsequence $\{H_{n_i}\}$ which is convergent in $L^2(K)$.

Note that no regularity is assumed on $\epsilon(x)$, $\mu(x)$.

We first prove these theorems under the additional assumption that

$$\Omega \text{ is bounded, simply connected and has no holes.} \qquad (7.18)$$

Recall Poincaré's inequality.

Lemma 7.3. *(1) Let T be a bounded linear functional on $H^1(\Omega)$ such that $T(1) \neq 0$. Then, there is a constant C such that*

$$\|u\| \leq C(\|\nabla u\| + |T(u)|), \quad \forall u \in H^1(\Omega).$$

(2) There is a constant C such that

$$\|u\| \leq C\|\nabla u\|, \quad \forall u \in H_0^1(\Omega).$$

Proof. If (1) does not hold, there is a sequence $\{u_n\} \subset H^1(\Omega)$ satsifying $\|u_n\| = 1$, $\|\nabla u_n\| \to 0$, $T(u_n) \to 0$. By Rellich's theorem, there is a subsequence $\{u_{n_i}\}$ convergent in $L^2(\Omega)$. Let $u_{n_i} \to v \in L^2(\Omega)$. Then, v satisfies $\|v\| = 1$. Letting $n_i \to \infty$ in $(\partial_j u_{n_i}, \varphi) = -(u_{n_i}, \partial_j \varphi)$, where $\varphi \in C_0^\infty(\Omega)$ and $\partial_j = \partial/\partial x_j$, we have $(v, \partial_j \varphi) = 0$. We use the well-known mollifier technique. Take $\rho \in C_0^\infty(\mathbf{R}^3)$ such that $\rho(x) = 0$ for $|x| > 1$, $\int_{\mathbf{R}^3} \rho(x) dx = 1$ and put $\rho_\kappa(x) = \kappa^{-3} \rho(x/\kappa)$. We let $v_\kappa(x) = \int_{\mathbf{R}^3} \rho_\kappa(x - y) v(y) dy$, where we define $v(x) = 0$, $x \notin \Omega$. Then, as is well-known, $v_\kappa(x) \to v(x)$ in $L^2(\Omega)$ as $\kappa \to 0$. Take a compact set $K \subset \Omega$. Then, for $x \in K$ and small $\kappa > 0$, $\rho_\kappa(x - \cdot) \in C_0^\infty(\Omega)$. By $(v, \partial_j \varphi) = 0$ for $\varphi = \rho_\kappa(x - \cdot)$, we have $\partial_j v_\kappa(x) = 0$ for $x \in K$. Letting $\kappa \to \infty$, we have $\nabla v(x) = 0$ for any $x \in \Omega$. Therefore v is a constant. However $T(u_{n_i}) \to T(v) = 0$, which implies $v = 0$, since $T(1) \neq 0$. We have thus arrived at a contradiction.

Take a large ball B such that $B \supset \Omega$. Then, by (1)

$$\|u\|_{L^2(B)} \leq C(\|\nabla u\|_{L^2(B)} + |\int_{B\setminus\Omega} u dx|), \quad \forall u \in H^1(B),$$

from which (2) follows. $\qquad\qquad\qquad\qquad\qquad\qquad\qquad\qquad \square$

We use a Helmholtz decomposition different from the standard one.

Lemma 7.4. *The orthogonal decomposition*

$$L^2(\Omega) = \nabla H_0^1(\Omega) \oplus \{E \in L^2(\Omega); \nabla \cdot (\epsilon E) = 0\}$$

holds with respect to the inner product

$$(E, F)_\epsilon = (\epsilon E, F)_{L^2(\Omega)}.$$

Proof.   Note that by Lemma 7.3 (2), $\|\nabla u\|$ defines a norm on $H_0^1(\Omega)$, hence $\nabla H_0^1(\Omega)$ is a closed subspace of $L^2(\Omega)$. For $f \in H_0^1(\Omega)$, we have

$$(\epsilon E, \operatorname{grad} f) = -(\operatorname{div}(\epsilon E), f).$$

Therefore, E is orthogonal to $\nabla H_0^1(\Omega)$ if and only if $\nabla \cdot (\epsilon E) = 0$. Let $H_1 = \{E \in L^2(\Omega)\,;\, \nabla \cdot (\epsilon E) = 0\}$. We have proven that

$$u \perp \nabla H_0^1(\Omega) \iff u \in H_1,$$

from which the lemma follows immediately.   □

Let us now prove Theorem 7.10 under the assumption (7.18). Using Lemma 7.4, we decompose E_n as

$$E_n = \nabla f_n + U_n, \quad f_n \in H_0^1(\Omega), \quad \operatorname{div}(\epsilon U_n) = 0.$$

By Lemma 7.3 (2), $\{f_n\}$ is a bounded set in $H^1(\Omega)$. Therefore, there exists a subsequence convergent in $L^2(\Omega)$, which is denoted by $\{f_n\}$ again. We then have, letting $f_{k\ell} = f_k - f_\ell$,

$$
\begin{aligned}
-(\operatorname{div}(\epsilon(E_k - E_\ell)), f_{k\ell}) &= -(\operatorname{div}(\epsilon \nabla f_{k\ell}), f_{k\ell}) \\
&= (\epsilon \nabla f_{k\ell}, \nabla f_{k\ell}) \\
&\geq C\|\nabla f_{k\ell}\|^2 = C\|\nabla f_k - \nabla f_\ell\|^2.
\end{aligned}
$$

Since the left-hand side tends to 0, $\{\nabla f_n\}$ is convergent in $L^2(\Omega)$.

Using the assumption (7.18) and Theorems 7.8, 7.9, we have $\epsilon U_n = \nabla p_n + \operatorname{curl} A_n$, where $p_n \in H_0^1(\Omega)$. Since $\operatorname{div}(\epsilon U_n) = \Delta p_n = 0$, we have $p_n = 0$. Therefore,

$$\epsilon U_n = \operatorname{curl} A_n, \quad \operatorname{div} A_n = 0, \quad \nu \cdot A_n = 0, \quad \|A_n\| \leq C\|\epsilon U_n\|.$$

Theorem 7.5 (2) shows that $\{A_n\}$ is bounded in $H^1(\Omega)$, hence it has a subsequence convergent in $L^2(\Omega)$, which is denoted by $\{A_n\}$ again. Letting $A_{k\ell} = A_k - A_\ell$, we now have

$$|(\operatorname{curl}(E_k - E_\ell), A_{k\ell})| \leq C\|A_{k\ell}\| \to 0.$$

On the other hand, using $\nu \times E_n = 0$, we have

$$
\begin{aligned}
(\operatorname{curl}(E_k - E_\ell), A_{k\ell}) &= (E_k - E_\ell, \operatorname{curl} A_{k\ell}) \\
&= (\nabla(f_k - f_\ell), \operatorname{curl} A_{k\ell}) + (\epsilon^{-1}\operatorname{curl} A_{k\ell}, \operatorname{curl} A_{k\ell}) \\
&\geq C\|\operatorname{curl} A_{k\ell}\|^2 - C'\|\nabla(f_k - f_\ell)\|^2.
\end{aligned}
$$

Therefore, $\{U_n\}$ is convergent in $L^2(\Omega)$, which proves Theorem 7.10.

We next prove Theorem 7.11 under the assumption (7.18). Applying Theorem 7.7 to μH_n, we obtain

$$\mu H_n = \nabla g_n + \operatorname{curl} B_n, \quad g_n \in H^1(\Omega), \quad \operatorname{div} B_n = 0, \quad \nu \times B_n = 0,$$

$$\|\nabla g_n\| + \|B_n\| \le C\|H_n\|.$$

Define a linear functional T by

$$T(f) = \int_\Omega f \, dx.$$

Then, by Lemma 7.3

$$\|f\|_{L^2(\Omega)} \le C(\|\nabla f\|_{L^2(\Omega)} + |T(f)|)$$

holds for any $f \in H^1(\Omega)$. Putting

$$c_n = T(g_n)/T(1),$$

we have $T(g_n - c_n) = 0$, hence $h_n = g_n - c_n$ satisfies

$$\|h_n\|_{L^2(\Omega)} \le C\|\nabla h_n\|_{L^2(\Omega)} = C\|\nabla g_n\|_{L^2(\Omega)}.$$

Therefore, denoting $g_n - c_n$ by g_n again, we see that

$$\sup_n \|g_n\|_{H^1(\Omega)} < \infty.$$

Then, by Rellich's theorem, there exists a subsequence of $\{g_n\}$ convergent in $L^2(\Omega)$, which is denoted by $\{g_n\}$ again.

Letting $g_{k\ell} = g_k - g_\ell$, and $B_{k\ell} = B_k - B_\ell$, we have

$$(\mu(H_k - H_\ell), \nabla g_{k\ell}) = \|\nabla g_{k\ell}\|^2 + (\operatorname{curl} B_{k\ell}, \nabla g_{k\ell}).$$

For $g \in H^2(\Omega)$, we have, using $\nu \times B_{k\ell} = 0$,

$$(\nabla \times B_{k\ell}, \nabla g) = (B_{k\ell}, \nabla \times \nabla g) = 0.$$

Hence $(\operatorname{curl} B_{k\ell}, \nabla g_{k\ell}) = 0$. We also have, using $\nu \cdot (\mu H_n) = 0$,

$$(\mu(H_k - H_\ell), \nabla g_{k\ell}) = -(\nabla \cdot (\mu(H_k - H_\ell)), g_{k\ell}).$$

We thus obtain

$$\|\nabla(g_k - g_\ell)\|^2 \le C\|g_k - g_\ell\|,$$

which proves that $\{\nabla g_n\}$ is also convergent in $L^2(\Omega)$.

By the equality $\mu H_n = \nabla g_n + \nabla \times B_n$, we have

$$(H_k - H_\ell, \nabla \times B_{k\ell}) = (\mu^{-1} \nabla g_{k\ell}, \nabla \times B_{k\ell}) + (\mu^{-1} \nabla \times B_{k\ell}, \nabla \times B_{k\ell}).$$

The right-hand side is estimated from below by

$$C\|\nabla \times B_{k\ell}\|^2 - C'\|\nabla g_{k\ell}\|^2,$$

while, by $\nu \times B_n = 0$, the left-hand side is equal to

$$(H_k - H_\ell, \nabla \times B_{k\ell}) = (\nabla \times (H_k - H_\ell), B_{k\ell}).$$

We then arrive at

$$\|\nabla \times B_{k\ell}\|^2 \le C(\|\nabla g_{k\ell}\|^2 + \|B_{k\ell}\|).$$

We apply Theorem 7.10 to $\{B_n\}$ (with $\epsilon = 1$) and find a subsequence convergent in $L^2(\Omega)^3$, which we denote by $\{B_n\}$ again. Then, $\{\nabla \times B_n\}$ is convergent, which proves that $\{H_n\}$ is convergent.

We remove the assumption (7.18). We take $\{\chi_j\} \in C_0^\infty(\mathbf{R}^3)$ such that $\sum_j \chi_j = 1$ in a neigborhood of $\partial\Omega$ and on each support of χ_j, the assumption (7.18) is satisfied. Then, the above arguments work for $\{\chi_j E_n\}$ and $\{\chi_j H_n\}$ on supp χ_j. Let $\chi = 1 - \sum_j \chi_j$. Then, $\{\chi E_n\}$ and $\{\chi H_n\}$ on $\mathbf{R}^3$ satisfy the assumptions in Theorems 7.10 and 7.11. The inequality (7.5) then completes the proof.

7.3   Self-adjointness

Let Ω be an interior or exterior domain in $\mathbf{R}^3$ with C^∞-boundary $\partial\Omega$. Let $\epsilon(x)$, $\mu(x)$ be as in (7.13). We put

$$\mathcal{M}(x) = \begin{pmatrix} \epsilon(x) & 0 \\ 0 & \mu(x) \end{pmatrix}, \quad \mathcal{L} = i \begin{pmatrix} 0 & \mathrm{curl} \\ -\mathrm{curl} & 0 \end{pmatrix} \tag{7.19}$$

and define a self-adjoint realization of the operator

$$\mathcal{H} = \mathcal{M}(x)^{-1}\mathcal{L} \tag{7.20}$$

on the Hilbert space $\mathbf{H} = (L^2(\Omega))^6$ equipped with the inner product

$$(f,g) = (\epsilon(x)f_E(x), g_E(x)) + (\mu(x)f_H(x), g_H(x)), \tag{7.21}$$

where $f = (f_E, f_H)$, $g = (g_E, g_H)$ and (E, F) is the standard inner product of $L^2(\Omega)^3$:

$$(E, F) = \int_\Omega E(x) \cdot \overline{F(x)}dx.$$

Lemma 7.5. *Assume (7.13). Then the operator $\mathcal{H}$ with domain*

$$D(\mathcal{H}) = H_0(\mathrm{curl}; \Omega) \times H(\mathrm{curl}\,\Omega)$$

is self-adjoint.

Proof. For $f = (f_E, f_H), g = (g_E, g_H) \in D(\mathcal{H})$, we have

$$(\mathcal{H}f, g) = i\,(\operatorname{curl} f_H, g_E) - i\,(\operatorname{curl} f_E, g_H).$$

For $U, V \in H(\operatorname{curl}; \Omega)$, by integration by parts and Theorem 7.4

$$(\operatorname{curl} U, V) = (U, \operatorname{curl} V) + (\nu \times U, V)_{\partial\Omega}.$$

Since $(\nu \times U) \cdot \overline{V} = -(\nu \times \overline{V}) \cdot U$, we have $(\nu \times U, V)_{\partial\Omega} = 0$ if $\nu \times U = 0$ or $\nu \times V = 0$ on $\partial\Omega$. This shows that $\mathcal{H}$ is symmetric.

To prove the self-adjointness, assume that

$$i\,(\operatorname{curl} H, E') - i\,(\operatorname{curl} E, H') = (E, A) - (H, B), \quad \forall(E, H) \in D(\mathcal{H}).$$

Then, taking $H = 0$ and $E \in C_0^\infty(\Omega)$, we have $-i\,(\operatorname{curl} E, H') = (E, A)$. This implies $-i\operatorname{curl} H' = A \in L^2(\Omega)$, hence $H' \in H(\operatorname{curl}; \Omega)$. Taking $E = 0$ and $H \in C^\infty(\overline{\Omega})$, we have $i\,(\operatorname{curl} H, E') = -(H, B)$. This implies $-i\operatorname{curl} E' = B \in L^2(\Omega)$, hence $E' \in H(\operatorname{curl}; \Omega)$, moreover, by integration by parts in $(H, \operatorname{curl} E') = (\operatorname{curl} H, E')$, we have $\nu \times E' = 0$. Therefore, $E' \in H_0(\operatorname{curl}; \Omega)$. $\qquad\square$

Let A be a self-adjoint operator on a Hilbert space $\mathbf{H}$. Its resolvent set and the specrum are denoted by $\rho(A)$ and $\sigma(A)$. For the self-adjoint operator, $\sigma(A)$ is a closed set in $\mathbf{R}$.

We say that an operator $A \in \mathbf{B}(\mathbf{H}_\Omega)$ has a *local compactness* property if for any $R > 0$, the operator $\chi_R A$ is compact, where χ_R is the cut-off function to the domain $\Omega \cap \{|x| < R\}$.

As is clear from the definition, the null space of $\mathcal{H}$ is given by

$$\begin{aligned} N(\mathcal{H}) &= \{f \in D(\mathcal{H}); \mathcal{H}f = 0\} \\ &= \{(E, H) \in D(\mathcal{H}); \operatorname{curl} E = \operatorname{curl} H = 0\}. \end{aligned} \tag{7.22}$$

Let $\mathcal{P}_0$ be the projection onto $N(\mathcal{H})$ and $\mathcal{P} = 1 - \mathcal{P}_0$. The local compactness of the resolvent $\mathcal{P}(\mathcal{H} - z)^{-1}$ is crucial in the study of spectral properties of $\mathcal{H}$.

Lemma 7.6. *Assume (7.13). Then $\mathcal{P}(\mathcal{H} - z)^{-1}$ has a local compactness property for any $z \in \rho(\mathcal{H})$.*

Proof. Fix $z \notin \mathbf{R}$ and let $u_n = (E_n, H_n) = \mathcal{P}(\mathcal{L} - z)^{-1} f_n$, where $\sup \|f_n\| < \infty$. Then $\|u_n\| + \|\mathcal{L}u_n\| \le C$, where C is a constant independent of n. This yields $\|E_n\| + \|\operatorname{curl} E_n\| + \|H_n\| + \|\operatorname{curl} H_n\| \le C$. Since u_n is orthogonal to $N(\mathcal{L})$, we have

$$(\epsilon E_n, E') + (\mu H_n, H') = 0, \quad \forall(E', H') \in N(\mathcal{L}).$$

Taking $E' = \nabla\psi$, $\psi \in C_0^\infty(\Omega)$ and $H' = 0$, we have $(\epsilon E_n, \nabla\psi) = -(\nabla \cdot (\epsilon E_n), \psi) = 0$, which implies $\nabla \cdot (\epsilon E_n) = 0$. Taking $E' = 0$ and $H' = \nabla\psi$, $\psi \in C^\infty(\overline{\Omega})$, we have $(\mu H_n, \nabla\psi) = (\mu H_n, \nu\psi)_{\partial\Omega} - (\nabla \cdot (\mu E_n), \psi) = 0$, which implies $\nu \cdot (\mu H_n) = 0$ on $\partial\Omega$ and $\nabla \cdot (\mu H_n) = 0$ in Ω. The lemma then follows from Theorems 7.10 and 7.11. $\qquad\square$

7.4 Spectral properties of self-adjoint operators

In this section, we summarize basic spectral properties for self-adjoint operators without proof. For the details, see e.g. [51], [102], [85].

7.4.1 *Spectral measure*

Let $\mathbf{H}$ be a Hilbert space. The *spectral decomposition* or the *spectral measure* is a family of bounded operators $E(\lambda)$ depending on a parameter $\lambda \in \mathbf{R}$ having the following propeties.

$$E(\lambda)^* = E(\lambda) = E(\lambda)^2, \tag{7.23}$$

$$E(\lambda)E(\mu) = E(\mu)E(\lambda) = E(\mu), \quad \text{if} \quad \lambda \geq \mu, \tag{7.24}$$

$$\text{s} - \lim_{\lambda\to\infty} E(\lambda) = I, \quad \text{s} - \lim_{\lambda\to-\infty} E(\lambda) = 0, \tag{7.25}$$

$$\text{s} - \lim_{\epsilon\downarrow 0} E(\lambda + \epsilon) = E(\lambda). \tag{7.26}$$

These properties imply that any $\lambda \in \mathbf{R}$, $E(\lambda - 0) = \text{s} - \lim_{\epsilon\downarrow 0} E(\lambda - \epsilon)$ exists. One can then associate operator-valued Stieltjes integrals in the following way. For a semi-open interval $(a, b]$

$$\int_{(a,b]} dE(\lambda) = E(b) - E(a),$$

and for the others

$$\int_{[a,b]} dE(\lambda) = E(b) - E(a - 0),$$

$$\int_{(a,b)} dE(\lambda) = E(b - 0) - E(a),$$

$$\int_{[a,b)} dE(\lambda) = E(b) - E(a - 0),$$

$$\int_{\{a\}} dE(\lambda) = E(a) - E(a - 0).$$

For a step function $f(\lambda) = \sum_i c_i \chi_{I_i}(\lambda)$, we define

$$\int_{-\infty}^{\infty} f(\lambda)dE(\lambda) = \sum_i c_i \int_{I_i} dE(\lambda).$$

For a continuous function $f(\lambda)$ defined on a compact interval I, the integral $\int_I f(\lambda)dE(\lambda)$ is defined as the limit of integrals for the step functions converging to $f(\lambda)$. Finally, for a bounded continuous function on $\mathbf{R}$, the integral $\int_{-\infty}^{\infty} f(\lambda)dE(\lambda)$ is defined as the strong limit $\text{s} - \lim_{n\to\infty} \int_{-n}^{n} f(\lambda)dE(\lambda)$. We put

$$Op(f) = \int_{-\infty}^{\infty} f(\lambda)dE(\lambda). \tag{7.27}$$

It has the following properties

$$Op(f)Op(g) = Op(fg), \quad Op(f)^* = Op(\overline{f}). \tag{7.28}$$

Moreover, for any $u \in \mathbf{H}$,

$$\|Op(f)u\|^2 = \int_{-\infty}^{\infty} |f(\lambda)|^2 d(E(\lambda)u, u). \tag{7.29}$$

This implies

$$\|Op(f)\| \leq \sup|f(\lambda)|. \tag{7.30}$$

If f is unbounded, let D be the set of $u \in \mathbf{H}$ satisfying

$$\int_{-\infty}^{\infty} |f(\lambda)|^2 d(E(\lambda)u, u) < \infty,$$

and define $\int_{-\infty}^{\infty} f(\lambda)dE(\lambda)u$ for $u \in D$ as the $\mathbf{H}$-valued integral. If $f(\lambda)$ is real-valued, the operator

$$Op(f) : u \to \int_{-\infty}^{\infty} f(\lambda)dE(\lambda)u$$

is self-adjoint with domain D.

Conversely, for a self-adjoint operator A on a Hilbert space $\mathbf{H}$, there exists a unique spectral measure $E(\lambda)$ such that A is written as

$$A = \int_{-\infty}^{\infty} \lambda dE(\lambda),$$

$$D(A) = \left\{ u \in \mathbf{H}; \int_{-\infty}^{\infty} \lambda^2 d(E(\lambda)u, u) < \infty \right\}. \tag{7.31}$$

In fact, $E(\lambda)$ is computed from the resolvent $R(z) = (A - z)^{-1}$ by the following *Stone's formula*: for any $u, v \in \mathbf{H}$,

$$\frac{1}{2}\Big((E((a, b)) + E([a, b]))u, v \Big) = \lim_{\epsilon \downarrow 0} \frac{1}{2\pi i} \int_a^b \Big((R(\lambda + i\epsilon) - R(\lambda - i\epsilon))u, v \Big) d\lambda. \tag{7.32}$$

In particular, if a, b are not the eigenvalues of A,

$$(E((a,b))u, v) = \lim_{\epsilon \downarrow 0} \frac{1}{2\pi i} \int_a^b \Big((R(\lambda + i\epsilon) - R(\lambda - i\epsilon))u, v \Big) d\lambda. \qquad (7.33)$$

Using the spectral measure of A, we define an operator

$$f(A) = \int_{-\infty}^{\infty} f(\lambda) dE(\lambda)$$

with domain

$$D(f(A)) = \Big\{ u \in \mathbf{H} \, ; \int_{-\infty}^{\infty} |f(\lambda)|^2 D(E(\lambda)u, u) < \infty \Big\}.$$

We then have

$$\|f(A)\| \leq \sup_{\lambda} |f(\lambda)|.$$

7.4.2 *Classification of the spectrum*

Let A be a self-adjoint operator with the spectral decomposition $E(\lambda)$. Let $\rho(A)$, $\sigma(A)$ and $\sigma_p(A)$ be the resolvent set, spectrum, and the set of all eigenvalues of A. They are characterized as follows.

Lemma 7.7. *(1)* $\lambda \in \rho(A) \cap \mathbf{R} \Longleftrightarrow E((\lambda - \epsilon, \lambda + \epsilon)) = 0$ *for some* $\epsilon > 0$.
(2) $\lambda \in \sigma(A) \Longleftrightarrow E((\lambda - \epsilon, \lambda + \epsilon)) \neq 0$ *for any* $\epsilon > 0$.
(3) $\lambda \in \sigma_p(A) \Longleftrightarrow E(\{\lambda\}) = E(\lambda) - E(\lambda - 0) \neq 0$.

For $\lambda \in \sigma_p(A)$, the dimension of $E(\{\lambda\})\mathbf{H} = \{u \in D(A) \, ; \, (A - \lambda)u = 0\}$ is said to be the multiplicity of λ. The set of all isolated eigenvalues with finite multiplicities is said to be the *discrete spectrum* of A and is denoted by $\sigma_d(A)$. $\sigma_e(A) = \sigma(A) \setminus \sigma_d(A)$ is called the *essential spectrum* of A.

Lemma 7.8. *(1)* $\lambda \in \sigma(A)$ *if and only if there exists a sequence* $u_n \in D(A)$ *such that* $\|u_n\| = 1$, $\|(A - \lambda)u_n\| \to 0$.
(2) $\lambda \in \sigma_e(A)$ *if and only if there exists a sequence* $u_n \in D(A)$ *such that* $\|u_n\| = 1$, $\|(A - \lambda)u_n\| \to 0$ *and* $u_n \to 0$ *weakly in* $\mathbf{H}$.

There is another classification of spectra using the measure theoretic properties of $E(\lambda)$. For $u \in \mathbf{H}$, we put $f(\lambda; u) = (E(\lambda)u, u)$, which is a bounded monotone increasing function of $\lambda \in \mathbf{R}$. Let $\mathbf{H}_{cont}(A)$, $\mathbf{H}_{ac}(A)$ and $\mathbf{H}_{sc}(A)$ be the set of all $u \in \mathbf{H}$ such that $f(\lambda; u)$ is continuous, absolutely continuous with respect to the Lebeasgue measure $d\lambda$, singular continuous with respect to the Lebesgue measure $d\lambda$, respectively. They are closed subspaces of $\mathbf{H}$ and reduce A. Here, a subspace $S \subset \mathbf{H}$ *reduces A if*

$$u \in S \Longrightarrow \varphi(A)u \in S$$

holds for any bounded continuous function $\varphi(\lambda)$. Letting $\mathbf{H}_{pp}(A)$ be the closure of the linear hull of the eigenvectors of A, we then get the following orthogonal decomposition

$$
\begin{aligned}
\mathbf{H} &= \mathbf{H}_{pp}(A) \oplus \mathbf{H}_{cont}(A), \\
\mathbf{H}_{cont}(A) &= \mathbf{H}_{ac}(A) \oplus \mathbf{H}_{sc}(A).
\end{aligned}
\tag{7.34}
$$

The *continuous spectrum* of A is defined as the spectrum of A restricted to $\mathbf{H}_{cont}(A)$, and the *absolutely continuous spectrum* and the *singular continuous spectrum* of A are defined in the same way:

$$
\begin{aligned}
\sigma_c(A) &= \sigma(A\big|_{\mathbf{H}_{cont}(A)}), \\
\sigma_{ac}(A) &= \sigma(A\big|_{\mathbf{H}_{ac}(A)}), \\
\sigma_{sc}(A) &= \sigma(A\big|_{\mathbf{H}_{sc}(A)}).
\end{aligned}
\tag{7.35}
$$

7.4.3   *Local decay of scattering solutions*

For a self-adjoint operator A, $u(t) = e^{-itA}u_0$ represents the solution of the equation

$$i\partial_t u = Au, \quad u(0) = u_0.$$

If u_0 is an eigenvector of A with eigenvalue λ, $u(t) = e^{-it\lambda}u_0$ is simply time-periodic. If $u_0 \in \mathbf{H}_{cont}(A)$, in many physical situations, $u(t)$ has the local decay property and scatters to the space infinity. Mathematically, it is based on the following lemma due to Wiener.

Lemma 7.9. *Let $\varphi(\lambda)$ be continuous and of bounded variation on $\mathbf{R}$. Then,*

$$\lim_{T \to \infty} \frac{1}{T} \int_0^T \left| \int_\infty^\infty e^{-i\lambda t} d\varphi(\lambda) \right|^2 dt = 0.$$

Proof. We have only to consider the case in which $\varphi(\lambda)$ is a bounded, continuous and monotone increasing function on $\mathbf{R}$. Let $d\Phi$ be the product measure $d\varphi(\lambda)d\varphi(\mu)$ on $\mathbf{R}^2$, and $D = \{(\lambda, \lambda)\,;\, \lambda \in \mathbf{R}\}$. Since the right-hand side of

$$\int_D d\Phi \le \int_{|\lambda - \mu| < \epsilon} d\Phi = \int_{-\infty}^\infty \big(\varphi(\lambda + \epsilon) - \varphi(\lambda - \epsilon)\big) d\varphi(\lambda)$$

tends to 0 as $\epsilon \to 0$, D has 0 measure. Using

$$\frac{1}{T}\int_0^T \left|\int_{-\infty}^{\infty} e^{-it\lambda}d\varphi(\lambda)\right|^2 dt = \iint_{\mathbf{R}^2 \setminus \mathbf{D}} f_T(\lambda, \mu)d\varphi(\lambda)d\lambda(\mu),$$

$$f_T(\lambda, \mu) = \frac{e^{-i(\lambda-\mu)T}-1}{-i(\lambda-\mu)T}, \quad \lambda \ne \mu,$$

and Lebesgue's theorem, we obtain the lemma.   □

Now let us consider the case in which $\mathbf{H} = L^2(\Omega)$, where Ω is an exterior domain in $\mathbf{R}^n$. For $R > 0$, let $\chi_R(x)$ be the characteristic function of the set $\{x \in \Omega\,;\,|x| < R\}$. Then, the following lemma holds.

Lemma 7.10. *If A has the local compactness property, for $u \in \mathbf{H}_{cont}(A)$,*

$$\frac{1}{T}\int_0^T \|\chi_R e^{-itA}u\|^2 dt \to 0, \quad T \to \infty. \tag{7.36}$$

Moreover, if $u \in \mathbf{H}_{ac}(A)$,

$$\|\chi_R e^{-itA}u\| \to 0, \quad t \to \infty. \tag{7.37}$$

Proof. Since u is approximated by $E(I)u$, where I is a bounded interval on $\mathbf{R}$, we have only to consider the case that $u = E(I)u$. Then, $e^{itA}u = E(I)e^{itA}u$, and $\chi_R E(I)$ is compact. To see it, we have only to note

$$\chi_R E(I) = \chi_R (A+i)^{-1}(A+i)E(I),$$

and $(A+i)E(I)$ is bounded. Since any compact operator is approximated by finite rank operators, for any $\epsilon > 0$, we can find $v_j, w_j \in \mathbf{H}$ such that $\|\chi_R E(I) - \sum_{j=1}^N (\cdot, v_j)w_j\| < \epsilon$. Therefore, to prove (7.36), we have only to show

$$\frac{1}{T}\int_0^T \left|(e^{-itA}u, v_j)\right|^2 dt \to 0.$$

However, by the spectral decomposition

$$(e^{-itA}u, v_j) = \int_{-\infty}^{\infty} e^{-it\lambda}d(E(\lambda)u, v_j),$$

and letting $P_c(A)$ be the projection onto $\mathbf{H}_{cont}(A)$, we have

$$(E(\lambda)u, v_j) = (E(\lambda)u, P_c(A)v_j),$$

which is continuous with respect to λ. Lemma 7.9 then proves (7.36). If $u \in \mathbf{H}_{ac}(H)$, we replace $P_c(A)$ by $P_{ac}(A)$, the projection onto $\mathbf{H}_{ac}(A)$. Then

$$(E(\lambda)u, v_j) = (E(\lambda)u_j, P_{ac}(A)v_j)$$

is absolutely continuous. Therefore, Riemann-Lebesgue's lemma implies

$$\int_{-\infty}^{\infty} e^{-it\lambda}d(E(\lambda)u, v_j) = \int_{-\infty}^{\infty} e^{-it\lambda}\frac{d}{d\lambda}(E(\lambda)u, v_j)d\lambda \to 0.$$

This proves (7.37).   □

7.5 Spectrum of the Maxwell operator

7.5.1 *Interior problem*

Let Ω be a bounded domain in $\mathbf{R}^3$ with C^∞ boundary. We put

$$\mathcal{L}(\xi) = \begin{pmatrix} 0 & -\xi\times \\ \xi\times & 0 \end{pmatrix}, \tag{7.38}$$

$$\mathcal{M}(x) = \begin{pmatrix} \epsilon(x) & 0 \\ 0 & \mu(x) \end{pmatrix}. \tag{7.39}$$

Assume (7.13), and define $\mathcal{H}_{int}$ by

$$\mathcal{H}_{int} = \mathcal{M}(x)^{-1}\mathcal{L}(-i\nabla), \tag{7.40}$$

$$D(\mathcal{H}_{int}) = H_0(\mathrm{curl}\,;\Omega) \times H(\mathrm{curl}\,;\Omega). \tag{7.41}$$

Then, $\mathcal{H}_{int}$ is self-adjoint by virtue of Lemma 7.5.

Theorem 7.12. $\sigma(\mathcal{H}_{int}) = \sigma_e(\mathcal{H}_{int}) \cup \sigma_d(\mathcal{H}_{int})$, where $\sigma_e(\mathcal{H}_{int}) = \{0\}$ *is an isolated eigenvalue of infinite multiplicity, while* $\sigma_d(\mathcal{H}_{int}) = \{\lambda_1^{(\pm)}, \lambda_2^{(\pm)}, \cdots\}$ *and* $0 < \lambda_1^{(+)} < \lambda_2^{(+)} <\to \infty,\ 0 > \lambda_1^{(-)} > \lambda_2^{(-)} >\to -\infty.$

Proof. Let $\mathcal{P}_0$ be the projection onto the null space of $\mathcal{H}_{int}$ and $\mathcal{P} = 1 - \mathcal{P}_0$. The range of $\mathcal{P}_0$ is the eigenspace of $\mathcal{H}_{int}$ with 0 eigenvalue, which is easily seen to be infinite dimensional. By Lemma 7.6, $\mathcal{H}_{int}\big|_{\mathcal{P}\mathbf{H}}$ has only a discrete spectrum whose absolute value tends to infinity. If there exists $u = (E, H) \in D(\mathcal{H}_{int})$ such that $\|u\| = 1$ and $(\mathcal{H}_{int}u, u) > n$, letting $v = (E, -H)$, we have $(\mathcal{H}_{int}v, v) = -(\mathcal{H}_{int}u, u) < -n$. Hence

$$\sigma(\mathcal{H}_{int}) \cap \{\lambda\,;\,\lambda > n\} \neq \emptyset \iff \sigma(\mathcal{H}_{int}) \cap \{\lambda\,;\,\lambda < -n\} \neq \emptyset.$$

Therefore, the eigenvalues of $\mathcal{H}_{int}$ have the properties in the lemma. $\square$

For general knowledge of elliptic boundary value problems, see e.g. [2].

7.5.2 *Free Maxwell operator*

The free Maxwell operator $\mathcal{H}_0$ on $\mathbf{R}^3$ is defined by

$$\mathcal{H}_0 = \mathcal{M}_0^{-1}\mathcal{L}(-i\nabla), \quad \mathcal{M}_0 = \begin{pmatrix} \epsilon_0 I_3 & 0 \\ 0 & \mu_0 I_3 \end{pmatrix}, \tag{7.42}$$

where ϵ_0, μ_0 are positive constants, with domain

$$D(\mathcal{H}_0) = H_0(\mathrm{curl}\,;\mathbf{R}^3) \times H(\mathrm{curl}\,;\mathbf{R}^3). \tag{7.43}$$

Let $\mathbf{H}_0$ be $L^2(\mathbf{R}^3)^6$ equipped with the inner product

$$(u, v) = (\epsilon_0 u_E, v_E)_{L^2(\mathbf{R}^3)^3} + (\mu_0 u_H, v_H)_{L^2(\mathbf{R})^3},$$

where $u = (u_E, u_H)$, $v = (v_E, v_H)$. Then $\mathcal{H}_0$ is self-adjoint on $\mathbf{H}_0$.

Let us compute the spectral decomposition of $\mathcal{H}_0$. In Section 6.4, we have seen that for $\xi \neq 0$, the matrix $\mathcal{M}_0^{-1}\mathcal{L}(\xi)$ has eigenvalues

$$\lambda_0(\xi) = 0, \quad \lambda_\pm(\xi) = \pm\frac{|\xi|}{\sqrt{\epsilon_0\mu_0}},$$

each of which has multiplicity 2. Let $\mathcal{P}_0^{(0)}(\xi), \mathcal{P}_\pm^{(0)}(\xi)$ be the associated eigenprojections. Note that they depend only on $\xi/|\xi|$. Although the eigenvectors cannot be chosen globally smooth for $\xi \neq 0$, the eigenprojections $\mathcal{P}_\pm^{(0)}(\xi)$ are smooth. This can be seen from the well-known formula

$$\mathcal{P}_\pm^{(0)}(\xi) = -\frac{1}{2\pi i}\int_{C_\pm}(\mathcal{M}_0^{-1}\mathcal{L}(\xi) - z)^{-1}dz,$$

where $C_\pm$ is a contour enclosing only $\lambda_\pm(\xi)$. (See e.g. [51], p. 67.)

We can rewrite $\mathcal{H}_0$ and its resolvent $R_0(z)$ as

$$\mathcal{H}_0 = \mathcal{F}_0^*\big(\lambda_+(\xi)\mathcal{P}_+^{(0)}(\xi) + \lambda_-(\xi)\mathcal{P}_-^{(0)}(\xi)\big)\mathcal{F}_0,$$

$$R_0(z) = \mathcal{F}_0^*\Big(\frac{\mathcal{P}_0^{(0)}(\xi)}{-z} + \frac{\mathcal{P}_+^{(0)}(\xi)}{\lambda_+(\xi) - z} + \frac{\mathcal{P}_-^{(0)}(\xi)}{\lambda_-(\xi) - z}\Big)\mathcal{F}_0,$$

where $\mathcal{F}_0$ is defined by (6.53). We put

$$\mathcal{P}_0^{(0)} = \mathcal{F}_0^*\mathcal{P}_0^{(0)}(\xi)\mathcal{F}_0, \quad \mathcal{P}_\pm^{(0)} = \mathcal{F}_0^*\mathcal{P}_\pm^{(0)}(\xi)\mathcal{F}_0.$$

Lemma 7.11. $\mathcal{H}_0$ *has only one eigenvalue 0, which is of infinite multiplicity.*

Proof.   Obviouly, 0 is an eigenvalue with eigenprojection $\mathcal{P}_0^{(0)}$. If $\lambda \neq 0$ is an eigenvalue with eigenvector u, we have

$$-\lambda\mathcal{P}_0^{(0)}(\xi)\widehat{u}(\xi) = 0, \quad (\lambda_\pm(\xi) - \lambda)\mathcal{P}_\pm^{(0)}(\xi)\widehat{u}(\xi) = 0,$$

which implies $\widehat{u}(\xi) = 0$ for $|\xi| \neq |\lambda|\sqrt{\epsilon_0\mu_0}$, hence $u = 0$. $\square$

Let $E_0(\lambda)$ be the spectral measure for $\mathcal{H}_0$. Parseval's formula implies

$$(R_0(z)u, v) = \Big(\frac{1}{-z}\mathcal{P}_0^{(0)}(\xi)\widehat{u}(\xi), \widehat{v}(\xi)\Big) + \Big(\frac{1}{\lambda_+(\xi) - z}\mathcal{P}_+^{(0)}(\xi)\widehat{u}(\xi), \widehat{v}(\xi)\Big)$$

$$+ \Big(\frac{1}{\lambda_-(\xi) - z}\mathcal{P}_-^{(0)}(\xi)\widehat{u}(\xi), \widehat{v}(\xi)\Big).$$

Using Stones' formula (7.32), we then have for $I = [a, b] \subset (0, \infty)$,

$$(E_0(I)u, v) = (\chi_I^{(+)}(\xi)\mathcal{P}_+^{(0)}(\xi)\widehat{u}(\xi), \widehat{v}(\xi)),$$

and for $I = [a, b] \subset (-\infty, 0)$,

$$(E_0(I)u, v) = (\chi_I^{(-)}(\xi)\mathcal{P}_-^{(0)}(\xi)\widehat{u}(\xi), \widehat{v}(\xi)),$$

where $\chi_I^{(\pm)}$ is the characteristic function of the set $\{\xi \, ; \lambda_\pm(\xi) \in I\}$. This implies

$$E_0(I) = \begin{cases} \mathcal{F}_0^* \chi_I^{(+)}(\xi)\mathcal{P}_+^{(0)}(\xi)\mathcal{F}_0, & \text{if} \quad I = [a, b] \subset (0, \infty) \\ \mathcal{F}_0^* \chi_I^{(-)}(\xi)\mathcal{P}_-^{(0)}(\xi)\mathcal{F}_0, & \text{if} \quad I = [a, b] \subset (-\infty, 0). \end{cases} \tag{7.44}$$

Moreover, for any bounded Borel function $f(\lambda)$

$$f(\mathcal{H}_0) = f(0)\mathcal{P}_0^{(0)} + \mathcal{F}_0^* f\left(\frac{|\xi|}{\sqrt{\epsilon_0 \mu_0}}\right)\mathcal{P}_+^{(0)}(\xi)\mathcal{F}_0$$

$$+ \mathcal{F}_0^* f\left(-\frac{|\xi|}{\sqrt{\epsilon_0 \mu_0}}\right)\mathcal{P}_-^{(0)}(\xi)\mathcal{F}_0. \tag{7.45}$$

Lemma 7.12. *(1)* $\sigma(\mathcal{H}_0) = \mathbf{R}$.
(2) $\mathcal{H}_0$ *has no singular continuous spectrum.*

Proof. Take $\lambda > 0$ and let $u_n \in \mathbf{H}_0$ be such that $\|u_n\| = 1$, $\widehat{u}_n(\xi) = \mathcal{P}_+^{(0)}(\xi)\widehat{u}_n(\xi)$ and

$$\operatorname{supp} \widehat{u}_n(\xi) \subset \{\lambda - \frac{1}{n} < \frac{|\xi|}{\sqrt{\epsilon_0 \mu_0}} < \lambda + \frac{1}{n}\}.$$

Then, $u_n \in D(\mathcal{H}_0)$ and $\|(\mathcal{H}_0 - \lambda)u_n\| \leq 1/n$. Lemma 7.8 implies $\lambda \in \sigma(\mathcal{H}_0)$. Similary $(-\infty, 0) \subset \sigma(\mathcal{H}_0)$.

To prove (2), let $\chi_I(k)$ be the charateristic function of an interval I, and $\chi_{I,+}(k) = \chi_I(k/\sqrt{\epsilon_0 \mu_0})$ if $I \subset (0, \infty)$ and $\chi_{I,-}(k) = \chi_I(-k/\sqrt{\epsilon_0 \mu_0})$ if $I \subset (-\infty, 0)$. Then, by (7.45), for any $u \in \mathbf{H}_{cont}(\mathcal{L}_0)$,

$$(E_0(I)u, u) = \|\chi_{I,+}(|\xi|)\mathcal{P}_+^{(0)}(\xi)\widehat{u}(\xi)\|^2 + \|\chi_{I,-}(|\xi|)\mathcal{P}_-^{(0)}(\xi)\widehat{u}(\xi)\|^2,$$

which generates an absolutely continuous measure. $\qquad\qquad \square$

7.5.3 Exterior problem

Let us turn to an exterior domain Ω in $\mathbf{R}^3$ with C^∞ boundary, and consider the Maxwell operator

$$\mathcal{H} = \mathcal{M}(x)^{-1}\mathcal{L}(-i\nabla), \quad \mathcal{M}(x) = \begin{pmatrix} \epsilon(x) & 0 \\ 0 & \mu(x) \end{pmatrix}, \tag{7.46}$$

$$D(\mathcal{H}) = H_0(\text{curl}\,;\,\Omega) \times H(\text{curl}\,;\,\Omega),$$

where we assume (7.13) on $\epsilon(x), \mu(x)$. Recall that our Hilbert space is $\mathbf{H} = L^2(\Omega)^6$ equipped with the inner product

$$(u,v) = (\epsilon(x)u_E, v_E)_{L^2(\Omega)^3} + (\mu(x)u_H, v_H)_{L^2(\Omega)^3}.$$

Then, $\mathcal{H}$ is self-adjoint on $\mathbf{H}$. Let $\mathcal{P}_0$ be the projection onto $N(\mathcal{H}) = \{u \in D(\mathcal{H})\,;\, \mathcal{H}u = 0\}$, and $\mathcal{P} = 1 - \mathcal{P}_0$.

We add a new assumption:

$$|\epsilon(x) - \epsilon_0 I_3| + |\mu(x) - \mu_0 I_3| \leq C(1 + |x|)^{-\kappa_0}, \tag{7.47}$$

for some constants $\kappa_0, \epsilon_0, \mu_0 > 0$.

Lemma 7.13. $\sigma(\mathcal{H}) = \mathbf{R}$, *and 0 is an eigenvalue of $\mathcal{H}$ with infinite multiplicity.*

Proof.   It is easy to see that

$$N(\mathcal{H}) = \{(E, H) \in D(\mathcal{H})\,;\, \text{curl}\, E = 0,\ \nu \times E = 0,\ \text{curl}\, H = 0\}$$

is infinite dimensional. Take $\lambda > 0$. Since $\lambda \in \sigma(\mathcal{H}_0)$, by using the spectral decomposition, one can construct a sequence $\{u_n\} \subset \mathbf{H}_0$ such that $u_n \in D(\mathcal{H}_0)$, $\|(\mathcal{L}_0 - \lambda)u_n\| \to 0$ and $u_n = \Pi_+^{(0)} u_n$, $(u_m, u_n) = \delta_{mn}$. In fact, for a sequence of disjoint compact intervals $\{I_n\}$ converging to λ, take u_n in such a way that $u_n = E_0(I_n)u_n$. Then, $u_n \to 0$ weakly on $\mathbf{H}_0$.

Take $R > 0$ so that $\Omega^c \subset \{|x| < R-1\}$ and let $\chi_R(x) \in C^\infty(\mathbf{R}^3)$ be such that $\chi_R(x) = 0$ for $|x| < R$, $\chi_R(x) = 1$ for $|x| > R+1$. Put $v_{R,n} = \chi_R u_n$. Then, we have

$$\begin{aligned}(\mathcal{H} - \lambda)v_{R,n} &= [\mathcal{H}, \chi_R]u_n + \chi_R \mathcal{M}^{-1}\mathcal{M}_0(\mathcal{H}_0 - \lambda)u_n \\ &\quad + \lambda\chi_R(\mathcal{M}^{-1}\mathcal{M}_0 - 1)u_n.\end{aligned} \tag{7.48}$$

By the assumption (7.47), for any $\epsilon > 0$, there exists $R = R_\epsilon > 0$ such that the 3rd term of the right-hand side of (7.48) is less than ϵ uniformly in n. Since $[\mathcal{H}, \chi_R]$ is a compactly supported function, by virtue of Lemma 7.6, $[\mathcal{H}, \chi_R]\mathcal{P}_+^{(0)}(\mathcal{H}_0 + i)^{-1}$ is a compact operator. Hence $\|[\mathcal{H}, \chi_R]u_n\| = \|[\mathcal{H}, \chi_R]\mathcal{P}_+^{(0)}(\mathcal{H}_0 + i)^{-1}(\mathcal{H}_0 + i)u_n\| \to 0$ as $n \to \infty$, since $(\mathcal{H}_0 + i)u_n \to 0$ weakly. Also, by the property of u_n, $\|(\mathcal{H}_0 - \lambda)u_n\| \to 0$. Therefore, there exists $n = n_\epsilon$ such that the 1st and the 2nd terms of the right-hand side of (7.48) is less than ϵ. This implies that $\lambda \in \sigma(\mathcal{H})$. Similiarly, one can prove $(-\infty, 0) \subset \sigma(\mathcal{H})$. $\qquad\square$

7.6   Time-dependent scattering theory

We are now in a position to start the study of scattering theory. Consider the electromagnetic wave $u = (E, H)$ in an exterior domain Ω governed by the Maxwell equation

$$\begin{cases} i\partial_t u = \mathcal{H}u, & \text{in} \quad \Omega, \\ u(0) = u_0, \\ \nu \times E = 0, & \text{on} \quad \partial\Omega. \end{cases} \tag{7.49}$$

Assume (7.13). If $u_0 \in \mathbf{H}_{cont}(\mathcal{H})$, by the local decay property (Lemma 7.10), the wave $u(t) = e^{-it\mathcal{H}}u_0$ escapes from any bounded part of the domain Ω as $t \to \pm\infty$. Since $e^{-it\mathcal{H}}$ is unitary, $u(t)$ propagates to infinity, i.e., it scatters. Assume that $\mathcal{M}(x) \to \mathcal{M}_0$ as $|x| \to \infty$. One may then infer that $u(t)$ will behave like a solution to the free Maxwell equation. Namely, for some $u_\pm \in \mathbf{H}_0$,

$$\|e^{-it\mathcal{H}}u_0 - e^{-it\mathcal{H}_0}u_\pm\| \to 0, \quad t \to \pm\infty. \tag{7.50}$$

Here, $\|\cdot\|$ is the norm of $\mathbf{H}$. Rewriting (7.50) as $\|u_0 - e^{it\mathcal{H}}\chi_\Omega e^{-it\mathcal{H}_0}u_\pm\| \to 0$, where χ_Ω is the characteristic function of Ω, we arrive at the definition of *wave operator*

$$W_\pm = \text{s} - \lim_{t\pm\infty} e^{it\mathcal{H}}\chi_\Omega\, e^{-it\mathcal{H}_0}\mathcal{P}_{ac}(\mathcal{H}_0), \tag{7.51}$$

where $\mathcal{P}_{ac}(\mathcal{H}_0)$ is the projection onto the absolutely continuous subspace for $\mathcal{H}_0$. Recall that $\mathcal{H}_0$ has no singular continuous spectrum by Lemma 7.12. From (7.50), we observe $u_0 = W_+u_+ = W_-u_-$. Noting that

$$\|e^{it\mathcal{H}}\chi_\Omega e^{-it\mathcal{H}_0}u\| = \|\chi_\Omega e^{-it\mathcal{H}_0}u\| \sim \|e^{-it\mathcal{H}_0}u\|_{\mathbf{H}_0}, \quad t \to \pm\infty,$$

$W_\pm$ is seen to be isometric on $\mathbf{H}_{ac}(\mathcal{H}_0)$. We then have

$$(W_+)^*W_-u_- = u_+. \tag{7.52}$$

We now define

$$S = (W_+)^*W_- : u_- \to u_+, \tag{7.53}$$

and call it the *scattering operator*. It assigns the outgoing data in the remote future to the incoming data in the remote past. The scattering operator is thus concerned with only the waves at infinity, however, it is expected to determine the whole physical system. Our issue below is centered around the scattering operator.

A fairly general framework has already been established for the mathematical scattering theory, which one can consult in e.g. [51], [85] or [100].

We do not dwell upon this general theory, and proceed directly to our problem. We impose a new condition:

$$|\epsilon(x) - \epsilon_0 I_3| + |\mu(x) - \mu_0 I_3| \le (1 + |x|)^{-1-\epsilon}, \quad \epsilon > 0. \qquad (7.54)$$

Namely, the constant κ_0 in (7.47) is assumed to be greater than 1, which is often called the *short-range* condition.

Theorem 7.13. *Assume (7.13) and (7.54). Then the strong limit (7.51) exists and has the following properties.*
(1) $\|W_\pm u\| = \|u\|$ for $u \in \mathbf{H}_{ac}(\mathcal{H}_0) = N(\mathcal{H}_0)^\perp$.
(2) For any bounded Borel function $\varphi(\lambda)$, $\varphi(\mathcal{H})W_\pm = W_\pm \varphi(\mathcal{H}_0)$.
(3) $\operatorname{Ran} W_\pm \subset \mathbf{H}_{ac}(\mathcal{H})$.

Proof.   The existence of $W_\pm$ is proven in the next section. The short-range assumption is crucial here, since for $\kappa_0 \le 1$, $W_\pm$ no longer exists and we need to modify the definition of wave operators. The assertion (1) is already proven above. Taking the limit as $t \to \pm\infty$ of

$$e^{is\mathcal{H}}e^{i(t-s)\mathcal{H}}\chi_\Omega e^{-i(t-s)\mathcal{H}_0}P_{ac}(\mathcal{H}_0) = e^{it\mathcal{H}}\chi_\Omega e^{-it\mathcal{H}_0}P_{ac}(\mathcal{H}_0)e^{is\mathcal{H}_0},$$

we have $e^{is\mathcal{H}}W_\pm = W_\pm e^{is\mathcal{H}_0}$. For $f \in \mathcal{S}$, we use the Fourier inversion formula $\varphi(\lambda) = (2\pi)^{-1/2}\int_{-\infty}^{\infty}e^{is\lambda}\widehat{\varphi}(s)ds$ to obtain $\varphi(\mathcal{H})W_\pm = W_\pm\varphi(\mathcal{H}_0)$, which can then be extended to any bounded Borel function $\varphi(\lambda)$. This proves (2). Let $E(\lambda)$ be the spectral measure for $\mathcal{H}$. Then by (2) for any interval I and $u \in \mathbf{H}_{ac}(\mathcal{H}_0)$,

$$(E(I)W_\pm u, W_\pm u) = (W_\pm E_0(I)u, W_\pm E_0(I)u) = (E_0(I)u, u),$$

which gives rise to an absolutely continuous measure. $\square$

The wave operator $W_\pm$ is said to be *asymptotically complete* if

$$\operatorname{Ran} W_\pm = \mathbf{H}_{ac}(\mathcal{H}). \qquad (7.55)$$

If this is true, for any $u_0 \in \mathbf{H}_{ac}(\mathcal{H})$, $e^{-it\mathcal{H}}u_0$ will behave like a solution to the free Maxwell equation as $t \to \pm\infty$. Moreover, it implies that

$$S \text{ is unitary on } \mathbf{H}_0. \qquad (7.56)$$

Namely, it assures a 1 to 1 correspondence between the incoming data and the outgoing data. If, furthermore, $\mathcal{H}$ has no singular continuous spectrum, i.e. if

$$\mathbf{H} = \mathbf{H}_{ac}(\mathcal{H}) \oplus \mathbf{H}_{pp}(\mathcal{H}), \qquad (7.57)$$

we have obtained a satisfactory classification of the space-time behavior of solutions to the equation $i\partial_t u = \mathcal{H}u$.

There are several ways for the proof of asymptotic completeness. Below, we show the method due to Enss [27], which agrees with our intuition of propagation of waves.

Theorem 7.14. *Assume (7.13) and (7.54). Then,* $\operatorname{Ran} W_\pm = \mathbf{H}_{ac}(\mathcal{H})$.

We explain the idea of the proof of Theorem 7.14 leaving technical details in the next section. Take $\varphi(\lambda) \in C_0^\infty((0,\infty))$ and consider $u \in \mathbf{H}_{ac}(\mathcal{H})$ satisfying $u = \varphi(\mathcal{H})u$. We observe the behavior of $e^{-it\mathcal{H}}\varphi(\mathcal{H})u$ as $t \to \infty$.

Step 1. The scattering waves disappear from a bounded part.

We use the following fact.

Lemma 7.14. *For* $\phi(\lambda) \in C_0^\infty(\mathbf{R} \setminus \{0\})$, $\phi(\mathcal{H}) - \phi(\mathcal{H}_0)$ *is compact on* $\mathbf{H}$.

Here and in the sequel, we extend $v \in \mathbf{H}$ to be 0 outside Ω.

Due to this lemma and the local decay property,

$$\|\chi_\Omega \varphi(\mathcal{H})e^{-it\mathcal{H}}u - \chi_\Omega \varphi(\mathcal{H}_0)e^{-it\mathcal{H}}u\| \to 0, \qquad (7.58)$$

as $t \to \infty$. Take $\chi(x) \in C^\infty(\mathbf{R}^3)$ such that $\chi(x) = 0$ for $|x| < 1$, $\chi(x) = 1$ for $|x| > 2$, and put $\chi_R(x) = \chi(x/R)$, where R is large enough so that $\Omega^c \subset \{|x| < R\}$. Since $(\chi_\Omega(x) - \chi_R(x))\varphi(\mathcal{H}_0) = (\chi_\Omega(x) - \chi_R(x))(\mathcal{H}_0 + i)^{-1}\mathcal{P}^{(0)}\varphi(\mathcal{H}_0)(\mathcal{H}_0 + i)$ is compact, again using the local decay property,

$$\|\chi_\Omega \varphi(\mathcal{H})e^{-it\mathcal{H}}u - \chi_R(x)\varphi(\mathcal{H}_0)e^{-it\mathcal{H}}u\| \to 0, \qquad (7.59)$$

for any fixed R. This means that, asymptotically, the wave $e^{-it\mathcal{H}}u$ lives only in the region $|x| > R$, and its frequency is restricted to a compact set in $(0, \infty)$.

Step 2. Only the outgoing waves survive.

Take $\chi_\pm(s) \in C^\infty(\mathbf{R})$ such that $\chi_+(s) + \chi_-(s) = 1$, $\chi_+(s) = 0$ for $s < -1/2$, $\chi_-(s) = 0$ for $s > 1/2$, and put

$$p_\pm(x,\xi) = \chi(x)\chi_\pm(\hat{x} \cdot \hat{\xi})\varphi(\lambda_+(\xi))\mathcal{P}_+^{(0)}(\xi), \quad \hat{x} = x/|x|. \qquad (7.60)$$

Consider the following operators $P_\pm$:

$$P_\pm u = (2\pi)^{-3/2} \int_{\mathbf{R}^3} e^{ix\cdot\xi} p_\pm(x,\xi)\hat{u}(\xi)d\xi. \qquad (7.61)$$

They are bounded operators on $L^2(\mathbf{R}^3)$. Note that

$$\chi_R(x)\varphi(\mathcal{H}_0) = \chi_R(x)P_+ + \chi_R(x)P_-.$$

On the support of $p_+(x,\xi)$, x,ξ with opposite directions are cut off, and for $p_-(x,\xi)$, x,ξ with same directins are cut off. Therefore, P_+ should be called an outgoing operator, and P_- an incoming operator. We then have:

Lemma 7.15. $\|P_-e^{-it\mathcal{H}}u\| \to 0, \quad t \to \infty.$

Step 3. For large time t, $e^{-it\mathcal{H}}$ is well approximated by $e^{-it\mathcal{H}_0}\chi_R P_+$.

The following lemma holds.

Lemma 7.16. *Let* $u_s = e^{-is\mathcal{H}}\varphi(\mathcal{H})u$. *Then, for any* $\epsilon > 0$, *there exist* $s > 0$ *and* $R > 0$ *such that*

$$\sup_{t \geq 0} \|e^{-it\mathcal{H}}u_s - e^{-it\mathcal{H}_0}\chi_R P_+ e^{-is\mathcal{H}}u\| < \epsilon.$$

Step 4. The proof of asymptotic completeness.

Suppose there exists $u \in \mathbf{H}_{ac}(\mathcal{H})$ such that $u \perp \operatorname{Ran} W_+$. Take $0 < a < c < d < b$ and $\varphi(\lambda) \in C_0^\infty((a,b))$ such that $\varphi(\lambda) = 1$ on (c,d). Let $u_s = e^{-is\mathcal{H}}\varphi(\mathcal{H})u$. Then, by Lemma 7.16, we have

$$\|u_s\|^2 = (e^{-it\mathcal{H}}u_s, e^{-it\mathcal{H}}u_s)$$
$$= (e^{-it\mathcal{H}_0}\mathcal{P}^{(0)}\chi_R P_+ e^{-is\mathcal{H}}u, e^{-it\mathcal{H}}u_s) + O(\epsilon)$$

holds uniformly for $t \geq 0$. Therefore, letting $t \to \infty$, we have

$$\|u_s\|^2 = (W\chi_R P_+ e^{-is\mathcal{P}}u, u_s) + O(\epsilon).$$

However, by using Theorem 7.13 (2), we have

$$(W_+\chi_R P_+ e^{-is\mathcal{H}}u, e^{-is\mathcal{H}}\varphi(\mathcal{H})u) = (\varphi(\mathcal{H})e^{is\mathcal{H}}W_+\chi_R P_+ e^{-is\mathcal{H}}u, u)$$
$$= (\varphi(\mathcal{H})W_+ e^{is\mathcal{H}_0}\chi_R P_+ e^{-is\mathcal{H}}u, u)$$
$$= 0.$$

This implies $\varphi(\mathcal{H})u = 0$ for any $\varphi \in C_0^\infty((0,\infty))$. Similarly, this holds for any $\varphi \in C_0^\infty((-\infty,0))$. Therefore, $u = 0$. This completes the proof of Theorem 7.14.

As has been done by Enss [27], a little modification of the arguments in the Step 1 allows us to take u from $\mathbf{H}_{cont}(\mathcal{H})$. We can then conclude the non-existence of singular continuous spectrum of $\mathcal{H}$.

Theorem 7.15. *Under the assumption (7.13) and (7.54), $\mathcal{H}$ has no singular continuous spectrum.*

7.7 Tools from scattering theory

We summarize mathematical tools needed for the proof of above facts.

7.7.1 *Functional calculus*

For $z = x + iy \in \mathbf{C}$, put

$$\overline{\partial_z} = \frac{1}{2}(\partial_x + i\partial_y), \quad dzd\overline{z} = -2idxdy.$$

Then, the following *generalized Cauchy formula* holds.

Lemma 7.17. *(1) Let $D \subset \mathbf{C}$ be a bounded domain with C^∞ boundary. Then, for $\lambda \in D$ and $\varphi(z) \in C^1(\overline{D})$,*

$$\varphi(\lambda) = \frac{1}{2\pi i} \int_{\partial D} \frac{\varphi(z)}{z - \lambda} dz + \frac{1}{2\pi i} \int_D \frac{\overline{\partial_z}\varphi(z)}{z - \lambda} dzd\overline{z}.$$

Here, $\overline{D}$ means the closure of D.
(2) If $F(z) \in C^1(\mathbf{C}) \cap L^\infty(\mathbf{C})$, $\lim_{|z| \to \infty} F(z) = 0$, and

$$\int_{\mathbf{C}} \left| \frac{\overline{\partial_z}F(z)}{z - \lambda} \right| dxdy < \infty,$$

we have

$$F(\lambda) = \frac{1}{2\pi i} \int_{\mathbf{C}} \frac{\overline{\partial_z}F(z)}{z - \lambda} dzd\overline{z}.$$

Proof. Let $D_\epsilon = D \setminus \{|z - \lambda| \leq \epsilon\}$. Then, by Gauss-Green's formula,

$$\frac{1}{2\pi i} \int_{\partial D_\epsilon} \frac{\varphi(z)}{z - \lambda} dz = -\frac{1}{2\pi i} \int_{D_\epsilon} \frac{\overline{\partial_z}\varphi(z)}{z - \lambda} dzd\overline{z}.$$

Letting $\epsilon \to 0$, we prove (1). Letting $D = \{|z - \lambda| < R\}$ and $R \to \infty$ in (1), we obtain (2). $\square$

The following useful formula is often refered to as *Helffer-Sjöstrand formula* ([39]).

Lemma 7.18. *Let $f(x) \in C(\mathbf{R}) \cap L^\infty(\mathbf{R})$ and $F(z) \in C^1(\mathbf{C})$ satisfy*

$$\begin{cases} F(x) = f(x), \quad x \in \mathbf{R}, \\ \lim_{|z| \to \infty} F(z) = 0, \\ |\overline{\partial_z}F(z)| \leq C|\mathrm{Im}\, z|(1 + |z|)^{-2-\epsilon}, \quad \epsilon > 0. \end{cases}$$

Then, for any self-adjoint operator A, we have

$$f(A) = \frac{1}{2\pi i} \int_{\mathbf{C}} \overline{\partial_z} F(z)(z - A)^{-1} dz d\bar{z},$$

where the integral of the right-hand side converges in the sense of operator norm.

Proof. Since $\|(z - A)^{-1}\| \leq 1/|\text{Im}\, z|$, we have $\|\overline{\partial_z} F(z)(z - A)^{-1}\| \in L^1(\mathbf{C})$. Let $E(\lambda)$ be the spectral decomposition of A, and put $\rho(\lambda) = (E(\lambda)u, v)$. Then

$$
\begin{aligned}
(f(A)u, v) &= \int_{-\infty}^{\infty} f(\lambda) d\rho(\lambda) \\
&= \frac{1}{2\pi i} \int_{-\infty}^{\infty} \int_{\mathbf{C}} \frac{\overline{\partial_z} F(z)}{z - \lambda} dz d\bar{z} d\rho(\lambda) \\
&= \frac{1}{2\pi i} \int_{\mathbf{C}} \overline{\partial_z} F(z)((z - A)^{-1} u, v) dz d\bar{z},
\end{aligned}
$$

which proves the lemma. $\square$

Let us construct $F(z)$ appearing in Lemma 7.18, which is called an *almost analytic extension* of f.

Lemma 7.19. *Let $N \geq 2, s \in \mathbf{R}$, and $f(x) \in C^N(\mathbf{R})$ satisfy*

$$|f^{(k)}(x)| \leq C_k(1 + |x|)^{s-k}, \quad 0 \leq k \leq N.$$

Then, there exists $F(z) \in C^1(\mathbf{C})$ such that

$$
\begin{cases}
F(x) = f(x), \quad x \in \mathbf{R}, \\
|F(z)| \leq C(1 + |z|)^s, \quad z \in \mathbf{C}, \\
|\overline{\partial_z} F(z)| \leq C|\text{Im}\, z|^n (1 + |z|)^{s-n-1}, \quad 1 \leq n \leq N - 1.
\end{cases}
$$

Moreover, if $\text{supp}\, f \subset [a, b]$, we can construct $F(z)$ so that

$$\text{supp}\, F(z) \subset \{z\,;\, \text{Re}\, z \in [a, b],\ \text{Im}\, z \in [-C, C]\}$$

for some $C > 0$.

Proof. We take $\chi \in C_0^\infty(\mathbf{R})$ satisfying $\chi(y) = 1$ $(|y| < 1)$, $\chi(y) = 0$ $(|y| > 2)$, and put

$$F(z) = \sum_{n=0}^{N-1} \frac{i^n}{n!} f^{(n)}(x) y^n \chi\left(\frac{y}{\langle x \rangle}\right), \quad \langle x \rangle = (1 + |x|^2)^{1/2}.$$

Then, we have

$$2\overline{\partial_z}F(z) = \frac{i^{N-1}}{(N-1)!}f^{(N)}(x)y^{N-1}\chi\left(\frac{y}{\langle x\rangle}\right)$$
$$+ \sum_{n=0}^{N-1}\frac{i^n}{n!}f^{(n)}(x)y^n\chi'\left(\frac{y}{\langle x\rangle}\right)\left(\frac{i}{\langle x\rangle} - \frac{xy}{\langle x\rangle^3}\right).$$

The 1st term of the right-hand side is estimated from above by

$$C\langle x\rangle^{s-N}|y|^{N-1} \le C|y|^n\langle z\rangle^{z-n-1},$$

since $|y| \le 2\langle x\rangle$. The 2nd term of the right-hand side is estimated from above by

$$C\sum_{n=0}^{N-1}\langle x\rangle^{s-n-1}|y|^n\left|\chi'\left(\frac{y}{\langle x\rangle}\right)\right| \le C\langle x\rangle^{s-1}\left|\chi'\left(\frac{y}{\langle x\rangle}\right)\right| \le C|y|^n\langle z\rangle^{s-n-1},$$

since $\langle x\rangle \le |y| \le 2\langle x\rangle$. By the similar computation, we can show $|F(z)| \le C\langle z\rangle^s$. $\qquad\square$

By the well-known Borel's procedure, putting

$$F(z) = \sum_{n=0}^{\infty}\frac{i^n}{n!}f^{(n)}(x)y^n\chi(t_n y),$$

and choosing $0 < t_1 < t_2 < \cdots$ suitably, one can take $N = \infty$ in the above lemma. However, we do not enter into the details.

Let us now prove Lemma 7.14. Letting $\mathcal{P}$ and $\mathcal{P}^{(0)}$ be the projections onto the orthogonal complements of $N(\mathcal{H})$ and $N(\mathcal{H}_0)$, we have $\phi(\mathcal{H}) = \phi(\mathcal{H})\mathcal{P}$, $\phi(\mathcal{H}_0) = \phi(\mathcal{H}_0)\mathcal{P}^{(0)}$. Therefore, by the local compactness, we have only to prove the compactness of $\chi(x)(\phi(\mathcal{H}) - \phi(\mathcal{H}_0))$ for $\chi(x) \in C^{\infty}(\mathbf{R}^3)$ such that $\chi(x) = 1$ for $|x| > R+1$, $\chi(x) = 0$ for $|x| < R$, where R is a large constant. Let $R(z)$ and $R_0(z)$ be the resolvent of $\mathcal{H}$ and $\mathcal{H}_0$, respectively. Using $\mathcal{H}_0 = \mathcal{M}_0^{-1}\mathcal{M}\mathcal{H}$, we have

$$(\mathcal{H}_0 - z)\chi R(z) = [\mathcal{H}_0, \chi]R(z) + \chi\mathcal{M}_0^{-1}\mathcal{M} + \chi z(\mathcal{M}_0^{-1}\mathcal{M} - 1)R(z).$$

Hence,

$$\chi R(z) = R_0(z)\chi\mathcal{M}_0^{-1}\mathcal{M} + R_0(z)V(z)R(z),$$

$$V(z) = [\mathcal{H}_0, \chi] + \chi z(\mathcal{M}_0^{-1}\mathcal{M} - 1).$$

Let $\Phi(z)$ be the almost analytic extension of $\phi(\lambda)$ constructed in Lemma 7.19. Then, by Lemma 7.18,

$$\chi\phi(\mathcal{H}) = \phi(\mathcal{H}_0)\chi\mathcal{M}_0^{-1}\mathcal{M} + K,$$

$$K = -\frac{1}{2\pi i} \int_{\mathbf{C}} \overline{\partial_z} \Phi(z) R_0(z) V(z) R(z) dz d\overline{z}.$$

Let $D = \operatorname{supp} \Phi(z)$, which is a compact set in $\mathbf{C}$ and $0 \notin D$. Then, letting $z = t + is$, we can find a constant $C > 0$ such that

$$\left|\frac{\lambda + i}{\lambda - z}\right| \le C \frac{|\lambda - t| + |t| + 1}{|\lambda - t| + |s|} \le \frac{C}{|s|}, \quad \forall z \in D, \quad \forall \lambda \in \mathbf{R}.$$

Therefore, by using the spectral decomposition, we have

$$\|R(z)(\mathcal{H} + i)\| \le C/|\operatorname{Im} z|, \quad \forall z \in D. \tag{7.62}$$

We decomose $R(z)$, $R_0(z)$ into

$$R(z) = \mathcal{P}(\mathcal{H} + i)^{-1}(\mathcal{H} + i)R(z) + \mathcal{P}_0 R(z),$$

$$R_0(z) = \mathcal{P}^{(0)}(\mathcal{H}_0 + i)^{-1}(\mathcal{H}_0 + i)R_0(z) + \mathcal{P}_0^{(0)} R_0(z),$$

Then, K is decomposed into

$$K = K_1 + K_0,$$

$$K_0 = -\frac{1}{2\pi i} \int_D \overline{\partial_z} \Phi(z) \mathcal{P}_0^{(0)} R_0(z) V(z) \mathcal{P}_0 R(z) dz d\overline{z}.$$

Since the integrand of K_1 is a sum of terms containing $\mathcal{P}$ or $\mathcal{P}^{(0)}$, by the local compactness and (7.62), K_1 is shown to be compact. Here, we use Lemma 7.19 with $n = 2$.

Since $\mathcal{P}_0 R(z) = \frac{1}{-z}\mathcal{P}_0$, $\mathcal{P}_0^{(0)} R_0(z) = \frac{1}{-z}\mathcal{P}_0^{(0)}$, we have

$$K_0 = -\frac{1}{2\pi i} \int_D \overline{\partial_z} \Phi(z) \frac{1}{z^2} \mathcal{P}_0^{(0)} V(z) \mathcal{P}_0 dz d\overline{z}.$$

Since $0 \notin D$, $K_0 = 0$ by integration by parts.

We have thus proven that $\chi\phi(\mathcal{H}) - \phi(\mathcal{H}_0)\chi\mathcal{M}_0^{-1}\mathcal{M}$ is compact. By the local compactness, $(\chi_\Omega - \chi)\phi(\mathcal{H})$ and $\phi(\mathcal{H}_0)(1 - \chi\mathcal{M}_0^{-1}\mathcal{M})$ are compact. This completes the proof of Lemma 7.14.

7.7.2 *Pseudo-differential calculus*

Let us first recall basic knowledge of pseudo-differential operators (called ΨDO's for brevity). For $p(x, \xi) \in C^\infty(\mathbf{R}^n \times \mathbf{R}^n)$ satisfying

$$\sup_{x, \xi \in \mathbf{R}^n} |\partial_x^\alpha \partial_\xi^\beta p(x, \xi)| < \infty, \quad \forall \alpha, \beta, \tag{7.63}$$

we associate an operator P by

$$Pu(x) = (2\pi)^{-n/2} \int_{\mathbf{R}^n} e^{ix \cdot \xi} p(x, \xi) \widehat{u}(\xi) d\xi, \tag{7.64}$$

for $u \in S$. By using $(1 + |x|^2)^{-1}(1 - \Delta_\xi)e^{ix \cdot \xi} = e^{ix \cdot \xi}$ and integration by parts, it is easy to show that $Pu \in S$ if $u \in S$. However, a much stronger fact holds.

Theorem 7.16. *The following inequality holds*

$$\|Pu\|_{L^2(\mathbf{R}^n)} \leq C_n \Big(\sup_{x, \xi \in \mathbf{R}^n} \sum_{|\alpha| + |\beta| \leq N} |\partial_x^\alpha \partial_\xi^\beta p(x, \xi)| \Big) \|u\|_{L^2(\mathbf{R}^n)},$$

where C_n and N are constants depending only on n.

For the proof see [16].

The function $p(x, \xi)$ is called the *symbol* of the operator P. A remarkable fact is that if p has a certain decay property, P's form an algebra of operators. For $m, s \in \mathbf{R}$, let $S^{m,s}$ be the set of functions $p(x, \xi)$ satisfying

$$|p|_{m,s,k} := \sup_{x, \xi \in \mathbf{R}^n} \sum_{|\alpha| + |\beta| \leq k} \langle x \rangle^{|\alpha| - m} \langle \xi \rangle^{|\beta| - s} |\partial_x^\alpha \partial_\xi^\beta p(x, \xi)| < \infty, \quad \forall k \geq 0.$$

$$(7.65)$$

The associated class of ΨDO's with symbol in $S^{m,s}$ is also denoted by $S^{m,s}$.

Theorem 7.17. *Let P_1, P_2 be ΨDO's with symbol $p_1 \in S^{m_1, s_1}, p_2 \in S^{m_2, s_2}$. Then, their product $P_1 P_2$ is a $\Psi DO \in S^{m_1 + m_2, s_1 + s_2}$, and its symbol $p(x, \xi)$ has the following property:*

$$p(x, \xi) = p_1(x, \xi) p_2(x, \xi) + r(x, \xi),$$

$$|\partial_x^\alpha \partial_\xi^\beta r(x, \xi)| \leq C_{\alpha\beta} \sum_{|\gamma| = 1} |p_1^{(\gamma)}|_{m_1, s_1 - 1, M + k} |p_{2(\gamma)}|_{m_2 - 1, s_2, M + k}$$

$$\times \langle x \rangle^{m_1 + m_2 - 1 - |\alpha|} \langle \xi \rangle^{s_1 + s_2 - 1 - |\beta|},$$

where $p_1^{(\gamma)} = \partial_\xi^\gamma p_1(x, \xi)$, $p_{2(\gamma)} = (-i\partial_x)^\gamma p_2(x, \xi)$, $k = |\alpha| + |\beta|$, and M is a constant.

Theorem 7.18. *Let P be a ΨDO with symbol $p \in S^{m,s}$. Then its adjoint P^* is also a ΨDO with symbol $p^* \in S^{m,s}$, and has the following property:*

$$p^*(x, \xi) = \overline{p(x, \xi)} + r(x, \xi),$$

$$|\partial_x^\alpha \partial_\xi^\beta r(x, \xi)| \leq C_{\alpha\beta} \sum_{|\gamma| = 1} |p_{(\gamma)}^{(\gamma)}|_{m-1, s-1, M+k} \langle x \rangle^{m-1-|\alpha|} \langle \xi \rangle^{s-1-|\beta|},$$

where $k = |\alpha| + |\beta|$, and M is a constant.

We omit the proof of these theorems. See e.g. [42], [103]. Note that in our applications below, we only consider the symbols compactly supported or rapidly decreasing in ξ, in which case the proof of L^2-boundedness of P is easy.

The following lemma, which can be easily proven by Theorem 7.17, is frequently used in the sequel.

Lemma 7.20. *Let P be a ΨDO with symbol $\in S^{0,0}$. Then for any $s \in \mathbf{R}$, $\langle x \rangle^s P \langle x \rangle^{-s}$ is a bounded operator on $L^2(\mathbf{R}^n)$.*

In particular, we can apply this lemma to
$$Pu(x) = (2\pi)^{-n/2} \int_{\mathbf{R}^n} e^{ix \cdot \xi} p(\xi) \widehat{u}(\xi) d\xi,$$
where $p(\xi) \in C_0^\infty(\mathbf{R}^n)$. In this case, the proof is done by elementary integration by parts.

Let us return to the operator $P_\pm$ in (7.61).

Lemma 7.21. *For $s \geq 0$,*
$$\sup_{\pm t \geq 0} \|(1 \pm t + |x|)^s P_\mp e^{-it\mathcal{H}_0} \langle x \rangle^{-s}\|_{\mathbf{B}(\mathbf{H}_0)} < \infty.$$

Proof.   We have only to prove the lemma for $s = 0, 1, 2, \cdots$. The general case follows from it by interpolation using Hadamard's three line theorem (see e.g. [85], Vol. 2, p. 33). We consider the case $t \geq 0$, and rewrite $P_- e^{-it\mathcal{H}_0} u$ as
$$(2\pi)^{-3/2} \int_{\mathbf{R}^3} e^{i(x \cdot \xi - tc_0 |\xi|)} p_-(x, \xi) \widehat{u}(\xi) d\xi, \tag{7.66}$$
where $c_0 = 1/\sqrt{\epsilon_0 \mu_0}$. On $\operatorname{supp} p_-(x, \xi)$, $\widehat{x} \cdot \widehat{\xi} < 1/2$. Therefore, for $t \geq 0$,
$$|\nabla_\xi (x \cdot \xi - tc_0 |\xi|)| = |x - tc_0 \widehat{\xi}| \geq C(|x| + t),$$
for a constant $C > 0$. Let L be the differential operator
$$L = -i |\nabla_\xi (x \cdot \xi - tc_0 |\xi|)|^{-2} \nabla_\xi (x \cdot \xi - tc_0 |\xi|) \cdot \nabla_\xi.$$
Noting that $L e^{i(x \cdot \xi - tc_0 |\xi|)} = e^{i(x \cdot \xi - tc_0 |\xi|)}$, we have by integration by parts
$$\int_{\mathbf{R}^3} e^{i(x \cdot \xi - tc_0 |\xi|)} p_-(x, \xi) \widehat{u}(\xi) d\xi = \int_{\mathbf{R}^3} e^{i(x \cdot \xi - tc_0 |\xi|)} (L^*)^s \big(p_-(x, \xi) \widehat{u}(\xi)\big) d\xi,$$
for any $s > 0$. We then have
$$(L^*)^s \big(p_-(x, \xi) \widehat{u}(\xi)\big) = \sum_{|\alpha| \leq s} p_\alpha(t, x, \xi) \partial_\xi^\alpha \widehat{u}(\xi),$$
where $p_\alpha(t, x, \xi)$ is compactly supported in ξ, and
$$|\partial_x^\beta \partial_\xi^\gamma p_\alpha(t, x, \xi)| \leq C_{\beta\gamma} (1 + t + |x|)^{-s}.$$
The lemma then follows from Theorem 7.16. $\square$

Let $P_\pm(R) = \chi(x/R)P_\pm$, where $\chi \in C^\infty(\mathbf{R}^n)$ satisfying $\chi(x) = 0$ for $|x| < 1$ and $\chi(x) = 1$ for $|x| > 2$.

Lemma 7.22. *As* $R \to \infty$,

$$\sup_{\pm t \geq 0} \|(\chi_\Omega e^{-it\mathcal{H}} - e^{-it\mathcal{H}_0})P_\pm(R)^*\|_{\mathbf{B}(\mathbf{H}_0)} \to 0. \tag{7.67}$$

Proof. Take $R_0 > 0$ large enough so that $\chi_{R_0}(x) = \chi(x/R_0) = 0$ on Ω^c. We also take $\psi \in C_0^\infty((0,\infty))$ such that $\psi = 1$ on $\operatorname{supp}\varphi$, where φ is from (7.60). Letting

$$V = \mathcal{H}\chi_{R_0} - \chi_{R_0}\mathcal{H}_0,$$

we have

$$e^{-it\mathcal{H}}\chi_{R_0}\psi(\mathcal{H}_0) - \chi_{R_0}e^{-it\mathcal{H}_0}\psi(\mathcal{H}_0) = -i\int_0^t e^{-i(t-s)\mathcal{H}}V\psi(\mathcal{H}_0)e^{-is\mathcal{H}_0}ds.$$

Therefore, for $t > 0$,

$$\|(e^{-it\mathcal{H}}\chi_{R_0} - \chi_{R_0}e^{-it\mathcal{H}_0})P_+(R)^*\| \leq \int_0^t \|V\psi(\mathcal{H}_0)e^{-is\mathcal{H}_0}P_+(R)^*\|ds.$$

Here the norm $\|\cdot\|$ is the norm in $\mathbf{B}(\mathbf{H}_0; \mathbf{H})$. Lemma 7.21 implies that

$$\sup_{t \geq 0} \|\langle x \rangle^{-s} e^{-it\mathcal{H}_0}P_+^*(1 + t + |x|)^s\| < \infty, \tag{7.68}$$

for any $s \geq 0$. Noting that

$$V = [\mathcal{H}_0, \chi_{R_0}] + \chi_{R_0}(\mathcal{M}^{-1}\mathcal{M}_0 - 1)\mathcal{H}_0,$$

and using (7.68) with $s = 1 + \epsilon$, we estimate $\|V\psi(\mathcal{H}_0)e^{-is\mathcal{H}_0}P_+(R)^*\|$ from above by

$$\|V\psi(\mathcal{H}_0)e^{-it\mathcal{H}_0}P_+^*(1+t+|x|)^{1+\epsilon}\| \times \|(1+t+|x|)^{-1-\epsilon}\chi_R\| \leq C(1+t+R)^{-1-\epsilon}.$$

This proves

$$\sup_{t \geq 0} \|(e^{-it\mathcal{H}}\chi_{R_0} - \chi_{R_0}e^{-it\mathcal{H}_0})P_+(R)^*\| \leq CR^{-\epsilon}. \tag{7.69}$$

Using (7.68) again, we have

$$\sup_{t \geq 0} \|(1 - \chi_{R_0})e^{-it\mathcal{H}_0}P_+(R)^*\| \leq CR^{-\epsilon},$$

so that one can replace $\chi_{R_0}e^{-it\mathcal{H}_0}$ by $e^{-it\mathcal{H}_0}$ in (7.69). In view of

$$P_+(R)\chi_{R_0} = \chi_R P_+ \chi_{R_0} = \chi_R P_+ + \chi_R[P_+, \chi_{R_0}],$$

and noting that the symbol of $[P_+, \chi_{R_0}]$ behaves like $O(|x|^{-1})$, we have

$$\|(\chi_{R_0} - 1)P_+(R)^*\| \to 0.$$

We can thus replace $e^{-it\mathcal{H}}\chi_{R_0}$ by $e^{-it\mathcal{H}}$ in (7.69). This proves Lemma 7.22 for the case $t \geq 0$. $\qquad\square$

Let us prove Lemma 7.15. Lemma 7.22 implies

$$\sup_{t\geq 0}\|P_-(R)\big(e^{-it\mathcal{H}}\chi_\Omega - e^{-it\mathcal{H}_0}\big)\| \to 0, \quad R \to \infty. \tag{7.70}$$

Therefore, for any $u \in \mathbf{H}_{ac}(\mathcal{H})$ and $\epsilon > 0$,

$$P_-(R)e^{-it\mathcal{H}}u = P_-(R)e^{-it\mathcal{H}_0}\chi_\Omega u + O(\epsilon),$$

as $R \to \infty$ uniformly in $t > 0$. Using

$$P_-(R)e^{-it\mathcal{H}_0}\chi_\Omega u = P_-(R)e^{-it\mathcal{H}_0}\mathcal{P}^{(0)}\chi_\Omega u + P_-(R)\mathcal{P}_0^{(0)}\chi_\Omega u,$$

Therefore, choosing R large enough, we have

$$P_-(R)e^{-it\mathcal{H}}u = P_-(R)e^{-it\mathcal{H}_0}\mathcal{P}^{(0)}\chi_\Omega u + O(\epsilon),$$

uniformly in $t > 0$. Using Lemma 7.21 and approximating u by rapidly decreasing functions, we then get $\|P_- e^{-it\mathcal{H}}u\| \to 0$.

Let us prove Lemma 7.16. We first show

$$\|P_+(R)^* - P_+(R)\| \leq C/R. \tag{7.71}$$

In fact, since the symbol $p_+(x,\xi;R)$ of $P_+(R)$ is real-valued, the symbol $q(x,\xi;R)$ of $Q_R = P_+(R)^* - P_+(R)$ is estimated by the derivatives $\partial_x^\gamma \partial_\xi^\gamma p_+(x,\xi;R)$, $(|\gamma| = 1)$, which behave like $O(1/R)$. Therefore $\|Q_R\| = O(1/R)$.

Then, by Lemma 7.22, for any $\epsilon > 0$, there exists $R > 0$ such that

$$\sup_{t\geq 0}\|\big(\chi_\Omega e^{-it\mathcal{H}} - e^{-it\mathcal{H}_0}\big)P_+(R)\|_{\mathbf{B}(\mathbf{H}_0)} < \epsilon.$$

For this R, by Lemma 7.15, there exists $s > 0$ such that

$$e^{-is\mathcal{H}}\varphi(\mathcal{H})u = P_+(R)e^{-is\mathcal{H}}u + O(\epsilon).$$

Since $u_s = e^{-it\mathcal{H}}\varphi(\mathcal{H})u = e^{-it\mathcal{H}}u$, we obtain Lemma 7.16.

We prove the existence of wave operators. We take $\varphi(\lambda) \in C_0^\infty((0,\infty))$, and show the existence of the limit $\lim_{t\to\infty} e^{it\mathcal{H}}\chi_R(x)e^{-it\mathcal{H}_0}\varphi(\mathcal{H}_0)u$ for $u \in C_0^\infty(\mathbf{R}^3)^6$, where $R > 0$ is a large constant. Computing $\frac{d}{dt}e^{it\mathcal{H}}\chi_R e^{-it\mathcal{H}_0}u$, we have only to show that

$$\int_0^\infty \|V e^{-it\mathcal{H}_0}\varphi(\mathcal{H}_0)u\|dt < \infty, \tag{7.72}$$

where $V = \mathcal{H}\chi_R - \chi_R\mathcal{H}_0$. Take $\chi_j(s) \in C^\infty(\mathbf{R})$ such that

$$\chi_1(s) + \chi_2(s) + \chi_3(s) = 1, \quad \text{on} \quad \mathbf{R},$$

$$\chi_1(s) = 1, \quad s < \frac{c_0}{4}, \quad \chi_1(s) = 0, \quad s > \frac{c_0}{2},$$

$$\chi_3(s) = 0, \quad s < 2c_0, \quad \chi_3(s) = 1, \quad s > 3c_0.$$

We insert $\chi_1(|x|/t) + \chi_2(|x|/t) + \chi_3(|x|/t) = 1$. Since V decays like $O(|x|^{-1-\epsilon})$, we easily have

$$\int_0^\infty \|\chi_2(|x|/t)Ve^{-it\mathcal{H}_0}\varphi(\mathcal{H}_0)u\|dt < \infty.$$

Noting that on the support of $\chi_1(|x|/t)$ and $\chi_3(|x|/t)$,

$$|\nabla_\xi(x \cdot \xi - tc_0|\xi|)| \geq C(t + |x|),$$

we have by integration by parts

$$\|\langle x\rangle^s \chi_j(|x|/t)Ve^{-it\mathcal{H}_0}\varphi(\mathcal{H}_0)\langle x\rangle^{-s}\| \leq C_N(1+t)^{-s}$$

for any $s > 0$. We have, therefore

$$\int_0^\infty \|\chi_j(|x|/t)Ve^{-it\mathcal{H}_0}\varphi(\mathcal{H}_0)u\|dt < \infty, \quad j = 1, 3,$$

which proves the existence of the wave operator.

Chapter 8

Stationary scattering theory

In this chapter, we study the properties of solutions to the boundary value problem of the stationary Maxwell equation $(\mathcal{H} - \lambda)u = 0$ in an exterior domain. Observing the behavior at infinity of the resolvent, we derive a generalized Fourier transformation for $\mathcal{H}$, and characterize the physically natural solution space. The main theorem is Theorem 8.14 in which the scattering operator is formulated in a stationary manner.

8.1 Resolvent estimates for the free Laplacian

For a self-adjoint operator A in a Hilbert space $\mathbf{H}$, the resolvent $(A - z)^{-1}$ does not exist when $z = \lambda \in \sigma(A)$. However, if $\lambda \in \sigma_c(A)$, it sometimes occurs that $\lim_{\epsilon \to 0}(A - \lambda \mp i\epsilon)^{-1} := (A - \lambda \mp i0)^{-1}$ exists as an operator from $\mathbf{X}$ to $\mathbf{Y}$, where $\mathbf{X}$ and $\mathbf{Y}$ are Banach spaces such that $\mathbf{X} \subset \mathbf{H} \subset \mathbf{Y}$ with continuous inclusion. This is called the *limiting absorption principle*. For example, the resolvent of the Laplacian in $\mathbf{R}^3$ has the limit

$$(-\Delta - \lambda \mp i0)^{-1}f = \frac{1}{4\pi} \int_{\mathbf{R}^3} \frac{e^{\pm i\sqrt{\lambda}|x-y|}}{|x-y|} f(y)dy,$$

if $f \in C_0^\infty(\mathbf{R}^3)$ and $\lambda > 0$. As $r = |x| \to \infty$, the right-hand side has the asymptotic expansion

$$\frac{1}{4\pi} \int_{\mathbf{R}^3} \frac{e^{\pm i\sqrt{\lambda}|x-y|}}{|x-y|} f(y)dy = \sqrt{\frac{\pi}{2}} \frac{e^{\pm i\sqrt{\lambda}r}}{r} \hat{f}(\pm\sqrt{\lambda}\hat{x}) + O(r^{-2}), \quad \hat{x} = x/r.$$

As is suggested by these formulas, the limiting absorption principle for the Laplacian, or Schrödinger operator on $\mathbf{R}^n$, or manifold plays a key role in stationary scattering theory, and plenty of works have been devoted to this subject. We review here the method due to Agmon-Hörmander [4]. See also Hörmander [42] and Yafaev [101].

For $s \in \mathbf{R}$, let $L^{2,s}$ be the weighted Hilbert space defined by

$$L^{2,s} \ni f \iff \|f\|_s = \|(1 + |x|)^s f(x)\|_{L^2(\mathbf{R}^n)} < \infty.$$

Let $R_j = 2^j$ $(j \geq 0)$, $\Omega_0 = \{x \in \mathbf{R}^n ; |x| < 1\}$, $\Omega_j = \{x \in \mathbf{R}^n ; 2^{j-1} < |x| < 2^j\}$ $(j \geq 1)$, and

$$\mathcal{B} \ni u \iff \|u\|_{\mathcal{B}} = \sum_{j=0}^{\infty} R_j^{1/2} \|u\|_{L^2(\Omega_j)} < \infty.$$

The dual space $\mathcal{B}^*$ is identified with the following space

$$\mathcal{B}^* \ni v \iff \|v\|_{\mathcal{B}^*} = \left(\sup_{R>1} \frac{1}{R} \int_{|x|<R} |v(x)|^2 dx \right)^{1/2} < \infty.$$

The following inclusion relation holds : For $s > 1/2$,

$$L^{2,s} \subset \mathcal{B} \subset L^{2,1/2} \subset L^2 \subset L^{2,-1/2} \subset \mathcal{B}^* \subset L^{2,-s}. \tag{8.1}$$

The space $\mathcal{B}_0^*$ is defined by

$$\mathcal{B}_0^* \ni v \iff \lim_{R \to \infty} \frac{1}{R} \int_{|x|<R} |v(x)|^2 dx = 0. \tag{8.2}$$

Lemma 8.1. *If $T \in \mathbf{B}(L^{2,s})$ for any $s \in \mathbf{R}$, then $T \in \mathbf{B}(\mathcal{B}) \cap \mathbf{B}(\mathcal{B}^*)$.*

For the proof, see [42], Theorem 14.1.4. Due to this lemma, bounded ΨDO's are also bounded on $\mathcal{B}$ and $\mathcal{B}^*$.

Now let us state the limiting absorption principle for $-\Delta$ in $\mathbf{R}^3$.

Theorem 8.1. *Let $r_0(z) = (-c_0^2 \Delta - z)^{-1}$ and $\lambda > 0$, where $c_0 > 0$ is a constant.*
(1) For $s > 1/2$, there exists a strong limit

$$\text{s} - \lim_{\epsilon \to 0} r_0(\lambda \pm i\epsilon) := r_0(\lambda \pm i0) \in \mathbf{B}(L^{2,s}; L^{2,-s}). \tag{8.3}$$

Moreover, for any $f \in L^{2,s}$, $r_0(\lambda \pm i0)f$ is an $L^{2,-s}$-valued continuous function of $\lambda > 0$.
(2) For any $f, g \in \mathcal{B}$, the limit

$$\lim_{\epsilon \to 0} (r_0(\lambda \pm i\epsilon)f, g) = (r_0(\lambda \pm i0)f, g)$$

exists, and the right-hand side is a continuous function of $\lambda \in \mathbf{R} \setminus \{0\}$. Moreover

$$\|r_0(\lambda \pm i0)f\|_{\mathcal{B}^*} \leq C\|f\|_{\mathcal{B}},$$

where C is independent of λ, when λ varies over a compact set in $(0, \infty)$.

(3) For any $f \in \mathcal{B}$ and $\lambda > 0$,

$$r_0(\lambda \pm i0)f - \frac{1}{c_0^2}\sqrt{\frac{\pi}{2}}\frac{e^{\pm i\frac{\sqrt{\lambda}}{c_0}r}}{r}\widehat{f}(\pm\frac{\sqrt{\lambda}}{c_0}\widehat{x}) \in \mathcal{B}_0^*, \quad r = |x|, \quad \widehat{x} = x/r.$$

(4) For any $f \in \mathcal{B}$,

$$\left(\frac{\partial}{\partial r} \mp i\frac{\sqrt{\lambda}}{c_0}\right)r_0(\lambda \pm i0)f \in \mathcal{B}_0^*.$$

A solution $u \in \mathcal{B}^*$ of the Helmholtz equation

$$(-c_0^2\Delta - \lambda)u = f$$

is said to satisfy the *radiation condition* if

$$\left(\frac{\partial}{\partial r} \mp i\frac{\sqrt{\lambda}}{c_0}\right)u \in \mathcal{B}_0^*,$$

(*outgoing* for $-$, and *incoming* for $+$). By Theorem 8.1 (4), $r_0(\lambda \pm i0)f$ satisfies the radiation condition. The converse is also true.

Theorem 8.2. *The solution of the Helmholtz equation satisfying the radiation condition is unique, and is given by $r_0(\lambda \pm i0)f$.*

We do not give the proof of these fundamental facts. However, we take notice of the following lemma, which is a crucial step of the proof of limiting absorption principle.

Lemma 8.2. *Let $\lambda > 0$ and $u \in \mathcal{B}_0^*$ satisfy the equation*

$$(-\Delta - \lambda)u = f.$$

If $f \in L^{2,s}$ with $s > 1/2$, then $u \in L^{2,s-1}$, and

$$\|u\|_{s-1} \leq C_s\|f\|_s,$$

where the constant C_s is independent of λ when λ varies over a compact set in $(0, \infty)$.

Since $u \in \mathcal{B}_0^*$, Theorem 8.1 (4) yields $\widehat{f}(\xi) = 0$ on $\{|\xi| = \sqrt{\lambda}\}$. Then, the lemma follows from [3] Theorem B.1, or [42] Theorem 14.2.4. As a matter of fact, this lemma should be proven without using the limiting absorption principle, since almost the same fact is used in its proof, and it can be done by checking the arguments in the proof of [42].

8.2   Resolvent estimates for the free Maxwell operator

We represent the resolvent of the free Maxwell operator by that of $-\Delta$. Let $\mathcal{P}_0^{(0)}$ be the projection onto the null space of $\mathcal{H}_0$, and $\mathcal{P}^{(0)} = 1 - \mathcal{P}_0^{(0)}$. Multiplying the free Maxwell equation

$$(\mathcal{H}_0 - z)u = f, \quad \text{in} \quad \mathbf{R}^3, \tag{8.4}$$

by $\mathcal{P}_0^{(0)}$, we have

$$\mathcal{P}_0^{(0)}u = -\frac{1}{z}\mathcal{P}_0^{(0)}f. \tag{8.5}$$

Therefore, we consider

$$(\mathcal{H}_0 - z)\mathcal{P}^{(0)}u = \mathcal{P}^{(0)}f. \tag{8.6}$$

Let $\mathcal{L} = \mathcal{L}(-i\nabla)$ with $\mathcal{L}(\xi)$ from (7.38), and $\mathcal{M}_0' = \begin{pmatrix} \mu_0 I_3 & 0 \\ 0 & \epsilon_0 I_3 \end{pmatrix}$. Then we have

$$\mathcal{L}\mathcal{M}_0 = \mathcal{M}_0'\mathcal{L}. \tag{8.7}$$

We multiply the equation (8.6) by $(\mathcal{L} + z\mathcal{M}_0')\mathcal{M}_0$ to get

$$\left(-\Delta - z^2\epsilon_0\mu_0\right)\mathcal{P}^{(0)}u = (\mathcal{L} + z\mathcal{M}_0')\mathcal{M}_0\mathcal{P}^{(0)}f,$$

where we have used (8.7), the fact $\mathcal{L}^2 u = (\text{curl}\,\text{curl}\,u_E, \text{curl}\,\text{curl}\,u_H)$ for $u = (u_E, u_H)$, and $\mathcal{P}^{(0)}u$ is divergence free. Again using (8.7), we have $(\mathcal{L} + z\mathcal{M}_0')\mathcal{M}_0 = \mathcal{M}_0'\mathcal{M}_0(\mathcal{H}_0 + z)$, hence

$$\left(-(\epsilon_0\mu_0)^{-1}\Delta - z^2\right)\mathcal{P}^{(0)}u = (\mathcal{H}_0 + z)\mathcal{P}^{(0)}f. \tag{8.8}$$

We have thus obtained the following expression of the resolvent $(\mathcal{H}_0 - z)^{-1}$.

Lemma 8.3. *Let* $c_0 = (\epsilon_0\mu_0)^{-1/2}$. *Then, for* $z \notin \mathbf{R}$,

$$(\mathcal{H}_0 - z)^{-1} = \left(-c_0^2\Delta - z^2\right)^{-1}(\mathcal{H}_0 + z)\mathcal{P}^{(0)} - \frac{1}{z}\mathcal{P}_0^{(0)}. \tag{8.9}$$

Let $\mathcal{P}_\pm^{(0)}(\xi)$ be the eigenprojection associated with the eigenvalue $\pm c_0|\xi|$ of $\mathcal{M}_0^{-1}\mathcal{L}(\xi)$. We also let

$$\mathcal{P}^{(0)}(\xi) = \mathcal{P}_+^{(0)}(\xi) + \mathcal{P}_-^{(0)}(\xi). \tag{8.10}$$

Then, passing to the Fourier transform, $(\mathcal{H}_0 + z)\mathcal{P}^{(0)}$ becomes the operator

$$(c_0|\xi| + z)\mathcal{P}_+^{(0)}(\xi) - (c_0|\xi| - z)\mathcal{P}_-^{(0)}(\xi). \tag{8.11}$$

We take bounded open intervals I, I_1, I_2 such that $I \subset\subset I_1 \subset\subset I_2 \subset\subset \mathbf{R} \setminus \{0\}$, and consider the limit $\lim_{\epsilon \to 0}(\mathcal{H}_0 - \lambda \mp i\epsilon)^{-1}$ for $\lambda \in I$. Take

$\varphi(k) \in C^\infty(\mathbf{R})$ such that $\varphi(k) = 0$ for $k \notin I_2$, $\varphi(k) = 1$ for $k \in I_1$. Since $\varphi(\mathcal{H}_0)$ restricts to the region $\pm c_0|\xi| \in I_2$, the limit

$$\lim_{\epsilon \to 0} (\mathcal{H}_0 - \lambda \mp i\epsilon)^{-1}(1 - \varphi(\mathcal{H}_0)) \in \mathbf{B}(\mathbf{H}_0) \qquad (8.12)$$

exists for $\lambda \in I$. Moreover, $\varphi(\mathcal{H}_0)(\mathcal{H}_0 + z)\mathcal{P}^{(0)}$ is a bounded ΨDO with smooth symbol. One can then transfer the limiting absorption principle for $(-c_0^2\Delta - z)^{-1}$ to the operator $(\mathcal{H}_0 - z)^{-1}\varphi(\mathcal{H}_0)$.

Theorem 8.3. *Let* $R_0(z) = (\mathcal{H}_0 - z)^{-1}$, *and* $\lambda \in \mathbf{R} \setminus \{0\}$.
(1) For $s > 1/2$, *there exists a strong limit*

$$\text{s} - \lim_{\epsilon \to 0} R_0(\lambda \pm i\epsilon) := R_0(\lambda \pm i0) \in \mathbf{B}(L^{2,s}; L^{2,-s}). \qquad (8.13)$$

Moreover, for any $f \in L^{2,s}$, $R_0(\lambda \pm i0)f$ *is an* $L^{2,-s}$-*valued continuous function of* $\lambda \neq 0$.
(2) For any $f, g \in \mathcal{B}$, *the limit*

$$\lim_{\epsilon \to 0} (R_0(\lambda \pm i\epsilon)f, g) = (R_0(\lambda \pm i0)f, g)$$

exists, where the right-hand side is a continuous function of $\lambda \in \mathbf{R} \setminus \{0\}$.
Moreover

$$\|R_0(\lambda \pm i0)f\|_{\mathcal{B}^*} \le C\|f\|_{\mathcal{B}},$$

where C *is independent of* λ, *when* λ *varies over a compact set in* $\mathbf{R} \setminus \{0\}$.
(3) For any $f \in \mathcal{B}$ *and* $\lambda \in \mathbf{R} \setminus \{0\}$,

$$R_0(\lambda \pm i0)f - \sqrt{\frac{2\pi}{c_0}} \frac{e^{\pm i\frac{\lambda}{c_0}r}}{r} \big(\mathcal{F}_\pm^{(0)}(\lambda)f\big)(\widehat{x}) \in \mathcal{B}_0^*, \quad \widehat{x} = x/|x|, \qquad (8.14)$$

where $\mathcal{F}_\pm^{(0)}(\lambda)$ *is defined by*

$$\big(\mathcal{F}_\pm^{(0)}(\lambda)f\big)(\omega) = \frac{|\lambda|}{c_0^{3/2}} \mathcal{P}_\pm^{(0)}(\omega)\widehat{f}(\pm\frac{\lambda}{c_0}\omega). \qquad (8.15)$$

Proof. All the assertions follow from Theorem 8.1 and Lemma 8.3. Here, to prove (3), note that

$$(\lambda \pm i0)^2 = \begin{cases} \lambda^2 + i0, & \text{if } \lambda > 0, \\ \lambda^2 - i0, & \text{if } \lambda < 0. \end{cases}$$

Therefore

$$R_0(\lambda \pm i0) = \big(- c_0^2\Delta - \lambda^2 \mp i\sigma(\lambda)0\big)^{-1}(\mathcal{H}_0 + \lambda)\mathcal{P}^{(0)} - \frac{1}{\lambda}\mathcal{P}_0^{(0)}, \qquad (8.16)$$

where $\sigma(\lambda) = 1$ for $\lambda > 0$, $\sigma(\lambda) = -1$ for $\lambda < 0$. Using

$$\mathcal{P}_{\pm}^{(0)}(\xi) = \frac{1}{2\pi i} \int_{C_{\pm}} (z - \mathcal{H}_0(\xi))^{-1} dz,$$

where $\mathcal{H}_0(\xi) = \mathcal{M}_0^{-1}\mathcal{L}(\xi)$ and $C_{\pm}$ are contours enclosing $\pm|\xi|$, we see that

$$\mathcal{P}_{+}^{(0)}(-\xi) = -\mathcal{P}_{-}^{(0)}(\xi). \tag{8.17}$$

$$\mathcal{P}_{\pm}^{(0)}(t\xi) = \mathcal{P}_{\pm}^{(0)}(\xi), \quad t > 0. \tag{8.18}$$

Let $g = (\mathcal{H}_0 + \lambda)\mathcal{P}^{(0)}f$. Then, we have

$$\widehat{g}(\xi) = \Big((c_0|\xi| + \lambda)\mathcal{P}_{+}^{(0)}(\xi) - (c_0|\xi| - \lambda)\mathcal{P}_{-}^{(0)}(\xi) \Big) \widehat{f}(\xi),$$

hence

$$\widehat{g}\Big(\pm \frac{|\lambda|}{c_0}\widehat{x} \Big) = \begin{cases} 2\lambda \mathcal{P}_{+}^{(0)}(\pm\widehat{x})\widehat{f}\Big(\pm \dfrac{|\lambda|}{c_0}\widehat{x} \Big), & \text{if } \lambda > 0, \\[4mm] 2\lambda \mathcal{P}_{-}^{(0)}(\pm\widehat{x})\widehat{f}\Big(\pm \dfrac{|\lambda|}{c_0}\widehat{x} \Big), & \text{if } \lambda < 0. \end{cases} \tag{8.19}$$

These properties together with Theorem 8.1 (3) imply (8.15).   □

8.3   Radiation condition

The uniqueness of the solution of the Maxwell equation in $\mathbf{R}^3$

$$\operatorname{curl} E - ikH = 0, \quad \operatorname{curl} H + ikE = 0$$

is not guaranteed unless we impose some boundary condition at infinity. Three formulations are known. Commonly used condition is the following one

$$H(x) \times \frac{x}{|x|} - E(x) = o(|x|^{-1}), \quad |x| \to \infty, \tag{8.20}$$

called *Silver-Müller radiation condition*.

As has been explained in the previous section, the radiation condition of Sommerfeld is usually imposed on the Helmholtz equation in $\mathbf{R}^n$

$$(-\Delta - k^2)u = 0, \tag{8.21}$$

in the following form

$$u(x) = O(r^{-(n-1)/2}), \quad \frac{\partial u}{\partial r} - iku = o(r^{-(n-1)/2}), \quad r = |x| \to \infty. \tag{8.22}$$

If $k > 0$, u is called outgoing, and for $k < 0$, u is called incoming. It is known that the latter condition in (8.22) is replaced by

$$\frac{\partial u}{\partial x_j} - ik\hat{x}_j u = o(r^{-(n-1)/2}), \quad \hat{x}_j = x_j/r, \quad 1 \leq j \leq n, \qquad (8.23)$$

which seems to be stronger, however, it is equivalent.

In the case of the vacuum, E and H must satisfy

$$\operatorname{div} E = \operatorname{div} H = 0. \qquad (8.24)$$

The formula $\operatorname{curl}\operatorname{curl} = \operatorname{grad}\operatorname{div} - \Delta$ implies

$$(-\Delta - k^2)E = (-\Delta - k^2)H = 0. \qquad (8.25)$$

We compute

$$\begin{aligned}
ik\,(H \times \hat{x} - E) &= \operatorname{curl} E \times \hat{x} - ikE \\
&= (\partial_r - ik)E - (\hat{x} \cdot \partial_1 E, \hat{x} \cdot \partial_2 E, \hat{x} \cdot \partial_3 E).
\end{aligned} \qquad (8.26)$$

Let $\Omega_R = \{x \in \mathbf{R}^3\,;\, |x| > R\}$, $R > 0$, and $\mathcal{V}$ be the set of C^∞-functions $u(x)$ on Ω_R satisfying

$$\partial_x^\alpha u(x) = O(r^{-1}), \quad \forall \alpha. \qquad (8.27)$$

For $A(x) \in \mathcal{V}$, we put

$$A_r(x) = \hat{x} \cdot A(x), \quad A_t(x) = A(x) - A_r(x)\hat{x}. \qquad (8.28)$$

Then from (8.26), we see that if $E \in \mathcal{V}$

$$\begin{aligned}
ik\,(H \times \hat{x} - E) &= (\partial_r - ik)E - \nabla E_r + 0(r^{-2}) \\
&= (\partial_r - ik)E - (\nabla - ik\hat{x})E_r - ikE_r\hat{x} + O(r^{-2}) \quad (8.29) \\
&= (\partial_r - ik)E_t - (\nabla_{tan} + ik\hat{x})E_r + O(r^{-2}),
\end{aligned}$$

where

$$\nabla_{tan} = \nabla - \hat{x}\partial_r. \qquad (8.30)$$

Since E and H satisfy the Helmholtz equation, it is natural to impose the Sommerfeld radiation condition on them. The formula (8.29) means that in the Silver-Müller radiation condition, we impose furthermore that

$$E_r = o(r^{-1}). \qquad (8.31)$$

If Ee^{-ikt} represents the outgoing spherical wave of the form $ae^{ik(|x|-t)}$, the condition (8.31) means that $\hat{x} \cdot a = o(r^{-1})$, i.e. $Ee^{ik(|x|-t)}$ must be a transverse wave with respect to the direction of propagation. This is a natural property for electromagnetic waves.

From the view point of differential equation, an imporatnt role of the radiation condition is to distinguish between $R_0(\lambda + i0)$ and $R_0(\lambda - i0)$. Theorem 8.3 suggests that, unlike the case of the Helmholtz equation, in which one observes the phase $e^{\pm i\lambda r}$, one can make use of the amplitude $\mathcal{F}_\pm^{(0)}(\lambda)f$. We thus define the radiation condition for the Maxwell operator as follows.

Definition 8.1. A solution $u \in \mathcal{B}^*$ of the Maxwell equation $(\mathcal{H}_0 - \lambda)u = f$, $\lambda \neq 0$, is said to satisfy the *radiation condition* if

$$\left(1 - \mathcal{P}_\pm^{(0)}(\widehat{x})\right)u \in \mathcal{B}_0^*.$$

For the $+$ case, it is said to be the outgoing radiation condition, and for the $-$ case, the incoming radiation condition. This type of radiation condition was introduced by Mochizuki [73].

Theorem 8.4. *For $\lambda \neq 0$ and $f \in \mathcal{B}$, $R_0(\lambda \pm i0)f$ satisfies the radiation condition.*

Proof.   This follows from Theorem 8.3.   □

The converse is also true.

Theorem 8.5. *The solution in $\mathcal{B}^*$ of the equation $(\mathcal{H}_0 - \lambda)u = f$, $\lambda \in \mathbf{R} \setminus \{0\}$ satisfying the radiation condition is unique.*

We prove this theorem in the next section (Lemma 8.7) in a more general situation.

8.4   Resolvent estimates in an exterior domain

Let Ω be a connected domain exterior to a bounded open set in $\mathbf{R}^3$ with C^∞ boundary, and $\mathbf{H} = L^2(\Omega)^6$ with the inner product

$$(u, v) = \int_\Omega \langle \mathcal{M}(x)u(x), v(x)\rangle dx,$$

where $\langle \, , \, \rangle$ is the standard inner product of $\mathbf{C}^6$. We consider the Maxwell operator $\mathcal{H} = \mathcal{M}(x)^{-1}\mathcal{L}(-i\nabla)$ equipped with the domain

$$D(\mathcal{H}) = H_0(\mathrm{curl};\, \Omega) \times H(\mathrm{curl};\, \Omega)$$

under the assumptions (7.13) and (7.54), i.e.

$$CI_3 \leq \epsilon(x) \leq C^{-1}I_3, \quad CI_3 \leq \mu(x) \leq C^{-1}I_3,$$

$$|\epsilon(x) - \epsilon_0 I_3| + |\mu(x) - \mu_0 I_3| \le C(1 + |x|)^{-1-\epsilon}, \quad \epsilon > 0,$$

where ϵ_0, μ_0 are positive constants. The spaces $L^{2,s}$, $\mathcal{B}$, $\mathcal{B}^*$ and $\mathcal{B}_0^*$ are naturally extended to Ω. By a solution $u \in \mathcal{B}^*$ of $(\mathcal{H} - z)u = f$, we mean $u \in \mathcal{B}^*$ satisfying

$$\chi(x)u \in D(\mathcal{H}), \tag{8.32}$$

for any $\chi \in C_0^\infty(\overline{\Omega})$ such that $\chi(x) = 1$ near $\partial\Omega$, and

$$(u, (\mathcal{H} - \overline{z})\psi) = (f, \psi), \tag{8.33}$$

for any $\psi \in C_0^\infty(\Omega)$. Hence $u = (u_E, u_H)$ satisfies $\nu \times u_E = 0$ on $H^{-1/2}(\partial\Omega)$.

Lemma 8.4. *For any* $\phi(\lambda) \in C_0^\infty(\mathbf{R})$ *and* $s \in \mathbf{R}$, $\langle x \rangle^s \phi(\mathcal{H}) \langle x \rangle^{-s} \in \mathbf{B}(\mathbf{H})$.

Proof. Let $\chi(x) \in C^\infty(\mathbf{R}^3)$ be such that $\chi(x) = 0$ near $\partial\Omega$ and $\chi(x) = 1$ near infinity. Then,

$$\chi(x)\langle x \rangle^s R(z) \langle x \rangle^{-s} = R(z)\chi + R(z)[\mathcal{H}, \chi(x)\langle x \rangle^s]R(z)\langle x \rangle^{-s}.$$

For $0 \le s \le 1$, we have

$$\|R(z)[\mathcal{H}, \chi(x)\langle x \rangle^s]R(z)\langle x \rangle^{-s}\| \le C|\operatorname{Im} z|^{-2}.$$

Using the almost analytic extension of $\phi(\lambda)$ and Lemma 7.18, we then see that the lemma holds for $0 \le s \le 1$. Next we take $\widetilde{\chi}(x)$ similar to $\chi(x)$ satisfying $\widetilde{\chi}(x) = 1$ on $\operatorname{supp}\chi$, and insert it between $[\mathcal{H}, \chi(x)\langle x \rangle^s]$ and $R(z)$. Doing the similar computation to $\widetilde{\chi}(x)\langle x \rangle^{s-1}R(z)\langle x \rangle^{s-1}$, we prove the lemma for $1 \le s \le 2$. Repeating this procedure, we prove the lemma for any $s \ge 0$, and, taking the adjoint, for the case $s \le 0$. $\square$

Lemma 8.5. *If* $u \in \mathcal{B}_0^*$ *satisfies* $(\mathcal{H} - \lambda)u = f \in L^{2,s}$ *with* $s > 1/2$ *and* $\lambda \in \mathbf{R} \setminus \{0\}$, *we have*

$$\|u\|_{s-1} \le C_s(\|u\|_{\mathcal{B}^*} + \|f\|_s),$$

where the constant C_s *is independent of* λ *when* λ *varies over a compact set in* $\mathbf{R} \setminus \{0\}$.

Proof. Take $\chi(x) \in G^\infty(\mathbf{R}^3)$ such that $\chi(x) = 0$ near $\partial\Omega$ and $\chi(x) = 1$ near infinity. Then, we have

$$(\mathcal{H}_0 - \lambda)\chi u = g,$$

$$g = \lambda\chi(\mathcal{M}_0^{-1}\mathcal{M} - 1)u + [\mathcal{H}_0, \chi]u + \mathcal{M}_0^{-1}\mathcal{M}\chi f.$$

Take $\varphi(k) \in C_0^\infty(\mathbf{R})$ such that $\varphi(k) = 1$ on a small neighborhood of λ and $\operatorname{supp}\varphi \subset \mathbf{R} \setminus \{0\}$. Then, noting that $|\mathcal{M}_0^{-1}\mathcal{M}(x) - 1| \le C(1 + |x|)^{-1-\epsilon}$ and

$u \in L^{2,-s'}$ for any $s' > 1/2$, we have $g \in L^{2,t}$ for any $t < \min\{\frac{1}{2} + \epsilon, s\}$. Therefore $(1 - \varphi(\mathcal{H}_0))\chi u = (1 - \varphi(\mathcal{H}_0))(\mathcal{H}_0 - \lambda)^{-1}g \in L^{2,t}$, and

$$\|(1 - \varphi(\mathcal{H}_0))\chi u\|_t \le C_s(\|u\|_{\mathcal{B}^*} + \|f\|_s).$$

The part $\varphi(\mathcal{H}_0)\chi u$ satisfies

$$(\mathcal{H}_0 - \lambda)\varphi(\mathcal{H}_0)\chi u = \varphi(\mathcal{H}_0)g.$$

Since $\varphi(\mathcal{H}_0) = \varphi(\mathcal{H}_0)\mathcal{P}^{(0)}$, we have, in view of (8.8),

$$(-c_0^2\Delta - \lambda^2)\varphi(\mathcal{H}_0)\chi u = (\mathcal{H}_0 + \lambda)\varphi(\mathcal{H}_0)g.$$

Since $\varphi(\mathcal{H}_0)\chi u \in \mathcal{B}_0^*$, we apply Lemma 8.2 to see that $\varphi(\mathcal{H}_0)\chi u \in L^{2,t-1}$, hence $\chi u \in L^{2,t-1}$, and

$$\|\chi u\|_{t-1} \le C(\|u\|_{\mathcal{B}^*} + \|f\|_s).$$

Since $t - 1 > -1/2$, we have improved the decay of u. We use this fact to g, and apply the same arguments as above to obtain more decay of u. Repeating this procedure, we finally arrive at $u \in L^{2,s-1}$, and prove the lemma. □

Lemma 8.6. *Let I be a compact interval in $\mathbf{R} \setminus \{0\}$. Then, $\sigma_p(\mathcal{H}) \cap I$ is a finite set, and each eigenvalue has a finite multiplicity.*

Proof. Suppose there exists an infinite orthonormal system of eigenvectors $\{u_n\}_{n=1}^{\infty}$ with eigenvalues in I. Then, by Lemma 8.5, for any $\epsilon > 0$, there exists $R > 0$ such that $\sup_n \|u_n\|_{L^2(B_R{}^c)} < \epsilon$, where $B_R{}^c = \{|x| > R\}$. By the local compactness of $\mathcal{P}(\mathcal{H}+i)^{-1}$, $\{u_n\}$ has a subsequence which converges in $L^2(B_R \cap \Omega)$ for any $R > 0$. Therefore, there exists a subsequence $\{u_{n_i}\}$ convergent in $L^2(\Omega)$. This is a contradiction. □

We define the radiation condtion for the exterior problem in the same way as in Definition 8.1.

Lemma 8.7. *Let $\lambda \in \mathbf{R} \setminus \sigma_p(\mathcal{H})$. Suppose $u = (E, H) \in \mathcal{B}^*$ satisfies $(\mathcal{H}-\lambda)u = 0$, the boundary condition $\nu \times E = 0$, and the radiation condition. Then $u = 0$.*

Proof. Integrating $\langle \mathcal{MH}u, u \rangle$ over $\Omega \cap B_R$ and taking the imaginary part, we have

$$\text{Re} \int_{|x|=R} (\widehat{x} \times E \cdot \overline{H} - \widehat{x} \times H \cdot \overline{E})dS = 0,$$

which entails, by (6.30),

$$\text{Re} \int_{|x|=R} \langle \mathcal{M}_0^{-1} \mathcal{L}(\widehat{x}) u, u \rangle_{\mathcal{M}_0} dS = 0.$$

Putting $r = R$ and integrating with respect to r, we then obtain

$$\lim_{R \to \infty} \frac{1}{R} \int_{\Omega_R} \langle \mathcal{M}_0^{-1} \mathcal{L}(\widehat{x}) u, u \rangle_{\mathcal{M}_0} dx = 0,$$

where $\Omega_R = \Omega \cap \{|x| < R\}$. Assuming that u satisfies the outgoing radiation condition, we can replace u by $\mathcal{P}^{(0)}(\widehat{x}) u$, and obtain

$$\lim_{R \to \infty} \frac{1}{R} \int_{\Omega_R} \langle \mathcal{P}^{(0)}(\widehat{x}) u, \mathcal{P}^{(0)}(\widehat{x}) u \rangle_{\mathcal{M}_0} dx = 0.$$

Using the radiation condition again, we see that $u \in \mathcal{B}_0^*$. Then, $u \in L^2(\Omega)$ by virtue of Lemma 8.5. Since $\lambda \notin \sigma_p(\mathcal{H})$, we have $u = 0$. $\square$

We can now prove the limiting absorption principle for $\mathcal{H}$. Take bounded intervals I, I_1, I_2 such that $I \subset\subset I_1 \subset\subset I_2 \subset\subset \mathbf{R} \setminus \sigma_p(\mathcal{H})$, and let

$$J_+ = \{z \in \mathbf{C} \setminus \mathbf{R} \,;\, \text{Re}\, z \in I, \, 0 < \text{Im}\, z \le 1\}.$$

We consider the behavior of $R(z) = (\mathcal{H} - z)^{-1}$ for $z \in J_+$.

Take $\varphi(\lambda) \in C_0^\infty(\mathbf{R})$ such that $\varphi(\lambda) = 1$ for $\lambda \in I_1$, and $\varphi(\lambda) = 0$ for $\lambda \notin I_2$. Then, there exists a limit

$$\lim_{z \to \lambda \in I} R(z)(1 - \varphi(\mathcal{H})) \in \mathbf{B(H)}. \tag{8.34}$$

We consider $u(z) = R(z)\varphi(\mathcal{H}) f$ for $z \in J$.

Lemma 8.8. *For any $s > 1/2$, there exists a constant $C_s > 0$ such that*

$$\|u(z)\|_{-s} \le C_s \|f\|_s, \quad \forall z \in J_+.$$

Proof. Without loss of generality, we assume that $1/2 < s < (1 + \epsilon)/2$, where ϵ is the exponent in (7.54). Suppose this lemma does not hold. Then, there exist $z_n \in J_n$, $f_n \in L^{2,s}$, $n = 1, 2, \cdots$ such that

$$\|f_n\|_s \to 0, \quad \|u_n\|_{-s} = 1,$$

where $u_n = R(z_n)\varphi(\mathcal{H}) f_n$. We can assume that $z_n \to \lambda$. If $\text{Im}\, \lambda > 0$, we easily obtain a contradiction. Therefore, we consider the case $\lambda \in I$. By the local compactness of $\mathcal{P}(\mathcal{H} + i)^{-1}$, there exists a subsequence of $\{u_n\}$, which is denoted by $\{u_n\}$ again, such that $\{u_n\}$ converges in $L^2(\Omega_R)$ for any $R > 0$, where $\Omega_R = \Omega \cap \{|x| < R\}$. Take $\chi \in C^\infty(\mathbf{R}^3)$ such that $\chi(x) = 0$ near $\partial\Omega$ and $\chi(x) = 1$ near infinity. Then, we have

$$(\mathcal{H}_0 - z_n)\chi u_n = g_n,$$

$$g_n = z_n(\mathcal{M}_0^{-1}\mathcal{M} - 1)\chi u_n + [\mathcal{H}_0, \chi]u_n + \chi\mathcal{M}_0^{-1}\mathcal{M}\varphi(\mathcal{H})f_n.$$

Taking note of $|\mathcal{M}_0^{-1}\mathcal{M}(x) - 1| \leq C(1+|x|)^{-1-\epsilon}$, the 1st term of the right-hand side is seen to converge in $L^{2,s}$. By Theorem 8.3, $\chi u_n = (\mathcal{H}_0 - z_n)^{-1}g_n$ converges in $L^{2,-s}$, hence so does u_n. Let $u_n \to u$ in $L^{2,-s}$. Then, u satisfies $(\mathcal{H} - \lambda)u = 0$. Moreover, it satisfies the radiation condition, since

$$\chi u = (\mathcal{H}_0 - \lambda - i0)^{-1}\big(\lambda(\mathcal{M}_0^{-1}\mathcal{M} - 1)\chi u + [\mathcal{H}_0, \chi]u\big).$$

Lemma 8.7 then implies $u = 0$, which is a contradiction.   □

Note that we have proven a uniform bound

$$\|R(z)f\|_{-s} \leq C_s\|f\|_s, \quad \forall s > 1/2, \quad \forall z \in J_+. \tag{8.35}$$

In the course of the above proof, we have obtained a formula

$$\chi R(z) = R_0(z)\chi\mathcal{M}_0^{-1}\mathcal{M} + R_0(z)\big(z\chi(\mathcal{M}_0^{-1}\mathcal{M} - 1) + [\mathcal{H}_0, \chi]\big)R(z). \tag{8.36}$$

This and (8.35) imply

$$\|R(z)f\|_{\mathcal{B}^*} \leq C\|f\|_s, \quad \forall s > 1/2, \quad \forall f \in L^{2,s}.$$

Taking the adjoint, we have

$$\|R(z)f\|_{-s} \leq C\|f\|_{\mathcal{B}}, \quad \forall s > 1/2, \quad \forall f \in \mathcal{B}.$$

We use this to the right-hand side of (8.36) to obtain

$$\|R(z)f\|_{\mathcal{B}^*} \leq C\|f\|_{\mathcal{B}}, \quad \forall f \in \mathcal{B}. \tag{8.37}$$

Lemma 8.9. *Let $\lambda \in I$, $s > 1/2$ and $f \in L^{2,s}$. Then, there exists a strong limit* s $- \lim_{z \to \lambda} R(z)f$ *in $L^{2,-s}$.*

Proof. Let $u_n = R(z_n)f$, where $z_n \in J$ and $z_n \to \lambda$. Arguing as in the previous lemma, $\{u_n\}$ contains a subsequence $\{u'_n\}$ which converges to $u \in L^{2,-s}$. This u solves $(\mathcal{H} - \lambda)u = f$ and satisfies the radiation condition. Such u is unique. Starting this argument from a subsequence of $\{u_n\}$, we then see that any subsequence of $\{u_n\}$ contains a sub-subsequence which converges to one and the same $u \in L^{2,-s}$. This means that $\{u_n\}$ itself converges to u.   □

By the similar argument, one can show that the convergence of s $- \lim_{\epsilon \to 0} R(\lambda + i\epsilon)$ is uniform in $\lambda \in I$. This proves that $R(\lambda + i0)f$ is $L^{2,-s}$-valued continuous with respect to $\lambda \in I$.

Since $L^{2,s}$ is dense in $\mathcal{B}$, we can show the existence of the limit

$$\lim_{\epsilon \to 0}(R(\lambda + i\epsilon)f, g), \quad \forall f, g \in \mathcal{B}, \quad \forall \lambda \in I. \tag{8.38}$$

Moreover, $(R(\lambda + i0)f, g)$ is continuous with respect to $\lambda \in I$.

Let us summarize what we have proven.

Theorem 8.6. *(1) Let I be a compact interval in $\mathbf{R} \setminus \sigma_p(\mathcal{H})$. Then, there exists a constant $C > 0$ such that*

$$\|R(z)f\|_{\mathcal{B}^*} \le C\|f\|_{\mathcal{B}}, \quad \operatorname{Re} z \in I, \quad f \in \mathcal{B}.$$

(2) For $\lambda \in \mathbf{R} \setminus \sigma_p(\mathcal{H})$ and $f \in L^{2,s}$ with $s > 1/2$, there exists a strong limit $\mathrm{s} - \lim_{\epsilon \to 0} R(\lambda \pm i\epsilon)f := R(\lambda \pm i0)f$ in $L^{2,-s}$. Moreover, $R(\lambda \pm i0)f$ is an $L^{2,-s}$-valued continuous function of $\lambda \in \mathbf{R} \setminus \sigma_p(\mathcal{H})$.
(3) For $f, g \in \mathcal{B}$ and $\lambda \in \mathbf{R} \setminus \sigma_p(\mathcal{H})$, there exists a limit

$$\lim_{\epsilon \to 0} (R(\lambda \pm i\epsilon)f, g) := (R(\lambda \pm i0)f, g),$$

and the right-hand side is a continuous function of $\lambda \in \mathbf{R} \setminus \sigma_p(\mathcal{H})$. Moreover, we have

$$\|R(\lambda \pm i0)f\|_{\mathcal{B}^*} \le C\|f\|_{\mathcal{B}},$$

where the constant C is independent of λ in a compact set in $\mathbf{R} \setminus \sigma_p(\mathcal{H})$.
(4) For $\lambda \in \mathbf{R} \setminus \sigma_p(\mathcal{H})$ and $f \in \mathcal{B}$, $R(\lambda \pm i0)f$ satisfies the radiation condition.
(5) For $\lambda \in \mathbf{R} \setminus \sigma_p(\mathcal{H})$, the solution $u \in \mathcal{B}^*$ of the equation $(\mathcal{H} - \lambda)u = f \in \mathcal{B}$ satisfying the boundary condition $\nu \times E = 0$ and the radiation condition is unique, and is given by $u = R(\lambda \pm i0)f$.

Limiting absorption principle yields the absolute continuity of the spectrum.

Theorem 8.7. $\mathbf{H} = \mathbf{H}_{ac}(\mathcal{H}) \oplus \mathbf{H}_{pp}(\mathcal{H})$.

Proof. We have only to show that $\mathbf{H}_{pp}(\mathcal{H})^{\perp} \subset \mathbf{H}_{ac}(\mathcal{H})$. Take $f \in \mathbf{H}_{pp}(\mathcal{H})$. Decompose $\mathbf{R} \setminus \sigma_p(\mathcal{H})$ into a disjoint union of open intervals I_n, $n = 1, 2, \cdots$. Since $\mathbf{H}_{ac}(\mathcal{H})$ is a closed subspace, we have only to show that $E(I_n)f \in \mathbf{H}_{ac}(\mathcal{H})$. This is proven if we show that for any $g \in \mathbf{H}$, $(E(\lambda)g, g)$ is absolutely continuous with respect to $\lambda \in I_n$. Now, for any $h \in L^{2,s}$, $s > 1/2$, and $[a, b] \subset I_n$, we have, by Stone's formula

$$(E((b, a))h, h) = \frac{1}{2\pi i} \int_a^b ((R(\lambda + i0) - R(\lambda - i0))h, h)d\lambda.$$

Therefore, $E(I_n)h \in \mathbf{H}_{ac}(\mathcal{H})$. Approximating g by elements of $L^{2,s}$, we see that $E(I_n)g \in \mathbf{H}_{ac}(\mathcal{H})$. This completes the proof. $\qquad\square$

8.5 Spectral representation

Let us construct a spectral representation of $\mathcal{H}_0 = \mathcal{M}_0^{-1}\mathcal{L}(-i\nabla_x)$ on $\mathbf{H}_0 = (L^2(\mathbf{R}^3))^6$. Let

$$\mathbf{h} = \left(L^2(S^2)\right)^6 \tag{8.39}$$

equipped with the inner product

$$(\phi, \psi)_{\mathbf{h}} = \int_{S^2} \langle \phi, \psi \rangle_{\mathcal{M}_0} dS.$$

Let $\widehat{\mathbf{H}}_\pm$ be defined by

$$\widehat{\mathbf{H}}_\pm = \{h(\lambda, \omega) \in L^2(\mathbf{R}, \mathbf{h}, \lambda^2 d\lambda) \,;\, h(\lambda, \omega) = P_\pm^{(0)}(\omega)h(\lambda, \omega)\} \tag{8.40}$$

equipped with the inner product

$$(g, h)_{\widehat{\mathbf{H}}_\pm} = \int_{-\infty}^{\infty} (g(\lambda, \cdot), h(\lambda, \cdot))_{\mathbf{h}} \, \lambda^2 d\lambda.$$

In view of (8.17), we have

$$\mathbf{H}_+ \ni h(\lambda, \omega) \Longleftrightarrow h(\lambda, -\omega) \in \mathbf{H}_-. \tag{8.41}$$

Using $\mathcal{F}_\pm^{(0)}(\lambda)$ in (8.15), we put

$$\left(\mathcal{F}_\pm^{(0)} f\right)(\lambda) = \mathcal{F}_\pm^{(0)}(\lambda)f.$$

Then, $\mathcal{F}_\pm^{(0)} f$, originally defined for $f \in \mathcal{B}$, is uniquely extended to a unitary operator from $\mathbf{H}_{ac}(\mathcal{H}_0) = \mathbf{H}_0 \ominus \mathbf{H}_{pp}(\mathcal{H}_0)$ onto $\widehat{\mathbf{H}}_\pm$, and satisfies

$$\left(\mathcal{F}_\pm^{(0)} \mathcal{H}_0 f\right)(\lambda) = \lambda\left(\mathcal{F}_\pm^{(0)} f\right)(\lambda).$$

We construct a similar operator for the perturbed Maxwell operator $\mathcal{H}$ on $\mathbf{H} = (L^2(\Omega))^6$. Let $\chi \in C^\infty(\mathbf{R}^3)$ be such that $\chi = 0$ near $\partial\Omega$ and $\chi = 1$ near infinity. Define $Q_\pm(\lambda)$ by

$$Q_\pm(\lambda) = (\mathcal{H}_0 - \lambda)\chi R(\lambda \pm i0) \tag{8.42}$$

It is rewritten as follows

$$Q_\pm(\lambda) = \chi \mathcal{M}_0^{-1}\mathcal{M} + \left(\lambda\chi(\mathcal{M}_0^{-1}\mathcal{M} - 1) + \mathcal{M}_0^{-1}\mathcal{L}(\nabla\chi)\right)R(\lambda \pm i0), \tag{8.43}$$

where we used $[\mathcal{L}, \chi] = \mathcal{L}(\nabla\chi)$. In view of (8.36), we have

$$\chi R(\lambda \pm i0) = R_0(\lambda \pm i0)Q_\pm(\lambda). \tag{8.44}$$

We put

$$\mathcal{F}_\pm(\lambda) = \mathcal{F}_\pm^{(0)}(\lambda)Q_\pm(\lambda). \tag{8.45}$$

Theorem 8.6 (3) implies that for any compact interval $I \subset \mathbf{R} \setminus \sigma_p(\mathcal{H})$, there exists a constant $C > 0$ such that

$$\|\mathcal{F}_\pm(\lambda)f\|_\mathbf{h} \leq C\|f\|_\mathcal{B}, \quad \lambda \in I, \tag{8.46}$$

which yields

$$\|\mathcal{F}_\pm(\lambda)^*\phi\|_{\mathcal{B}^*} \leq C\|\phi\|_\mathbf{h}, \quad \lambda \in I. \tag{8.47}$$

The following Theorem is a direct consequence of Theorem 8.3 (3) and (8.44), (8.45).

Theorem 8.8. *For any* $f \in \mathcal{B}$ *and* $\lambda \in \mathbf{R} \setminus \sigma_p(\mathcal{H})$, *we have the asymptotic expansion:*

$$R(\lambda \pm i0)f - \sqrt{\frac{2\pi}{c_0}} \frac{e^{\pm i\frac{\lambda}{c_0}r}}{r} (\mathcal{F}_\pm(\lambda)f)(\widehat{x}) \in \mathcal{B}_0^*. \tag{8.48}$$

Lemma 8.10. *For* $f, g \in \mathcal{B}$ *and* $\lambda \in \mathbf{R} \setminus \sigma_p(\mathcal{H})$,

$$\frac{1}{2\pi i}\big(R(\lambda + i0)f - R(\lambda - i0)f, g\big) = \big(\mathcal{F}_\pm(\lambda)f, \mathcal{F}_\pm(\lambda)g\big)_\mathbf{h}. \tag{8.49}$$

Proof. Let

$$\langle u, v \rangle_\mathcal{M} = \langle \mathcal{M}u, v \rangle.$$

Letting $u = R(\lambda + i0)f$, $v = R(\lambda + i0)g$ and integrating $\langle (\mathcal{H} - \lambda)u, v \rangle_\mathcal{M} - \langle u, (\mathcal{H} - \lambda)v \rangle_\mathcal{M}$ over the region $\Omega_r = \Omega \cap \{|x| < r\}$, we have

$$((\mathcal{H} - \lambda)u, v)_{L^2(\Omega_r)} - (u, (\mathcal{H} - \lambda)v)_{L^2(\Omega_r)} = -i \int_{|x|=r} \langle \mathcal{M}_0^{-1}\mathcal{L}(\widehat{x})u, v \rangle_{\mathcal{M}_0} dS.$$

Let us note that

$$\mathcal{M}_0^{-1}\mathcal{L}(\widehat{x}) = c_0\mathcal{P}_+^{(0)}(\widehat{x}) - c_0\mathcal{P}_-^{(0)}(\widehat{x}).$$

Averaging with respect to r and using the radiation condition,

$$(f, v) - (u, g) = \lim_{R \to \infty} \frac{-i}{R} \int_{\Omega_R} \langle \mathcal{M}_0^{-1}\mathcal{L}(\widehat{x})u, v \rangle_{\mathcal{M}_0} dx$$

$$= \lim_{R \to \infty} \frac{-c_0 i}{R} \int_{\Omega_R} \langle \mathcal{P}_+^{(0)}(\widehat{x})u, \mathcal{P}_+^{(0)}(\widehat{x})v \rangle_{\mathcal{M}_0} dx.$$

We then have

$$\frac{1}{2\pi i}\big(R(\lambda + i0)f - R(\lambda - i0)f, g\big) = \lim_{R \to \infty} \frac{c_0}{2\pi R} \int_{\Omega_R} \langle \mathcal{P}_+^{(0)}(\widehat{x})u, \mathcal{P}_+^{(0)}(\widehat{x})v \rangle dx.$$

In view of Theorem 8.8, we obtain the lemma for $\mathcal{F}_+(\lambda)$. The proof for $\mathcal{F}_-(\lambda)$ is almost the same. $\qquad\square$

Notice that $\mathbf{R} \setminus \sigma_p(\mathcal{H}) = \cup_{i=1}^{\infty} I_i$, I_i being disjoint open intervals. Integrating (8.49) with respect to $\lambda \in \cup_{i=1}^{\infty} I_i$, we have Parseval's formula

$$\|P_{ac}(\mathcal{H})f\|^2 = \int_{\mathbf{R}} \|\mathcal{F}_{\pm}(\lambda)f\|_{\mathbf{h}}^2 \, d\lambda. \tag{8.50}$$

Recall the orthogonal decomposition in Theorem 8.7:

$$\mathbf{H} = \mathbf{H}_{ac}(\mathcal{H}) \oplus \mathbf{H}_{pp}(\mathcal{H}).$$

We define

$$(\mathcal{F}_{\pm}f)(\lambda) = \mathcal{F}_{\pm}(\lambda)f$$

for $f \in \mathcal{B}$. Then, (8.50) shows that $\mathcal{F}_{\pm}$ is uniquely extended to a bounded operator from $\mathbf{H}$ to $\widehat{\mathbf{H}}_{\pm}$. Moreover, $\mathcal{F}_{\pm}$ is isometric on $\mathbf{H}_{ac}(\mathcal{H})$, and $\mathcal{F}_{\pm}$: $\mathbf{H}_{pp}(\mathcal{H}) \to \{0\}$. Such an operator $\mathcal{F}_{\pm}$ is called a *partial isometry* on $\mathbf{H}$, and $\mathbf{H}_{ac}(\mathcal{H})$, $\mathrm{Ran}\,(\mathcal{F}_{\pm})$ are said to be the *initial set* and *final set* of $\mathcal{F}_{\pm}$, respectively.

Theorem 8.9. *(1) $\mathcal{F}_{\pm}$ is a partial isometry with initial set $\mathbf{H}_{ac}(\mathcal{H})$ and final set $\widehat{\mathbf{H}}_{\pm}$.*
(2) For $f \in D(\mathcal{H})$, we have

$$(\mathcal{F}_{\pm}\mathcal{H}f)(\lambda) = \lambda(\mathcal{F}_{\pm}f)(\lambda).$$

(3) $\mathcal{F}_{\pm}(\lambda)^* \in \mathbf{B}(\mathbf{h}; \mathcal{B}^*)$ is an eigenoperator of $\mathcal{H}$ in the sense that

$$(\mathcal{H} - \lambda)\mathcal{F}_{\pm}(\lambda)^*\phi = 0, \quad \phi \in \mathbf{h}. \tag{8.51}$$

(4) Let K_n be a finite union of compact intervals in $\mathbf{R} \setminus \sigma_p(\mathcal{H})$, $n = 1, 2, \cdots$. Then for any $\widehat{g} \in \widehat{\mathbf{H}}_{\pm}$

$$\int_{K_n} \mathcal{F}_{\pm}(\lambda)^*\widehat{g}(\lambda)d\lambda \in \mathbf{H}.$$

Moreover, if $K_n \to \mathbf{R} \setminus \sigma_p(\mathcal{H})$, for any $f \in \mathbf{H}_{ac}(\mathcal{H})$, the inversion formula holds in the sense of strong limit in $\mathbf{H}$:

$$f = \mathrm{s} - \lim_{n \to \infty} \int_{K_n} \mathcal{F}_{\pm}(\lambda)^* (\mathcal{F}_{\pm}f)(\lambda)d\lambda.$$

Note that the equation (8.51) is understood in the weak sense, i.e.

$$(\mathcal{F}_{\pm}(\lambda)^*\phi, (\mathcal{H} - \lambda)f) = 0, \quad \forall f \in D(\mathcal{H}) \cap C_0^{\infty}(\overline{\Omega}).$$

In particular, $\mathcal{F}_{\pm}(\lambda)^*\phi$ satisfies the boundary condition $\nu \times E = 0$ on $\partial\Omega$.

Proof. We have already proved the assertion (1) except that $\mathrm{Ran}\,(\mathcal{F}_{\pm}) = \widehat{\mathbf{H}}_{\pm}$, which will be shown in the next section.

To prove (2), first note that if $f \in D(\mathcal{H}) \cap \mathcal{B}$, we have $R(z)\mathcal{H}f = R(z)(\mathcal{H} - z)f + zR(z)f$. Letting $z \to \lambda \pm i0$, we then have $R(\lambda \pm i0)\mathcal{H}f = f + \lambda R(\lambda \pm i0)f$, which implies

$$(\mathcal{F}_{\pm}\mathcal{H}f)(\lambda) = \lambda(\mathcal{F}_{\pm}f)(\lambda), \quad \forall f \in D(\mathcal{H}) \cap \mathcal{B}.$$

Take $\chi \in C_0^{\infty}(\mathbf{R}^3)$ such that $\chi(x) = 1$ for $|x| < 1$, $\chi(x) = 0$ for $|x| > 2$, and put $\chi_j(x) = \chi(x/j)$. For $f \in D(\mathcal{H})$, put $f_j = \chi_j f$. Then, $f_j \in D(\mathcal{H}) \cap \mathcal{B}$ and

$$\|f_j - f\| \to 0, \quad \|\mathcal{H}f_j - \mathcal{H}f\| \to 0, \quad \text{in } \mathbf{H}.$$

Applying Parseval's formula (8.50) to $\|P_{ac}(\mathcal{H})(f_j - f)\|$ and $\|P_{ac}(\mathcal{H})\mathcal{H}(f_j - f)\|$, we have

$$\|\mathcal{F}_{\pm}f_j - \mathcal{F}_{\pm}f\| \to 0, \quad \|\lambda\mathcal{F}_{\pm}f_j - \mathcal{F}_{\pm}\mathcal{H}f\| \to 0, \quad \text{in } \widehat{\mathbf{H}},$$

which implies (2).

Take $\phi \in \mathbf{h}$. Then, for any $f \in D(\mathcal{H}) \cap C_0^{\infty}(\overline{\Omega})$,

$$(\mathcal{F}_{\pm}(\lambda)^*\phi, (\mathcal{H} - \lambda)f) = (\phi, \mathcal{F}_{\pm}(\lambda)(\mathcal{H} - \lambda)f) = 0,$$

since $\mathcal{F}_{\pm}(\lambda)(\mathcal{H} - \lambda)f = 0$ by (2). This proves (3).

For $\widehat{g} \in \widehat{\mathbf{H}}_{\pm}$, $\mathcal{F}_{\pm}(\lambda)^*\widehat{g}(\lambda)$ is a $\mathcal{B}^*$-valued locally bounded function of $\lambda \in \mathbf{R} \setminus \sigma_p(\mathcal{H})$. We put

$$u_n = \int_{K_n} \mathcal{F}_{\pm}(\lambda)^*\widehat{g}(\lambda)d\lambda.$$

Then, $u_n \in \mathcal{B}^*$, and for any $h \in \mathcal{B}$,

$$(u_n, h) = \int_{K_n} (\mathcal{F}_{\pm}(\lambda)^*\widehat{g}(\lambda), h)d\lambda = \int_{K_n} (\widehat{g}(\lambda), \mathcal{F}_{\pm}(\lambda)h)_{\mathbf{h}}d\lambda. \tag{8.52}$$

This implies

$$|(u_n, h)| \leq \|\widehat{g}\|_{L^2(K_n;\mathbf{h})}\|\mathcal{F}_{\pm}h\|_{L^2(K_n;\mathbf{h})} \leq \|\widehat{g}\|_{L^2(K_n;\mathbf{h})}\|h\|_{\mathbf{H}}.$$

By Riesz' theorem, $u_n \in \mathbf{H}$ and $\|u_n\|_{\mathbf{H}} \leq \|\widehat{g}\|_{L^2(K_n;\mathbf{h})}$.

Now, let $\widehat{g} = \mathcal{F}_{\pm}f$ for $f \in \mathbf{H}_{ac}(\mathcal{H})$, and take an increasing sequence $K_1 \subset K_2 \subset \cdots \to \mathbf{R} \setminus \sigma_p(\mathcal{H})$. Arguing as above, we have for $n < m$

$$\|u_m - u_n\|_{\mathbf{H}} \leq \|\widehat{g}\|_{L^2(K_m - K_n;\mathbf{h})}.$$

This implies $u_n \to u$ in $\mathbf{H}$ for some $u \in \mathbf{H}$. Letting $n \to \infty$ in (8.52),

$$(u, h) = (\mathcal{F}_{\pm}f, \mathcal{F}_{\pm}h) = (P_{ac}(\mathcal{H})f, h).$$

Then, we have $u = f$, which completes the proof of (4). $\qquad \square$

8.6 The surjectivity of $\mathcal{F}_\pm$

Let us first recall a relation between the unitary group e^{-itH} and the resolvent $R(z) = (H - z)^{-1}$ of a self-adjoint operator H on a Hilbert space. Let $\chi_\pm(t) = 1$ for $\pm t \geq 0$, $\chi_\pm(t) = 0$ for $\pm t < 0$. We put

$$U_\pm^\epsilon(t) = \pm i\chi_\pm(t)e^{-itH - \epsilon|t|}, \quad \epsilon > 0.$$

Integrating the equation

$$R(\lambda \pm i\epsilon)\frac{d}{dt}e^{it(\lambda \pm i\epsilon - H)} = -ie^{it(\lambda \pm i\epsilon - H)},$$

we have

$$R(\lambda \pm i\epsilon) = \int_{-\infty}^{\infty} e^{it\lambda}U_\pm^\epsilon(t)dt,$$

hence

$$\frac{1}{2\pi}\int_{-\infty}^{\infty} e^{-it\lambda}R(\lambda \pm i\epsilon)d\lambda = \pm i\chi_\pm(t)e^{-itH - \epsilon|t|}.$$

Given another self-adjoint operator H_0 and letting $R_0(z) = (H_0 - z)^{-1}$, by the Plancherel-Parseval formula, we have for any bounded operator A

$$\int_{-\infty}^{\infty} \big(R_0(\lambda \pm i\epsilon)f, AR(\lambda \pm i\epsilon)g\big)d\lambda = 2\pi\int_0^{\infty} e^{-2\epsilon t}\big(e^{\mp itH_0}f, Ae^{\mp itH}g\big)dt,$$

$$\tag{8.53}$$

$$\int_{-\infty}^{\infty} \big(R_0(\lambda - i\epsilon)f, AR(\lambda + i\epsilon)g\big)d\lambda = 0. \tag{8.54}$$

We now derive a relation between wave operators and spectral representations.

Theorem 8.10. $W_\pm = (\mathcal{F}_\pm)^* \mathcal{F}_\pm^{(0)}$.

Proof. We must be careful about the difference of the Hilbert space structures of $\mathbf{H}_0$ and $\mathbf{H}$. Take $\chi \in C^\infty(\mathbf{R}^3)$ such that $\chi = 1$ near infinity, and $\chi = 0$ near $\partial\Omega$. Define the operator of extension and that of restriction by

$$\chi_{ext} : \mathbf{H} \ni f(x) \to \chi(x)f(x) \in \mathbf{H}_0, \tag{8.55}$$

$$\chi_{res} : \mathbf{H}_0 \ni f(x) \to \chi(x)f(x) \in \mathbf{H}. \tag{8.56}$$

As can be checked easily,

$$(\chi_{ext})^* = \mathcal{M}^{-1}\chi_{res}\mathcal{M}_0, \tag{8.57}$$

$$(\chi_{res})^* = \mathcal{M}_0^{-1}\chi_{ext}\mathcal{M}. \tag{8.58}$$

Then, by the local decay property, the wave operator is rewritten as

$$W_\pm = \text{s} - \lim_{t \to \pm\infty} e^{it\mathcal{H}} \mathcal{M}^{-1} \chi_{res} \mathcal{M}_0 e^{-it\mathcal{H}_0} P_{ac}(\mathcal{H}_0)$$
$$= \text{s} - \lim_{t \to \pm\infty} e^{it\mathcal{H}} (\chi_{ext})^* e^{-it\mathcal{H}_0} P_{ac}(\mathcal{H}_0). \tag{8.59}$$

Take $f \in \mathbf{H}_0$ such that $\widehat{f}(\xi) \in \left(C_0^\infty(\mathbf{R}^3 \setminus \{0\})\right)^6$, and $g \in D(\mathcal{H})$. Then, by integrating $\frac{d}{dt}(e^{it\mathcal{H}}(\chi_{ext})^* e^{-it\mathcal{H}_0} P_{ac}(\mathcal{H}_0)f, g)$, we have

$$(W_+f, g) = (P_{ac}(\mathcal{H}_0)f, \chi_{ext}g)$$
$$- i \int_0^\infty (e^{-it\mathcal{H}_0} P_{ac}(\mathcal{H}_0)f, (\mathcal{H}_0\chi_{ext} - \chi_{ext}\mathcal{H})e^{-it\mathcal{H}}g)dt.$$

Here, note that the 2nd term of the right-hand side is absolutely convergent by virtue of (7.72). In view of (8.53), (8.54), with $A = (\mathcal{H}_0\chi_{ext} - \chi_{ext}\mathcal{H})(\mathcal{H} + i)^{-1}$, and g replaced by $(\mathcal{H} + i)g$, we have

$$\frac{1}{2\pi i} \int_{-\infty}^\infty \left([R_0(\lambda + i\epsilon) - R_0(\lambda - i\epsilon)]P_{ac}(\mathcal{H}_0)f, (\mathcal{H}_0\chi_{ext} - \chi_{ext}\mathcal{H})R(\lambda + i\epsilon)g\right)d\lambda$$
$$= -i \int_0^\infty e^{-2\epsilon t}\left(e^{-it\mathcal{H}_0}f, (\mathcal{H}_0\chi_{ext} - \chi_{ext}\mathcal{H})e^{-it\mathcal{H}}g\right)dt.$$

Letting $\epsilon \to 0$, and passing to the spectral representation, we have

$$(W_+f, g)_\mathbf{H} = \int_{-\infty}^\infty \left(\mathcal{F}_+^{(0)}(\lambda)f, \mathcal{F}_+^{(0)}(\lambda)Q_+(\lambda)g\right)_\mathbf{h} d\lambda = (\mathcal{F}_+^{(0)}f, \mathcal{F}_+g)_{\widehat{\mathbf{H}}},$$

which proves the theorem for W_+. The proof for W_- is similar. $\square$

Theorem 8.10 yields $\mathcal{F}_\pm = \mathcal{F}_\pm^{(0)}(W_\pm)^*$. Since $W_\pm$ is unitary from $\mathbf{H}_0$ onto $\mathbf{H}_{ac}(\mathcal{H})$ and $\mathcal{F}_\pm^{(0)}$ is onto, we have $\text{Ran}\,\mathcal{F}_\pm = \widehat{\mathbf{H}}_\pm$. This proves Theorem 8.9 (1).

Let us also give an alternative proof of the surjectivity of $\mathcal{F}_\pm$, which do not use the asymptotic completeness of time-dependent wave operators. It is based on the following lemma.

Lemma 8.11. *Let*

$$\mathbf{h}_\pm = \{\mathcal{P}_\pm^{(0)}(\omega)h(\omega)\,;\, h \in \mathbf{h}\}.$$

Then, for $\lambda \in \mathbf{R} \setminus \sigma_p(\mathcal{H})$,

$$\mathcal{F}_\pm(\lambda)\mathcal{B} = \mathbf{h}_\pm.$$

We prove this lemma in §8.8 (see Corollary 8.1).

Now, suppose $\widehat{g} \in \widehat{\mathbf{H}}_{\pm}$ is orthogonal to $\mathcal{F}_{\pm}\mathcal{B}$. Take any open interval $I \subset \mathbf{R} \backslash \sigma_p(\mathcal{H})$. For $f \in \mathcal{B}$, let $f_I = \mathcal{F}_{\pm}^* \chi_I \mathcal{F}_{\pm} f$, where χ_I is the characteristic function of I. Then, we have

$$(\mathcal{F}_{\pm} f_I, \widehat{g}) = \int_I (\mathcal{F}_{\pm}(\lambda)f, \widehat{g}(\lambda))_{\mathbf{h}} d\lambda = 0.$$

Since this holds for any interval I, there is a set e_f of measure 0 such that $(\mathcal{F}_{\pm}(\lambda)f, \widehat{g}(\lambda))_{\mathbf{h}} = 0$, if $\lambda \notin e_f$. By the definition of $\mathcal{B}$, there exists a countable dense subset $\{f_\ell\}_{\ell=1}^\infty$ of $\mathcal{B}$, where each f_ℓ is compactly supported. Then, letting $e = \cup_{\ell=1}^\infty e_{f_\ell}$, we have

$$\mathcal{F}_{\pm}(\lambda)^* \widehat{g}(\lambda) = 0, \quad \text{if} \quad \lambda \notin e.$$

By Lemma 8.11, for $\lambda \notin e$, there exists $f \in \mathcal{B}$ such that $\mathcal{F}_{\pm}(\lambda)f = \widehat{g}(\lambda)$. Then, we have

$$(\mathcal{F}_{\pm}(\lambda)f, \widehat{g}(\lambda))_{\mathbf{h}} = \|\widehat{g}(\lambda)\|_{\mathbf{h}}^2 = 0.$$

Hence, $\widehat{g} = 0$, a.e., which proves that $\text{Ran}\, W_{\pm} = \widehat{\mathbf{H}}_{\pm}$.

8.7   S-matrix

Let us rewrite the scattering operator $S = (W_+)^* W_-$ by the spectral representation. For $\lambda \in \mathbf{R} \setminus \sigma_p(\mathcal{H})$, we define an operator on $\mathbf{h}$ by

$$\widehat{S}(\lambda) = 1 - 2\pi i A(\lambda), \tag{8.60}$$

$$A(\lambda) = \mathcal{F}_+(\lambda)\big(\mathcal{H}\chi_{res} - \chi_{res}\mathcal{H}_0\big)\mathcal{F}_+^{(0)}(\lambda)^*, \tag{8.61}$$

where $\chi \in C^\infty(\mathbf{R}^3)$, and $\chi = 1$ near infinity, $\chi = 0$ near $\partial\Omega$. In quantum mechanical potential scattering, $\widehat{S}(\lambda)$ is called the *S-matrix*, and $A(\lambda)$ the *scattering amplitude*. We use the same terminologies for the scattering of electromagnetic waves.

Theorem 8.11. *Let* $\widehat{S} = \mathcal{F}_+^{(0)} S (\mathcal{F}_+^{(0)})^*$. *Then, for* $u \in \widehat{\mathbf{H}}_0$

$$(\widehat{S}u)(\lambda) = \widehat{S}(\lambda)u(\lambda), \quad a.e. \ \lambda \in \mathbf{R} \setminus \sigma_p(\mathcal{H}). \tag{8.62}$$

$\widehat{S}(\lambda)$ *is unitary on* $\mathbf{h}_+$ *and strongly continuous with respect to* $\lambda \in \mathbf{R} \backslash \sigma_p(\mathcal{H})$. *Moreover,* $A(\lambda)$ *is a compact operator on* $\mathbf{h}_+$.

Proof. The unitarity of $\widehat{S}(\lambda)$ follows from that of S and (8.62), and the continuity in λ is a consequence of the formula (8.61). Since $\mathcal{F}_+^{(0)}(\lambda)^*\phi$ is a divergence free solution to the Maxwell equation, Ran $\mathcal{F}_+^{(0)}(\lambda)^*$ is contained in $H_{loc}^1(\Omega)$. Then, $\left(\mathcal{H}\chi - \chi\mathcal{H}_0\right)\mathcal{F}_+^{(0)}(\lambda)^*$ is a compact operator from $\mathbf{h}_+$ to $\mathcal{B}$, from which follows the compactness of $A(\lambda)$.

Noting that $S - P_{ac}(\mathcal{H}_0) = -(W_+)^*(W_+ - W_-)$, we consider $((W_+ - W_-)f, W_+g)$. Recalling (7.72), we compute

$$(W_+ - W_-)f = i\int_{-\infty}^{\infty} e^{it\mathcal{H}}Ve^{-it\mathcal{H}_0}P_{ac}(\mathcal{H}_0)f\,dt,$$

for $\widehat{f}(\xi), \widehat{g}(\xi) \in \left(C_0^\infty(\mathbf{R}^3 \setminus \{0\})\right)^6$, where $V = \mathcal{H}\chi_{res} - \chi_{res}\mathcal{H}_0$. This yields

$$((W_+ - W_-)f, W_+g) = i\int_{-\infty}^{\infty} (Ve^{-it\mathcal{H}_0}P_{ac}(\mathcal{H}_0)f, W_+e^{-it\mathcal{H}_0}g)\,dt,$$

where we have used $e^{-it\mathcal{H}}W_+ = W_+e^{-it\mathcal{H}_0}$ (see Theorem 7.13 (2)). Then,

$$
\begin{aligned}
&((W_+ - W_-)f, W_+g)\\
&= i\int_{-\infty}^{\infty} dt\,(Ve^{-it\mathcal{H}_0}P_{ac}(\mathcal{H}_0)f, \chi_{ext}e^{-it\mathcal{H}_0}P_{ac}(\mathcal{H}_0)g)\\
&\quad + \int_{-\infty}^{\infty} dt\int_0^{\infty} ds\,(Ve^{-it\mathcal{H}_0}P_{ac}(\mathcal{H}_0)f, e^{is\mathcal{H}}Ve^{-i(s+t)\mathcal{H}_0}P_{ac}(\mathcal{H}_0)g).
\end{aligned}
\tag{8.63}
$$

Note that the integrals in the right-hand side are absolutely convergent. To compute the 2nd term, we insert $e^{-\epsilon|t|}e^{-\epsilon s}$ and take the limit $\epsilon \to 0$. Passing to the spectral representation, we have

$$
\begin{aligned}
&\int_{-\infty}^{\infty} dt\,(Ve^{-it\mathcal{H}_0-\epsilon|t|}P_{ac}(\mathcal{H}_0)f, e^{is\mathcal{H}-\epsilon s}Ve^{-i(s+t)\mathcal{H}_0}P_{ac}(\mathcal{H}_0)g)\\
&= \int_{-\infty}^{\infty} dt\,(V^*e^{-is\mathcal{H}-\epsilon s}Ve^{-it\mathcal{H}_0-\epsilon|t|}P_{ac}(\mathcal{H}_0)f, e^{-i(s+t)\mathcal{H}_0}P_{ac}(\mathcal{H}_0)g)\\
&= \int_{-\infty}^{\infty} dt\int_{-\infty}^{\infty} d\lambda\,(\mathcal{F}_+^{(0)}(\lambda)V^*e^{-is\mathcal{H}-\epsilon s}Ve^{-it\mathcal{H}_0-\epsilon|t|}P_{ac}(\mathcal{H}_0)f, e^{-i(s+t)\lambda}\mathcal{F}_+^{(0)}(\lambda)g)\\
&= \int_{-\infty}^{\infty} dt\int_{-\infty}^{\infty} d\lambda\,(\mathcal{F}_+^{(0)}(\lambda)V^*e^{-is(\mathcal{H}-\lambda)-\epsilon s}Ve^{-it(\mathcal{H}_0-\lambda)-\epsilon|t|}P_{ac}(\mathcal{H}_0)f, \mathcal{F}_+^{(0)}(\lambda)g).
\end{aligned}
$$

Performing the integration with respect to t, we see that this is equal to

$$2\pi\int_{-\infty}^{\infty} (\mathcal{F}_+^{(0)}(\lambda)V^*e^{-is(\mathcal{H}-\lambda)-\epsilon s}V\frac{1}{2\pi i}[R_0(\lambda+i\epsilon)-R_0(\lambda-i\epsilon)]f, \mathcal{F}_+^{(0)}(\lambda)g)\,d\lambda.$$

Integrating with respect to s, we then have

$$-2\pi i\int_{-\infty}^{\infty} (\mathcal{F}_+^{(0)}(\lambda)V^*R(\lambda+i\epsilon)V\frac{1}{2\pi i}[R_0(\lambda+i\epsilon)-R_0(\lambda-i\epsilon)]f, \mathcal{F}_+^{(0)}(\lambda)g)\,d\lambda.$$

Letting $\epsilon \to 0$, we obtain

$$-2\pi i \int_{-\infty}^{\infty} (\mathcal{F}_+^{(0)}(\lambda) V^* R(\lambda + i0) V \mathcal{F}_+^{(0)}(\lambda)^* \mathcal{F}_+^{(0)}(\lambda) f, \mathcal{F}_+^{(0)}(\lambda) g) \, d\lambda.$$

Here, we have used the fact that

$$\frac{1}{2\pi i} \left(R_0(\lambda + i0) - R_0(\lambda - i0) \right) = (\mathcal{F}_+^{(0)}(\lambda))^* \mathcal{F}_+^{(0)}(\lambda),$$

which follows from Parseval's formula (Lemma 8.10). In order to guarantee the existence of $R(\lambda + i0)$, we need to assume that the support with respect to λ of $\mathcal{F}_+^{(0)}(\lambda) g$ is contained in $\mathbf{R} \setminus \sigma_p(\mathcal{H})$.

By the similar argument, the 1st term of the right-hand side of (8.63) is equal to

$$2\pi i \int_{-\infty}^{\infty} (\mathcal{F}_+^{(0)}(\lambda) \chi_{ext} V \mathcal{F}_+^{(0)}(\lambda)^* \mathcal{F}_+^{(0)}(\lambda) f, \mathcal{F}_+^{(0)}(\lambda) g) \, d\lambda.$$

Summing up, we obtain

$$((S - P_{ac}(\mathcal{H}_0)) f, g) = -2\pi i \int_{-\infty}^{\infty} (A(\lambda) \mathcal{F}_+^{(0)}(\lambda) f, \mathcal{F}_+^{(0)}(\lambda) g) \, d\lambda,$$

where $A(\lambda) \in \mathbf{B(h)}$ is defined by

$$A(\lambda) = \mathcal{F}_+^{(0)}(\lambda) \chi_{ext} V \mathcal{F}_+^{(0)}(\lambda)^* - \mathcal{F}_+^{(0)}(\lambda) V^* R(\lambda + i0) V \mathcal{F}_+^{(0)}(\lambda)^*.$$

This is rewritten as

$$A(\lambda) = \mathcal{F}_+^{(0)}(\lambda) Q_+(\lambda + i0) V \mathcal{F}_+^{(0)}(\lambda)^* = \mathcal{F}_+(\lambda) V \mathcal{F}_+^{(0)}(\lambda)^*.$$

We have thus completed the proof of the theorem. $\square$

8.8   Homogeneous equation

The aim of this section is to characterize the solution space of the homogeneous equation $\{u \in \mathcal{B} ; (\mathcal{H} - \lambda) u = 0\}$ by the spectral representation.

Let us begin with preparing a lemma for the Fourier transformation. We put $r_0(z) = (-\Delta - z)^{-1}$, and

$$F_0(k) f(\theta) = \widehat{f}(k\theta) = (2\pi)^{-n/2} \int_{\mathbf{R}^n} e^{-ik\theta \cdot x} f(x) dx.$$

We then have

$$\frac{1}{2\pi} ([r_0(\lambda + i0) - r_0(\lambda - i0)] f, g) = \frac{\lambda^{(n-2)/2}}{2} (F_0(\sqrt{\lambda}) f, F_0(\sqrt{\lambda}) g)_{L^2(S^{n-1})}.$$

This implies for $k > 0$

$$F_0(k) \in \mathbf{B}(\mathcal{B} ; L^2(S^{n-1})), \quad F_0(k)^* \in \mathbf{B}(L^2(S^{n-1}) ; \mathcal{B}^*).$$

We introduce a notation

$$u(x) \simeq v(x) \iff u - v \in \mathcal{B}_0^*. \tag{8.64}$$

Then, one can derive the following asymptotic expansion by applying the stationary phase method. We prove it in Appendix C.

Lemma 8.12. *For $\varphi \in L^2(S^{n-1})$, we have for $k > 0$*

$$\int_{S^{n-1}} e^{ik\omega \cdot x} \varphi(\omega) \, d\omega \simeq C(k) r^{-(n-1)/2} e^{ikr} \varphi(\widehat{x}) \\ + \overline{C(k)} r^{-(n-1)/2} e^{-ikr} \varphi(-\widehat{x}), \tag{8.65}$$

where $r = |x|$, $\widehat{x} = x/r$, $C(k) = e^{-(n-1)\pi i/4} (2\pi)^{(n-1)/2} k^{-(n-1)/2}$.

By definition

$$\left(\mathcal{F}_\pm^{(0)}(\lambda)^* \varphi\right)(x) = \lambda (2\pi c_0)^{-3/2} \int_{S^2} e^{\pm i \frac{\lambda}{c_0} \omega \cdot x} \mathcal{P}_\pm^{(0)}(\omega) \varphi(\omega) \, d\omega. \tag{8.66}$$

This and Lemma 8.12 imply the following lemma.

Lemma 8.13. *For $\varphi \in \mathbf{h}$,*

$$\left(\mathcal{F}_\pm^{(0)}(\lambda)^* \varphi\right)(x) \\ \simeq \mp i \, (2\pi c_0)^{-1/2} r^{-1} \left(e^{\pm i \frac{\lambda}{c_0} r} \mathcal{P}_\pm^{(0)}(\widehat{x}) \varphi(\widehat{x}) - e^{\mp i \frac{\lambda}{c_0} r} \mathcal{P}_\pm^{(0)}(-\widehat{x}) \varphi(-\widehat{x}) \right).$$

For the free case $\mathcal{H}_0$, the outgoing and incoming states are related by the space inversion $x \to -x$. Introducing a bounded operator J on $\mathbf{h}$ by

$$(J\varphi)(\omega) = -\varphi(-\omega), \tag{8.67}$$

and using (8.15), (8.17), we have

$$\mathcal{F}_+^{(0)}(\lambda) = J \mathcal{F}_-^{(0)}(\lambda). \tag{8.68}$$

It is convenient to put

$$A_J(\lambda) = A(\lambda) J, \tag{8.69}$$

$$\widehat{S}_J(\lambda) = J - 2\pi i \, A_J(\lambda). \tag{8.70}$$

By (8.61) and (8.68), we have

$$A_J(\lambda) = \mathcal{F}_+(\lambda)(\mathcal{H}\chi_{res} - \chi_{res}\mathcal{H}_0)\mathcal{F}_-^{(0)}(\lambda)^*, \tag{8.71}$$

$$\widehat{S}_J(\lambda) = \widehat{S}(\lambda) J. \tag{8.72}$$

Therefore, $\widehat{S}_J(\lambda)$ is a unitary operator from $\mathbf{h}_-$ onto $\mathbf{h}_+$. Recall that $\mathcal{F}_\pm(\lambda)^*$ is an eigenoperator of $\mathcal{H}$ in the sense that

$$(\mathcal{H} - \lambda)\mathcal{F}_\pm(\lambda)^*\varphi = 0, \quad \varphi \in \mathbf{h}_\pm.$$

It has the following representation.

Lemma 8.14.

$$\mathcal{F}_-(\lambda)^* = \big(\chi_{res} - R(\lambda + i0)(\mathcal{H}\chi_{res} - \chi_{res}\mathcal{H}_0)\big)\mathcal{F}_-^{(0)}(\lambda)^*.$$

Proof. By a direct computation, using (8.43), we have

$$Q_-(\lambda)^* = \chi_{res} + R(\lambda + i0)\big(\lambda\chi(1 - \mathcal{M}^{-1}\mathcal{M}_0) - [\mathcal{H}, \chi]\big).$$

Using $[\mathcal{H}, \chi] = \mathcal{M}^{-1}\mathcal{M}_0[\mathcal{H}_0, \chi]$, we have

$$[\mathcal{H}, \chi]\mathcal{F}_-^{(0)}(\lambda)^*\varphi = (\mathcal{H}\chi - \lambda\chi\mathcal{M}^{-1}\mathcal{M}_0)\mathcal{F}_-^{(0)}(\lambda)^*\varphi.$$

This and the formula $\mathcal{F}_-(\lambda)^* = Q_-(\lambda)^*\mathcal{F}_-^{(0)}(\lambda)^*$ prove the lemma. $\square$

As will be shown in the next lemma, $\mathcal{F}_-(\lambda)^*$ admits an asymptotic expansion similar to the one in Lemma 8.13, in which $\widehat{S}_J(\lambda)$ appears as a mapping from incoming waves to outgoing waves.

Lemma 8.15. *For* $\varphi \in \mathbf{h}_-$,

$$\big(\mathcal{F}_-(\lambda)^*\varphi\big)(x) \simeq -i\,(2\pi c_0)^{-1/2}r^{-1}\Big(e^{i\frac{\lambda}{c_0}r}(\widehat{S}_J(\lambda)\varphi)(\widehat{x}) - e^{-i\frac{\lambda}{c_0}r}\varphi(\widehat{x})\Big).$$

Proof. Lemma 8.13 implies

$$\mathcal{F}_-^{(0)}(\lambda)^*\varphi \simeq i(2\pi c_0)^{-1/2}r^{-1}\Big(e^{-i\frac{\lambda}{c_0}r}\varphi(\widehat{x}) - e^{i\frac{\lambda}{c_0}r}\varphi(-\widehat{x})\Big).$$

Theorem 8.8 implies

$$R(\lambda + i0)V\mathcal{F}_-^{(0)}(\lambda)^*\varphi \simeq \sqrt{\frac{2\pi}{c_0}}\frac{e^{i\frac{\lambda}{c_0}r}}{r}\mathcal{F}_+(\lambda)V\mathcal{F}_-^{(0)}(\lambda)^*\varphi,$$

where $V = \mathcal{H}\chi_{res} - \chi_{res}\mathcal{H}_0$. These two formulas, togethers with Lemma 8.14, prove the theorem. $\square$

Lemma 8.16. *For any* $\varphi \in \mathbf{h}_-$

$$\lim_{R \to \infty} \frac{1}{R}\int_{|x|<R}|\mathcal{F}_-(\lambda)^*\varphi|^2 dx = \frac{1}{\pi c_0}\|\varphi\|_{\mathbf{h}}^2.$$

Proof. In view of Lemma 8.15, we have

$$\lim_{R\to\infty}\frac{1}{R}\int_{|x|<R}|\mathcal{F}_-(\lambda)^*\varphi|^2dx$$

$$=\frac{1}{2\pi c_0}\lim_{R\to\infty}\frac{1}{R}\int_0^R\|e^{i\frac{\lambda}{c_0}r}\widehat{S}_J\varphi-e^{-i\frac{\lambda}{c_0}r}\varphi\|_{L^2(S^2)}^2dr.$$

Noting that

$$\|e^{i\frac{\lambda}{c_0}r}\widehat{S}_J\varphi-e^{-i\frac{\lambda}{c_0}r}\varphi\|_{L^2(S^2)}^2$$

$$=\|\widehat{S}_J\varphi\|^2+\|\varphi\|^2-e^{2i\frac{\lambda}{c_0}r}(\widehat{S}_J\varphi,\varphi)-e^{-2i\frac{\lambda}{c_0}r}(\varphi,\widehat{S}_J\varphi),$$

and the unitarity of $\widehat{S}_J$, we obtain the lemma. □

The definition of $\mathcal{B}^*$-norm and Lemma 8.16 imply the following lemma.

Lemma 8.17. *There exists a constant $C>0$ such that*

$$C^{-1}\|\varphi\|_{\mathbf{h}}\leq\|\mathcal{F}_-(\lambda)^*\varphi\|_{\mathcal{B}^*}\leq C\|\varphi\|_{\mathbf{h}}.$$

Lemma 8.18. *Suppose $u\in\mathcal{B}^*$ satisfies $(\mathcal{H}-\lambda)u=0$ and the boundary condition $E\cdot\nu=0$. Then, $(u,f)=0$ for any $f\in\mathcal{B}$ satisfying $\mathcal{F}_-(\lambda)f=0$.*

Proof. Let $\Omega_r=\Omega\cap\{|x|<r\}$, and $\langle\cdot,\cdot\rangle_{\Omega_r}$ the standard inner product of $L^2(\Omega_r)$. Letting $v=R(\lambda-i0)f$, and $\chi_r(x)$ be the characteristic function of Ω_r, we have

$$0=(\chi_r(\mathcal{H}-\lambda)u,v)=\langle(\mathcal{L}-\lambda\mathcal{M})u,v\rangle_{\Omega_r}$$

$$=-i\langle\mathcal{L}(\widehat{x})u,v\rangle_{S_r}+(\chi_ru,f),$$

where $S_r=\{|x|=r\}$, and $\langle\cdot,\cdot\rangle_{S_r}$ is the inner product of $L^2(S_r)$. Then, we have

$$(u,f)=i\lim_{R\to\infty}\frac{1}{R}\int_{r_0}^R\langle\mathcal{L}(\widehat{x})u,v\rangle_{S_r}dr,$$

where $r_0>0$ is chosen large enough. Since $u\in\mathcal{B}^*$, we have

$$\left|\frac{1}{R}\int_{r_0}^R\langle\mathcal{L}(\widehat{x})u,v\rangle_{S_r}dr\right|\leq C\|u\|_{\mathcal{B}^*}\frac{1}{R}\int_{r_0}^R|v|^2dx.$$

Since $\mathcal{F}_-(\lambda)f=0$, by Theorem 8.8, the right-hand side tends to 0 as $R\to\infty$. □

Let us recall Banach's closed range theorem.

Theorem 8.12. *Let X, Y be banach spaces, and $T \in \mathbf{B}(X; Y)$. Then, the following assertions are equivalent.*

(1) $R(T) = \overline{R(T)}$,

(2) $R(T^*) = \overline{R(T^*)}$,

(3) $R(T) = N(T^*)^\perp = \{y \in Y \, ; \, \langle y, y^* \rangle = 0 \,\, \forall y^* \in N(T^*)\}$,

(4) $R(T^*) = N(T)^\perp = \{x^* \in X^* \, ; \, \langle x, x^* \rangle = 0 \,\, \forall x \in N(T)\}$,

where $R(T) = \operatorname{Ran} T$, $N(T) = \operatorname{Ker} T$.

See [102], p. 205.

Let $\mathcal{N}_\lambda(\mathcal{H})$ be the set defined by

$$
u \in \mathcal{N}_\lambda(\mathcal{H}) \iff
\begin{cases}
u = (E, H) \in \mathcal{B}^*, \\
(\mathcal{H} - \lambda)u = 0, & \text{in} \quad \Omega, \\
\nu \times E = 0, & \text{on} \quad \partial\Omega.
\end{cases}
\tag{8.73}
$$

Theorem 8.13. $\mathcal{N}_\lambda(\mathcal{H}) = \mathcal{F}_-(\lambda)^* \mathbf{h}_-$, $\lambda \in \mathbf{R} \setminus \sigma_p(\mathcal{H})$.

Proof.   We know already that $\mathcal{N}_\lambda(\mathcal{H}) \supset \mathcal{F}_-(\lambda)^* \mathbf{h}_-$. Take $X = \mathcal{B}$, $Y = \mathbf{h}_-$, and $T = \mathcal{F}_-(\lambda)$ in Theorem 8.12. Lemma 8.17 shows that $R(T^*)$ is closed, and $N(T^*) = \{0\}$. Therefore, by virtue of Theorem 8.12, $R(T) = Y$, $R(T^*) = N(T)^\perp$. Lemma 8.18 implies that if $u \in \mathcal{B}^*$ satisfies $(\mathcal{H} - \lambda)u = 0$ and the boundary condition $E \cdot \nu = 0$, then, $u \in N(T)^\perp$. Hence, $u \in R(T^*)$. This proves $\mathcal{N}_\lambda(\mathcal{H}) \subset \mathcal{F}_-(\lambda)^* \mathbf{h}_-$. $\square$

The following fact is also proven.

Corollary 8.1. $\mathcal{F}_\pm(\lambda)\mathcal{B} = \mathbf{h}_\pm$.

By Lemma 8.15 and Theorem 8.13, any solution $u \in \mathcal{N}_\lambda(\mathcal{H})$ admits the asymptoic expansion in Lemma 8.15. It enables us to formulate the scattering in the stationary picture.

Theorem 8.14. *For any $\varphi_- \in \mathbf{h}_-$, there exist unique $u \in \mathcal{N}_\lambda(\mathcal{H})$ and $\varphi_+ \in \mathbf{h}_+$ having the asymptotic expansion*

$$
u \simeq -i \, (2\pi c_0)^{-1/2} r^{-1} \left(e^{i \frac{\lambda}{c_0} r} \varphi_+(\widehat{x}) - e^{-i \frac{\lambda}{c_0} r} \varphi_-(\widehat{x}) \right).
$$

Moreover, $\varphi_+ = S_J(\lambda)\varphi_-$.

Proof.   To prove the existence, put $u = \mathcal{F}_-(\lambda)\varphi_-$. To show the uniqueness, suppose $\varphi_- = 0$ and u, φ_+ satisfy the asymptotic expansion. Then, $u \in \mathcal{B}^*$ is a solution to the homogeneous equation satifying the outgoing radiation condition. Therefore, $u = 0$, and the theorem follows. $\square$

Chapter 9

Scattering and interior boundary value problem

In this and the next chapters, we study the inverse scattering for the Maxwell operator $\mathcal{H} = \mathcal{M}(x)^{-1}\mathcal{L}(-i\nabla_x)$ in an exterior domain $\Omega \subset \mathbf{R}^3$ under the following conditions.

(C-1) Ω is either $\mathbf{R}^3$ or a connected open set in $\mathbf{R}^3$ with C^∞ boundary $\partial\Omega$ such that Ω^c is compact.

(C-2) The matrix $\mathcal{M}(x) = \begin{pmatrix} \epsilon(x) & 0 \\ 0 & \mu(x) \end{pmatrix}$ is positive definite and

$$CI_6 \leq \mathcal{M}(x) \leq C^{-1}I_6, \quad on \quad \Omega$$

for a constant $C > 0$.

(C-3) There exist constants $\epsilon_0, \mu_0 > 0$ and $r_0 > 1$ such that

$$\epsilon(x) = \epsilon_0 I_3, \quad \mu(x) = \mu_0 I_3 \quad for \quad |x| > r_0 - 1.$$

Without loss of generality, we assume that

$$c_0 = (\epsilon_0\mu_0)^{-1/2} = 1. \tag{9.1}$$

We split Ω into two parts, and put

$$\Omega_{ext} = \{|x| > r_0\}, \quad \Omega_{int} = \Omega \cap \{|x| < r_0\}, \tag{9.2}$$

$$S = S_{ext} = \{|x| = r_0\}, \quad S_{int} = \partial\Omega. \tag{9.3}$$

In the *interior* domain Ω_{int}, one can solve the boundary value problem for the Maxwell equation and define the mapping from the electric field on S to the magnetic field on S. The aim of this chapter is to show that the S-matrix in an exterior domain Ω_{ext} for any fixed frequency $\lambda \neq 0$ and this electromagnetic map for the interior domain Ω_{int} determine each other. No regularity is assumed on $\mathcal{M}(x)$ in this chapter.

We use the notation $F_{x\to\xi}$ to denote the Fourier transform:

$$(F_{x\to\xi}f)(\xi) = (2\pi)^{-3/2}\int_{\mathbf{R}^3} e^{-ix\cdot\xi}f(x)\,dx. \tag{9.4}$$

For $m \in \mathbf{R}$, H^m_{loc} denotes the space of distributions $f \in \mathcal{S}'$ such that

$$H^m_{loc} \ni f \Longleftrightarrow \chi f \in H^m(\mathbf{R}^3), \quad \forall \chi \in C^\infty_0(\mathbf{R}^3),$$

and H^m_{comp} is the subspace of compactly supported distributions in H^m_{loc}:

$$H^m_{comp} \ni f \Longleftrightarrow f \in H^m_{loc}, \quad f = \chi f \quad \text{for some} \quad \chi \in C^\infty_0(\mathbf{R}^3).$$

The spaces $\mathbf{H}_0$ and $\mathbf{H}$ are defined by (6.40), (7.21), respectively.

9.1   Mapping properties of the resolvent

We extend the resolvent estimates on Sobolev spaces H^s for $-1 \le s \le 1$.

Lemma 9.1. *Let $\chi \in C^\infty(\mathbf{R}^3)$ be such that $\chi(x) = 1$ for $|x| > r_0 - 1/2$ and $\chi(x) = 0$ for $|x| < r_0 - 1$. Then, for $z \in \mathbf{C} \setminus \mathbf{R}$ and $-1 \le s \le 1$,*

$$\chi R(z)\chi \in \mathbf{B}\big(H^s(\mathbf{R}^3); H^s(\mathbf{R}^3)\big), \tag{9.5}$$

and for $\lambda \in \mathbf{R} \setminus \{0\}$, $t > 1/2$ and $-1 \le s \le 1$,

$$(1+|x|)^{-t}\chi R(\lambda \pm i0)\chi(1+|x|)^{-t} \in \mathbf{B}\big(H^s(\mathbf{R}^3); H^s(\mathbf{R}^3)\big). \tag{9.6}$$

Proof.   Let $u = (u_E, u_H) = R(z)\chi f$, $f = (f_E, f_H)$. Then, taking the divergence of $(\mathcal{L} - \lambda\mathcal{M})u = \mathcal{M}\chi f$, we have

$$-\lambda\,\mathrm{div}\,(\epsilon u_E) = \mathrm{div}\,(\epsilon\chi f_E), \quad -\lambda\,\mathrm{div}\,(\mu u_H) = \mathrm{div}\,(\mu\chi f_H).$$

Therefore,

$$\|\mathrm{div}\,\chi u_E\| + \|\mathrm{div}\,\chi u_H\| \le C\,(\|\chi f\|_{H^1} + \|u\|), \tag{9.7}$$

$\|\cdot\|$ being the norm of $L^2(\mathbf{R}^3)$. From the equation $(\mathcal{L} - z\mathcal{M})u = \mathcal{M}\chi f$, we have

$$\|\mathrm{curl}\,u_E\| + \|\mathrm{curl}\,u_H\| \le C\|\chi f\|. \tag{9.8}$$

Theorem 7.5 then yields

$$\chi R(z)\chi \in \mathbf{B}(H^1; H^1), \tag{9.9}$$

where $H^s = H^s(\mathbf{R}^3)$. To define $\chi R(z)\chi$ on H^{-1}, we consider

$$(\chi R(z)\chi f, g) = (f, \chi R(\bar{z})\chi g). \tag{9.10}$$

Note that the L^2-inner product $(\ ,\ )$ is uniquely extended to a coupling between H^{-1} and H^1 by passing to the Fourier transform

$$(f,g) = (\widehat{f},\widehat{g}) = (\langle\xi\rangle^{-1}\widehat{f}(\xi), \langle\xi\rangle\widehat{g}(\xi)) := {}_{H^{-1}}\langle f,g\rangle_{H^1},$$

where $\langle\xi\rangle = (1+|\xi|^2)^{1/2}$. Moreover, we have

$$\|f\|_{H^{-1}} = \sup_{\|g\|_{H^1}=1} |(f,g)_{L^2}|.$$

By (9.9), the right-hand side of (9.10) is extended to a continuous anti-linear functional on H^1:

$$H^1 \ni g \to {}_{H^{-1}}\langle f, \chi R(\bar{z})\chi g\rangle_{H^1} \in \mathbf{C},$$

where $\langle\xi\rangle = (1+|\xi^2|)^{1/2}$. We then have

$$\|\chi R(z)\chi f\|_{H^{-1}} = \sup_{\|g\|_{H^1}=1} |(f,\chi R(\bar{z})\chi g)| \le C\|\langle\xi\rangle^{-1}\widehat{f}(\xi)\| = \|f\|_{H^{-1}}.$$

Therefore,

$$\chi R(z)\chi \in \mathbf{B}(H^{-1}; H^{-1}). \tag{9.11}$$

Interpolating (9.9) and (9.11) (see Theorem D.2 in Appendix D), we obtain (9.5).

Applying Theorem 7.5 to $\chi R(z)\chi$, and letting $z \to \lambda \pm i0$, we get (9.6) for $s = 1$. Taking duality, we have (9.6) for $s = -1$, and finally for $-1 \le s \le 1$ by interpolation. $\qquad\square$

9.2 Layer potentials

In this and the next sections (subsections 9.3.1 and 9.3.2), we use the notation S, Ω_{ext} and Ω_{int} in a meaning different from the ones in (9.2) and (9.3). Let S be a compact C^∞-surface in $\mathbf{R}^3$. We put

$$g_\lambda(x) = \frac{e^{i\lambda|x|}}{4\pi|x|}, \tag{9.12}$$

and for $a(x) \in C^\infty(S;\mathbf{R}^3)$, define the layer potentials by

$$(G_S a)(x) = \int_S g_\lambda(x-y)a(y)dS_y, \tag{9.13}$$

$$(G_C a)(x) = \mathrm{curl}\int_S g_\lambda(x-y)a(y)\,dS_y, \tag{9.14}$$

$$(G_D a)(x) = \mathrm{grad}\,\mathrm{div}\int_S g_\lambda(x-y)a(y)\,dS_y. \tag{9.15}$$

We also put

$$\left(G_C^{(pv)}a\right)(x) = \text{p.v.} \int_S \text{curl}_x\left(g_\lambda(x-y)a(y)\right) dS_y, \tag{9.16}$$

$$\left(G_D^{(pv)}a\right)(x) = \text{p.v.} \int_S \text{grad}_x \text{div}_x\left(g_\lambda(x-y)a(y)\right) dS_y, \tag{9.17}$$

where p.v. means the principal value. To study the jump property across S of these operators, we need to decompose the vector fields $\nabla_x, \nabla_x \times, \nabla_x\cdot$ into normal and tangential components.

Let $\theta = (\theta^1, \theta^2) \in U$ be local coordinates on S, where U is a bounded open set in $\mathbf{R}^2$. Then, S is locally represented as $x = z(\theta)$, $\theta \in U$, $z(\theta)$ being a smooth map from U to $\mathbf{R}^3$. Let

$$g_{ij} = \frac{\partial z}{\partial \theta^i} \cdot \frac{\partial z}{\partial \theta^j}, \quad \left(g^{ij}\right) = \left(g_{ij}\right)^{-1}, \quad g = \det\left(g_{ij}\right).$$

Then the *surface gradient* grad_S is defined by

$$\text{grad}_S \varphi = g^{ij} \frac{\partial \varphi}{\partial \theta^i} \frac{\partial z}{\partial \theta^j}, \tag{9.18}$$

for $\varphi \in C^\infty(U; \mathbf{R})$, and the *surface divergence* div_S is defined by

$$\text{div}_S a = \frac{1}{\sqrt{g}} \frac{\partial}{\partial \theta^i} \left(\sqrt{g} g^{ij} \frac{\partial z}{\partial \theta^j} \cdot a\right), \tag{9.19}$$

for $a \in C^\infty(U; \mathbf{R}^3)$. Let ν be a unit normal field on S. Then, the *surface curl* is defined by

$$\text{curl}_S a = -\text{div}_S(\nu \times a). \tag{9.20}$$

Let us summarize the properties of these operators.

Lemma 9.2. *(1) Let* $(\,,\,)_S$ *be the inner product of* $L^2(S)$. *Then*

$$(\varphi, \text{div}_S a)_S = -(\text{grad}_S \varphi, a)_S. \tag{9.21}$$

(2) The gradient in the Euclidean metric is rewritten as

$$\nabla_x \varphi = \frac{\partial \varphi}{\partial \nu} \nu + \text{grad}_S \varphi \quad on \quad S. \tag{9.22}$$

(3) If a is tangential to S, we have

$$\text{curl}_S a = \nu \cdot \text{curl}\, a, \tag{9.23}$$

$$(\text{curl}_S a, \varphi)_S = (a, (\text{grad}_S \varphi) \times \nu)_S. \tag{9.24}$$

(4) Let $a = \alpha^j \dfrac{\partial z}{\partial \theta^j}$, *and put* $\alpha_i = g_{ij}\alpha^j$. *Then*

$$d\left(\alpha_i d\theta^i\right) = \left(\text{curl}_S a\right) \sqrt{g}\, d\theta^1 \wedge d\theta^2. \tag{9.25}$$

Proof.   A simple integration by parts yields (9.21). To prove (2), note that in any local coordinate $u = (u^1, u^2, u^3)$, the Euclidean metric in $\mathbf{R}^3$ is rewritten as

$$(dx)^2 = \sum_{i,j=1}^{3} \bar{g}_{ij} du^i du^j, \quad \bar{g}_{ij} = \frac{\partial x}{\partial u^i} \cdot \frac{\partial x}{\partial u^j}.$$

Then, the gradient in $\mathbf{R}^3$ is defined invariantly by

$$\operatorname{grad} \psi = \bar{g}^{ij} \frac{\partial \psi}{\partial u^i} \frac{\partial x}{\partial u^j}, \quad \psi \in C^\infty(\mathbf{R}^3). \tag{9.26}$$

Let $\nu(\theta)$ be the outer unit normal of S at $z(\theta) \in S$. For a small $\epsilon > 0$, the map

$$x(t, \theta) = t\nu(\theta) + z(\theta), \quad (t, \theta) \in (-\epsilon, \epsilon) \times U \tag{9.27}$$

is a diffeomorphism. Let $(t(x), \theta(x))$ be its inverse, and

$$S_t = \{x(t, \theta) \, ; \, \theta \in U\}.$$

Note that $\nu(\theta)$ is also the unit normal field on S_t at $x(t, \theta)$, which is easily seen by differentiating (9.27) by θ^i and taking the inner product with $\nu(\theta)$. We also see that

$$\frac{\partial t(x)}{\partial x^i} \Big|_S = \nu^i(\theta). \tag{9.28}$$

In fact, by differentiating $x^j = t\nu^j + z^j$ by x^i, multiplying by ν^j, summing up by j and letting $t = 0$, we have

$$\nu^i = \frac{\partial t}{\partial x^i} + \frac{\partial z}{\partial x^i} \cdot \nu.$$

Since $\frac{\partial z}{\partial x^i} = \frac{\partial \theta^j}{\partial x^i} \frac{\partial z}{\partial \theta^j}$, the 2nd term of the right-hand side vanishes. This yields (9.28).

In view of (9.27) and (9.28), we have

$$(dx)^2 = (dt)^2 + \sum_{i,j=1}^{2} g_{ij} d\theta^i d\theta^j$$

at any point on S. This and (9.26) imply (9.22).

We prove (3). Take $x_0 \in S$ arbitrarily, and consider the ball $B(x_0, \epsilon)$ of radius $\epsilon > 0$ centered at x_0. The surface S divides this ball into two parts. Let D be the part which lies in the oppsite side to the normal ν. Take any $\varphi \in C^\infty(\mathbf{R}^3)$. Then,

$$0 = \int_D (\operatorname{div} \operatorname{curl} a) \, \varphi \, dx = \int_{\partial D} (\nu \cdot \operatorname{curl} a) \varphi \, dS - \int_D \operatorname{curl} a \cdot \operatorname{grad} \varphi \, dx$$

$$= \int_{\partial D} (\nu \cdot \operatorname{curl} a) \varphi \, dS - \int_{\partial D} (\nu \times a) \cdot \operatorname{grad} \varphi \, dS.$$

Using (9.21), we have

$$\int_{\partial D} (\nu \times a) \cdot \operatorname{grad} \varphi \, dS = \int_{\partial D} (\nu \times a) \cdot \operatorname{grad}_S \varphi \, dS = - \int_{\partial D} (\operatorname{div}_S \nu \times a) \varphi \, dS,$$

which proves (9.23). Using (9.20), integration by parts and noting that $(\nu \times a) \cdot \operatorname{grad}_S \varphi = (\operatorname{grad}_S \varphi \times \nu) \cdot a$, we obtain (9.24).

If a is tangential to S, it is written as $a = \alpha^i \frac{\partial z}{\partial \theta^i}$. A direct computation then yields

$$g^{ij} \frac{\partial z}{\partial \theta^j} \cdot \left(\left(\frac{\partial z}{\partial \theta^1} \times \frac{\partial z}{\partial \theta^2} \right) \times a \right) = \delta_2^i g_{1j} \alpha^j - \delta_1^i g_{2j} \alpha^j.$$

Letting $\alpha_i = g_{ij} \alpha^j$, we then have

$$\operatorname{div}_S (\nu \times a) = \frac{1}{\sqrt{g}} \left(\frac{\partial \alpha_1}{\partial \theta^2} - \frac{\partial \alpha_2}{\partial \theta^1} \right). \tag{9.29}$$

We thus obtain (9.25). $\qquad\square$

We return to the layer potential. Using

$$\frac{\partial}{\partial \theta^j} \frac{e^{i\lambda|x-y|}}{|x-y|} = \frac{e^{i\lambda|x-y|}}{|x-y|} \left(i\lambda - \frac{1}{|x-y|} \right) \frac{x-y}{|x-y|} \cdot \frac{\partial x}{\partial \theta^j},$$

$$\frac{x-y}{|x-y|} \cdot \frac{\partial x}{\partial \theta^j} = \left(\frac{x-y}{|x-y|} - \left(\frac{x-y}{|x-y|} \cdot \nu(x) \right) \nu(x) \right) \cdot \frac{\partial x}{\partial \theta^j},$$

and Lemma 3.5 in Chap. 3, we see that the tangential derivative part of $G_C^{(pv)} a$ is continuous across S. By Theorem 3.3 in Chap. 3, the normal derivative part of $G_C^{(pv)} a$ has a jump across S. We have thus arrived at the following lemma.

Lemma 9.3. *For $a \in S^{0,\alpha}(S)$, $0 < \alpha \le 1$, the following jump formulas hold on S:*

$$(G_C a(x))_\pm = \mp \frac{1}{2} \nu \times a + G_C^{(pv)} a, \tag{9.30}$$

$$(G_C a(x))_+ - (G_C a(x))_- = -\nu \times a. \tag{9.31}$$

To study the operator G_D, we prepare the following lemma.

Lemma 9.4. *If a is tangent to S,*

$$\operatorname{div} \int_S g_\lambda(x-y) a(y) \, dy = \int_S g_\lambda(x-y) \operatorname{div}_S a \, dS.$$

Proof. We compute

$$\text{div} \int_S g_\lambda(x-y)a(y)dy = -\int_S (\nabla_y g_\lambda(x-y)) \cdot a(y)\, dS.$$

By (9.22) and (9.21), this is equal to

$$-\int_S \left((\partial_{\nu(y)} g_\lambda(x-y))\nu \cdot a + (\text{grad}_S\, g_\lambda(x-y)) \cdot a \right) dS$$

$$= -\int_S (\partial_{\nu(y)} g_\lambda(x-y))\nu \cdot a\, dS + \int_S g_\lambda(x-y)\, \text{div}_S\, a\, dS.$$

Since a is tangential to S, the 1st term of the right-hand side vanishes. This proves the lemma. $\qquad\square$

In view of Theorem 3.4 and Lemma 9.4, we obtain the following lemma.

Lemma 9.5. *Suppose* $a \in C^{1,\alpha}(S)$, $0 < \alpha \leq 1$, *is tangent to* S. *Then for* $x \in S$,

$$(G_D a(x))_\pm = \mp \frac{1}{2}(\text{div}_S\, a(x))\nu(x) + G_D^{(pv)} a(x), \tag{9.32}$$

$$(G_D a(x))_+ - (G_D a(x))_- = -(\text{div}_S\, a(x))\nu(x). \tag{9.33}$$

Let us recall the explicit form of the free resolvent. We put

$$\mathcal{H}_0(\xi) = \mathcal{M}_0^{-1}\mathcal{L}(\xi) = \begin{pmatrix} 0 & -\epsilon_0^{-1}\xi\times \\ \mu_0^{-1}\xi\times & 0 \end{pmatrix}. \tag{9.34}$$

Let $\mathcal{P}_0^{(0)}(\xi)$ be the projection onto the kernel of $\mathcal{H}_0(\xi)$, and $\mathcal{P}^{(0)}(\xi) = 1 - \mathcal{P}_0^{(0)}(\xi)$. Lemma 8.3 shows that $(\mathcal{H}_0 - \lambda \mp i0)^{-1}$ has the form

$$(F_{x\to\xi})^{-1}\left((\xi^2 - (\lambda \pm i0)^2)^{-1}(\mathcal{H}_0(\xi) + \lambda)\mathcal{P}^{(0)}(\xi) - \frac{1}{\lambda}\mathcal{P}_0^{(0)}(\xi) \right) F_{x\to\xi}.$$

Note that $\mathcal{P}_0^{(0)}(\xi)$ is rewritten as

$$\mathcal{P}_0^{(0)}(\xi) = |\xi|^{-2} \begin{pmatrix} |\xi\rangle\langle\xi| & 0 \\ 0 & |\xi\rangle\langle\xi| \end{pmatrix}, \tag{9.35}$$

where $|\xi\rangle\langle\xi|$ is the 3×3 matrix defined by

$$|\xi\rangle\langle\xi| = (\xi_i\xi_j).$$

Therefore, $(\mathcal{H}_0 - \lambda \mp i0)^{-1}$ has the form

$$(F_{x\to\xi})^{-1}(\xi^2 - (\lambda \pm i0)^2)^{-1}\left(\mathcal{H}_0(\xi) + \lambda - \frac{1}{\lambda}\mathcal{K}_0(\xi) \right) F_{x\to\xi}, \tag{9.36}$$

where

$$\mathcal{K}_0(\xi) = \begin{pmatrix} |\xi\rangle\langle\xi| & 0 \\ 0 & |\xi\rangle\langle\xi| \end{pmatrix}. \tag{9.37}$$

We put

$$\mathbf{G}_0(\lambda; x - y) = \begin{pmatrix} \lambda + \frac{1}{\lambda}\mathrm{grad}_x \, \mathrm{div}_x & -\frac{i}{\epsilon_0}\mathrm{curl}_x \\ \frac{i}{\mu_0}\mathrm{curl}_x & \lambda + \frac{1}{\lambda}\mathrm{grad}_x \, \mathrm{div}_x \end{pmatrix} g_\lambda(x - y) I_6, \tag{9.38}$$

where I_6 is the 6×6 identity matrix. This is the integral kernel of the resolvent $R_0(\lambda + i0)$. Taking account of (9.36), we define the potential operator for the free Maxwell equation by

$$(\mathcal{G}_0(\lambda)f)(x) = \int_S \mathbf{G}_0(\lambda; x - y)f(y)dS_y, \tag{9.39}$$

for $f \in L^2(S)$. We then have

$$\mathcal{G}_0(\lambda) = \begin{pmatrix} \lambda G_S + \frac{1}{\lambda}G_D & -\frac{i}{\epsilon_0}G_C \\ \frac{i}{\mu_0}G_C & \lambda G_S + \frac{1}{\lambda}G_D \end{pmatrix}. \tag{9.40}$$

We also define

$$\mathcal{G}_0^{(pv)}(\lambda) = \begin{pmatrix} \lambda G_S + \frac{1}{\lambda}G_D^{(pv)} & -\frac{i}{\epsilon_0}G_C^{(pv)} \\ \frac{i}{\mu_0}G_C^{(pv)} & \lambda G_S + \frac{1}{\lambda}G_D^{(pv)} \end{pmatrix}. \tag{9.41}$$

Lemmas 9.3 and 9.5 imply that they have the following jump property.

Lemma 9.6. *For* $f = (f_E, f_H) \in C^{1,\alpha}(S; \mathbf{R}^6), 0 < \alpha \leq 1,$ *and* $x \in S,$ *we have*

$$(\mathcal{G}_0(\lambda)f(x))_\pm = \mp \begin{pmatrix} \frac{1}{2\lambda}(\mathrm{div}_S f_E)\nu - \frac{i}{2\epsilon_0}\nu \times f_H \\ \frac{1}{2\lambda}(\mathrm{div}_S f_H)\nu + \frac{i}{2\mu_0}\nu \times f_E \end{pmatrix} + \mathcal{G}_0^{(pv)}(\lambda)f, \tag{9.42}$$

$$(\mathcal{G}_0(\lambda)f(x))_+ - (\mathcal{G}_0(\lambda)f(x))_- = \begin{pmatrix} -\frac{1}{\lambda}(\mathrm{div}_S f_E)\nu + \frac{i}{\epsilon_0}\nu \times f_H \\ -\frac{1}{\lambda}(\mathrm{div}_S f_H)\nu - \frac{i}{\mu_0}\nu \times f_E \end{pmatrix}. \tag{9.43}$$

For $f \in H^{-s}(S), s > 0$, we define a distribution $\delta_S f \in \mathcal{S}'$ by

$$\delta_S f : \mathcal{S} \ni \varphi \to \langle \delta_S f, \varphi \rangle = \int_S \varphi(x)f(x)dS \in \mathbf{C}. \tag{9.44}$$

The trace theorem implies

$$\left| \int_S \varphi(x)f(x)dS \right| \leq C_D \|f\|_{H^{-s}(S)} \|\varphi\|_H^{s+1/2}(D),$$

where D is any bounded open set containing S. Therefore,

$$\delta_S f \in H_{loc}^{-s-1/2}(\mathbf{R}^3), \tag{9.45}$$

which means that $\chi \delta_S f \in H^{-s-1/2}(\mathbf{R}^3)$ for any $\chi \in C_0^\infty(\mathbf{R}^3)$. Moreover, we have $\operatorname{supp} \delta_S f \subset S$.

Lemma 9.7. *For* $f \in H^{-s}(S)$, $s > 0$, *we have in the sense of distribution*

$$(\mathcal{H}_0 - \lambda)\mathcal{G}_0(\lambda)f = \delta_S f.$$

Proof. First we show that for $\varphi \in \mathcal{S}$,

$$\varphi = R_0(\lambda \pm i0)(\mathcal{H}_0 - \lambda)\varphi.$$

In fact, letting $u = \varphi - R_0(\lambda \pm i0)(\mathcal{H}_0 - \lambda)\varphi$, we have $(\mathcal{H}_0 - \lambda)u = 0$. Since u satisfies the radiation condition, $u = 0$. We then compute

$$((\mathcal{H}_0 - \lambda)\mathcal{G}_0(\lambda)f, \overline{\varphi}) = (f, R_0(\lambda - i0)(\mathcal{H}_0 - \lambda)\overline{\varphi})_{L^2(S)} = \langle \delta_S f, \varphi \rangle,$$

which proves the lemma. $\qquad\square$

9.3 Mapping properties on Sobolev spaces

9.3.1 *Function spaces*

Let S be a compact C^∞-surface in $\mathbf{R}^3$. We consider the Laplace-Beltrami operator $\Delta_S = \operatorname{div}_S \operatorname{grad}_S$ from the view point of quadratic forms (see Appendix E). Let $D = H^1(S)$, and define a quadratic form $a(\cdot, \cdot)$ with domain D by

$$a(u, v) = (u, v) + (\operatorname{grad}_S u, \operatorname{grad}_S v), \quad u, v \in D, \tag{9.46}$$

where $(\cdot, \cdot)$ is the inner product on $L^2(S)$, and

$$(\operatorname{grad}_S u, \operatorname{grad}_S v) = \int g^{ij} \frac{\partial u}{\partial \theta^i} \overline{\frac{\partial v}{\partial \theta^j}} \sqrt{g}\, d\theta.$$

Letting $\|u\|_a = \sqrt{a(u, u)}$, we show that for $u_n \in D$

$$\|u_n\|_{L^2(S)} \to 0, \quad \|u_n - u_m\|_a \to 0 \implies \|u_n\|_a \to 0.$$

In fact, letting $\operatorname{grad}_S u_n \to f$ in $L^2(S)$, we have for any $w \in H^1(S)$

$$(-u_n, \operatorname{div}_S w) = (\operatorname{grad}_S u_n, w) \to (f, w).$$

Since $\|u_n\|_{L^2(S)} \to 0$, we have $f = 0$, hence $\|u_n\|_a \to 0$. Therefore, $a(\cdot, \cdot)$ is a closable form. Let $\tilde{a}(\cdot, \cdot)$ be its closed extension. By Theorem E.1 in the Appendix, there exists a unique self-adjoint operator A satisfying

$$(Au, v) = (u, v) + (\operatorname{grad}_S u, \operatorname{grad}_S v), \quad u \in D(A), \quad v \in D(\tilde{a}). \tag{9.47}$$

Moreover,

$$u \in D(A^{1/2}) \iff \exists u_n \in H^1(S), \quad \exists f \in L^2(S) \quad \text{s.t.}$$
$$u_n \to u, \quad \text{grad}_S u_n \to f \quad \text{in} \quad L^2(S). \tag{9.48}$$

It is easy to see that

$$Au = (1 - \Delta_S) u, \quad u \in H^2(S). \tag{9.49}$$

This gives us an equivalent definition of the boundary Sobolev spaces: Let $D^\infty = \cap_{n \geq 1} D(A^n)$. For $s \in \mathbf{R}$, $H^s(S)$ is defined as the completion of D^∞ by the norm $\|A^{s/2}u\|$, i.e.

$$H^s(S) = D(A^{s/2}), \quad s \in \mathbf{R}. \tag{9.50}$$

A merit of this definition is that one can apply the interpolation theorem (Theorem D.1) to operators between these Sobolev spaces.

Now we assume that S is a boundary of a connected exterior domain Ω_{ext}. Let ν be the interior unit normal to $\partial\Omega_{ext}$. For $s \in \mathbf{R}$, let $H^s_{tan}(S)$ be the set defined by

$$H^s_{tan}(S) \ni u \iff u \in (H^s(S))^3, \quad \nu \cdot u = 0 \quad \text{on} \quad S. \tag{9.51}$$

We define $H^s_{loc}(\Omega_{ext})$ to be the Fréchet space of distributions u on Ω_{ext} equipped with seminorms $\|u\|_{H^s(B_n)}$, $n = 0, 1, 2, \cdots$, where $B_n = \{x \in \Omega_{ext}; |x| < n\}$. $H^s_{loc}(\mathbf{R}^3)$ is defined similarly. We also define

$$H^s_{tan}(\text{div}_S; S) \ni u \iff u \in H^s_{tan}(S), \ \text{div}_S u \in H^s(S), \tag{9.52}$$

$$H^s_{tan}(\text{curl}_S; S) \ni u \iff u \in H^s_{tan}(S), \ \text{div}_S(\nu \times u) \in H^s(S), \tag{9.53}$$

$$H^1(\text{curl}_S; \Omega_{int}) \ni u \iff u \in H^1(\Omega_{int})^3, \ \text{div}_S(\nu \times u) \in H^{1/2}(S), \tag{9.54}$$

$$H^1_{loc}(\text{curl}_S; \Omega_{ext}) \ni u \iff u \in H^1_{loc}(\Omega_{ext})^3, \ \text{div}_S(\nu \times u) \in H^{1/2}(S), \tag{9.55}$$

equipped with norms

$$\|u\|_{H^s_{tan}(\text{div}_S;S)} = \|u\|_{H^s(S)} + \|\text{div}_S u\|_{H^s(S)}, \tag{9.56}$$

$$\|u\|_{H^s_{tan}(\text{curl}_S;S)} = \|u\|_{H^s(S)} + \|\text{div}_S(\nu \times u)\|_{H^s(S)}, \tag{9.57}$$

$$\|u\|_{H^1(\text{curl}_S;\Omega_{int})} = \|u\|_{H^1(\Omega_{int})} + \|\text{div}_S(\nu \times u)\|_{H^{1/2}(S)}, \tag{9.58}$$

and semi-norms

$$\|u\|_{H^1(B_n)} + \|\text{div}_S(\nu \times u)\|_{H^{1/2}(S)}, \quad n = 1, 2, \cdots. \tag{9.59}$$

Here, $\Omega_{int} = \mathbf{R}^3 \setminus \overline{\Omega_{ext}}$. As can be checked easily

$$u \in H^s_{tan}(\text{div}_S; S) \implies \nu \times u \in H^s_{tan}(\text{curl}_S; S), \tag{9.60}$$

$$u \in H^s_{tan}(\text{curl}_S; S) \implies \nu \times u \in H^s_{tan}(\text{div}_S; S). \tag{9.61}$$

9.3.2 Layer potentials on Sobolev spaces

Let S_λ and D_λ be the single layer and double layer potentials:

$$S_\lambda \varphi(x) = \int_S g_\lambda(x - y)\varphi(y)dS_y,$$

$$D_\lambda \varphi(x) = \int_S \left(\frac{\partial}{\partial \nu(y)} g_\lambda(x - y)\right)\varphi(y)dS_y.$$

We define the *magnetic dipole operator* M_λ and the *electric dipole operator* N_λ by

$$M_\lambda = 2\nu \times G_C, \quad M^{(pv)} = 2\nu \times G_C^{(pv)}, \tag{9.62}$$

$$N_\lambda a = 2\nu \times G_D(\nu \times a). \tag{9.63}$$

Hölder estimates in Theorems 3.7, 3.8 and 3.9 are extended to Sobolev spaces in the following way.

Theorem 9.1. *For any $s \in \mathbf{R}$, the following operators are bounded:*

$$S_\lambda : H^s(S) \to H^{s+1}(S), \tag{9.64}$$

$$D_\lambda, D_\lambda^* : H^s(S) \to H^{s+1}(S), \tag{9.65}$$

$$M_\lambda, M_\lambda^* : H^s_{tan}(S) \to H^{s+1}_{tan}(S), \tag{9.66}$$

$$N_\lambda : H^s_{tan}(\mathrm{curl}_S; S) \to H^s_{tan}(\mathrm{div}_S; S). \tag{9.67}$$

This theorem is well-known (at least for experts), and can be proven by e.g. pseudo-differential calculus when S is C^∞. More general results and related real analysis estimates are given in [25], [29] and [92].

What is interesting is that, as has been noticed by [53], these continuity properties in Sobolev spaces can be derived from those in Hölder spaces and the following abstract theorem due to Lax [63].

Theorem 9.2. *Let X, Y be Banach spaces, and H_X, H_Y Hilbert spaces with inner product $(\, , \,)_{H_X}, (\, , \,)_{H_Y}$. Assume that*

$$X \subset H_X, \quad Y \subset H_Y,$$

where the inclusions are dense and continuous. Suppose there exist bounded operators $A : X \to Y$, $B : Y \to X$ satisfying

$$(A\varphi, \psi)_{H_Y} = (\varphi, B\psi)_{H_X}, \quad \forall \varphi \in X, \quad \forall \psi \in Y.$$

Then, we have

$$\|A\|_{\mathbf{B}(H_X; H_Y)} \le \left(\|A\|_{\mathbf{B}(X;Y)}\|B\|_{\mathbf{B}(Y;X)}\right)^{1/2}.$$

For the full proof of Theorem 9.1, see Theorem 4.3 in [53] and also Chapter 3 of [23].

Corollary 9.1. *For any $s \geq 0$, the following operators are bounded.*
(1) $\operatorname{grad} S_\lambda : H^s(S) \to H^s(S)$,
(2) $G_C : H^s(S) \to H^s(S)$,
(3) $\nu \times G_D : H^s_{tan}(\operatorname{div}_S ; S) \to H^s_{tan}(S)$.

Proof. By the orthogonal decomposition, $\operatorname{grad} S_\lambda$ is split into D_λ^* and $\operatorname{grad}_S S_\lambda$. Therefore, the assertion (1) follows from (9.64) and (9.65). The assertion (2) then follows from (1). If a is tangential to S, by Lemma 9.4, $G_D a = \operatorname{grad} S_\lambda(\operatorname{div} a)$, which implies

$$\nu \times G_D a = \nu \times \operatorname{grad}_S S_\lambda(\operatorname{div}_S a), \tag{9.68}$$

which implies (3) by virtue of (1). $\qquad \square$

The jump property is also extended on Sobolev spaces. We define the operator $(\operatorname{grad} S_\lambda)^{(pv)}$ by

$$(\operatorname{grad} S_\lambda)^{(pv)} \varphi(x) = \text{p.v.} \int_S \operatorname{grad}_x g_\lambda(x - y)\varphi(y) dS_y. \tag{9.69}$$

Lemma 9.8. *Let $s \geq 0$, and ν_x denote the unit normal to S at $x \in S$.*
(1) For $\varphi \in H^s(S)$,

$$\lim_{t \downarrow 0} (\operatorname{grad} S_\lambda \varphi)(x \pm t\nu_x) = (\operatorname{grad} S_\lambda)^{(pv)} \varphi(x) \pm \frac{1}{2}\varphi(x)\nu_x, \quad in \quad H^s(S).$$
$$\tag{9.70}$$

(2) For $a \in H^s(S)$, we have

$$\lim_{t \downarrow 0} (G_C a)(x \pm t\nu_x) = G_C^{(pv)} a(x) \pm \frac{1}{2}\nu_x \times a(x), \quad in \quad H^s(S). \tag{9.71}$$

(3) For $a \in H^s_{tan}(\operatorname{curl}_S; S)$, we have

$$\lim_{t \downarrow 0} \left((N_\lambda a)(x + t\nu_x) - (N_\lambda a)(x - t\nu_x) \right) = 0, \quad in \quad H^s(S). \tag{9.72}$$

(4) For $a \in H^s_{tan}(\operatorname{div}_S; S)$, we have

$$\lim_{t \downarrow 0} \left((\nu \times G_D a)(x + t\nu_x) - (\nu \times G_D a)(x - t\nu_x) \right) = 0, \quad in \quad H^s(S). \tag{9.73}$$

Proof. By Corollary 9.1, the operator $T_t\varphi(x) = (\operatorname{grad} S_\lambda \varphi)(x \pm t\nu_x) - (\operatorname{grad} S_\lambda)^{(pv)} \varphi(x) \mp \frac{1}{2}\varphi(x)\nu_x$ is uniformly bounded in $H^s(S)$ for small t. Since (9.70) holds for $\varphi \in C^\infty(S)$ (see the formulas (3.1) $\sim$ (3.4) in [53]), the assertion (1) is proved. The assertions (2) and (3) then follow from (1). If a is tangential to S, we have

$$-N_\lambda(\nu \times a) = 2\nu \times G_D(a).$$

By virtue of (9.65), (4) follows from (3). $\qquad \square$

We need a weak form of the above lemma. For the sake of simplicity, we prove it when S is a sphere. However, the result is true for general compact surfaces.

Lemma 9.9. *Assume that* $S = \{|x| = r\}$, *and let* $s \geq 0$.
(1) The operator: $a \to (G_C a)(x + t\nu_x)$ *is bounded on* $H^{-s}(S)$ *uniformly for small* $|t|$, *and*

$$\lim_{t\downarrow 0} \left((G_C a)(x + t\nu_x) - (G_C a)(x - t\nu_x), b\right)_{L^2(S)} = (\nu \times a, b)_{L^2(S)}, \quad (9.74)$$

holds for any $a \in H^{-s}(S)$ *and* $b \in H^s(S)$.
(2) The operator: $a \to (\nu \times G_D a)(x + t\nu_x)$ *is bounded on* $H^{-s}_{tan}(\mathrm{curl}_S ; S)$ *uniformly for small* $|t|$, *and*

$$\lim_{t\downarrow 0} \left((\nu \times G_D a)(x + t\nu_x) - (\nu \times G_D a)(x - t\nu_x), b\right)_{L^2(S)} = 0 \quad (9.75)$$

holds for any $a \in H^{-s}_{tan}(\mathrm{curl}_S ; S)$ *and* $b \in H^s_{tan}(\mathrm{curl}_S ; S)$.

Proof. Since $g_\lambda(x - y) = g_0(x - y) + i\lambda + O(|x - y|)$, we have only to prove this lemma for $\lambda = 0$. First we prove

$$\left((G_C a)(x + t\nu_x), b(x)\right)_{L^2(S)} = \left(\frac{r}{r+t}\right)^2 \left(a(y), (G_C b)(y - \frac{rt}{r+t}\nu_y)\right)_{L^2(S)} \tag{9.76}$$

for $a, b \in C^\infty(S)$. In fact, since $\nu_x = x/r$, noting that

$$(G_C a)(x) = -\int_S \frac{x - y}{|x - y|^3} \times a(y) dS_y,$$

we have by a simple computation

$$(G_C a)(x + t\nu_x) = -\left(\frac{r}{r+t}\right)^2 \int_S \frac{x - \frac{r}{r+t}y}{|x - \frac{r}{r+t}y|^3} \times a(y) dS_y.$$

We then have

$$\left((G_C a)(x + t\nu_x), b(x)\right)_{L^2(S)}$$

$$= \left(\frac{r}{r+t}\right)^2 \iint_{S \times S} a(y) \cdot \left(\frac{x - \frac{r}{r+t}y}{|x - \frac{r}{r+t}y|^3} \times \overline{b(x)}\right) dS_x dS_y$$

$$= \left(\frac{r}{r+t}\right)^2 \left(a(y), (G_C b)(\frac{r}{r+t}y)\right)_{L^2(S)},$$

which implies (9.76). By Corollary 9.1, G_C is bounded on $H^s(S)$, moreover the operator $b \to (G_C b)(\frac{ry}{r+t})$ is uniformly bounded on $H^s(S)$ for small $|t|$. Therefore, the operator: $a \to (G_C a)(x + t\nu_x)$ is bounded on $H^{-s}(S)$ uniformly for small $|t|$. In view of (9.71), we have

$$\lim_{t\downarrow 0} \left((G_C a)(x \pm t\nu_x), b(x)\right)_{L^2(S)} = \left(a, G_C^{(pv)} b \mp \frac{1}{2}\nu \times b\right)_{L^2(S)},$$

which yields (9.74).

To prove (9.75), we show

$$\left(\nu \times G_D a(x + t\nu_x), b(x)\right)_{L^2(S)}$$

$$= -\left(\frac{r}{r+t}\right)^3 \left(a(y), (G_D(\nu \times b))(y - \frac{rt}{r+t}\nu_y)\right)_{L^2(S)} \tag{9.77}$$

for $a, b \in H^s_{tan}(\mathrm{curl}_S\,;\,S)$. Putting

$$H(x) = \left(\frac{\partial^2}{\partial x_i \partial x_j} g_0(x)\right),$$

we have

$$G_D a(x) = \int_S H(x - y) a(y) dS_y.$$

Since $H(x)$ is homogeneous of degree -3, we have

$$G_D a(x + t\nu_x) = \left(\frac{r}{r+t}\right)^3 \int_S H(x - \frac{r}{r+t}y) a(y) dS_y.$$

Using the formula, we can obtain (9.77) by a direct computation.

By the same argument as above, we can show that the operator: $a \to (\nu \times G_D a)(x + t\nu_x)$ is bounded on $H^{-s}_{tan}(\mathrm{curl}_S\,;\,S)$ uniformly for small $|t|$. The assertion (9.75) then follows from (9.73). □

We use the following notation

$$\nu \times g = (\nu \times g_E, \nu \times g_H), \quad \text{for} \quad g = (g_E, g_H) \in \mathbf{C}^3 \times \mathbf{C}^3.$$

By virtue of (9.40) and Lemma 9.9, we obtain the following lemma.

Lemma 9.10. *Let $s \geq 0$. Then, for any $f = (f_E, f_H)$ with $f_E, f_H \in H^{-s}_{tan}(\mathrm{curl}_S\,;\,S)$ and $h = (h_E, h_H)$ with $h_E, h_H \in H^s_{tan}(\mathrm{curl}_S\,;\,S)$, we have*

$$\lim_{t \downarrow 0} \left((\nu \times \mathcal{G}_0(\lambda)f)(x + t\nu_x) - (\nu \times \mathcal{G}_0(\lambda)f)(x - t\nu_x), h(x)\right)_{L^2(S)}$$

$$= \frac{i}{\epsilon_0}(f_H, h_E) - \frac{i}{\mu_0}(f_E, h_H).$$

9.3.3 *Perturbed layer potentials*

Now, we return to the case of the domain Ω in (C-1) and use the notation $S, \Omega_{ext}, \Omega_{int}$ in the meaning in (9.2), (9.3). As has been proved in Lemma 9.1, $R(\lambda \pm i0) \in \mathbf{B}(H^s_{comp}; H^s_{loc})$ if $-1 \leq s \leq 1$. For $f \in H^{-1/2}(S)$, we have $\delta_S f \in H^{-1}(\mathbf{R}^3)$. We thus define

$$\mathcal{G}(\lambda)f = R(\lambda + i0)\delta_S f, \quad f \in H^{-1/2}(S). \tag{9.78}$$

It satisfies

$$(\mathcal{H} - \lambda)\mathcal{G}(\lambda)f = \delta_S f. \tag{9.79}$$

In fact, take $\varphi \in C_0^\infty(\mathbf{R}^3)$. Then,

$$(\mathcal{G}(\lambda)f, (H - \lambda)\varphi) = (\delta_S f, R(\lambda - i0)(H - \lambda)\varphi) = (\delta_S f, \varphi),$$

which proves (9.79).

Let $\chi \in C^\infty(\mathbf{R}^3)$ be such that $\chi(x) = 0$ for $|x| < r_0 - \frac{2}{3}$, $\chi(x) = 1$ for $|x| > r_0 - \frac{1}{3}$, and define $Q_\pm(\lambda)$ by (8.43). Since $\mathcal{M} = \mathcal{M}_0$ for $|x| > r_0 - 1$, we have

$$Q_\pm(\lambda) = \chi + \mathcal{M}_0^{-1}\mathcal{L}(\nabla\chi)R(\lambda \pm i0). \tag{9.80}$$

In view of (8.44), we then have

$$\chi\mathcal{G}(\lambda) = \chi\mathcal{G}_0(\lambda) + \chi R_0(\lambda + i0)\mathcal{M}_0^{-1}\mathcal{L}(\nabla\chi)R(\lambda + i0)\delta_S. \tag{9.81}$$

For $f \in H^{-1/2}(S)$, let $w = (w_E, w_H) = R(\lambda \pm i0)\delta_S f$. Noting the equation $(\mathcal{H} - \lambda)w = \delta_S f$, we see that w is C^∞ on $\{r_0 - 1 < |x| < r_0\} \cup \{r_0 < |x|\}$, since $\text{div}\, w_E = \text{div}\, w_H = 0$, and $(-\Delta - \lambda^2)w_E = (-\Delta - \lambda^2)w_H = 0$ hold there. Therefore, $\mathcal{M}_0^{-1}\mathcal{L}(\nabla\chi)R(\lambda + i0)\delta_S f \in C_0^\infty(\mathbf{R}^3)$. This implies that $\mathcal{G}(\lambda)f$ has the same jump property as $\mathcal{G}_0(\lambda)f$.

Since $\delta_S f$ is a compactly supported distribution, its Fourier transform is C^∞. By (8.15), $\mathcal{F}_\pm^{(0)}(\lambda)$ is extended to compactly supported distributions. We thus see that

$$\mathcal{F}_\pm(\lambda)\delta_S = \mathcal{F}_\pm^{(0)}(\lambda)Q_\pm(\lambda)\delta_S \tag{9.82}$$

is well-defined on $H^{-1/2}(S)$. By virtue of (9.81), $\mathcal{F}_\pm(\lambda)\delta_S f$ is observed in the asymptotic expansion of $\mathcal{G}(\lambda)f$ as $|x| \to \infty$. Let us summarize what we have seen.

Lemma 9.11. *For $f \in H^{-1/2}(S)$, we have*

$$(\mathcal{H} - \lambda)\mathcal{G}(\lambda)f = \delta_S f, \tag{9.83}$$

and $\mathcal{G}(\lambda)f$ has the same jump property as $\mathcal{G}_0(\lambda)f$ at S. Moreover, it has the following asymptotic expansion:

$$\mathcal{G}(\lambda)f \simeq \sqrt{\frac{2\pi}{c_0}}\frac{e^{i\frac{\lambda}{c_0}r}}{r}(\mathcal{F}_+(\lambda)\delta_S f)(\hat{x}), \quad \hat{x} = x/r. \tag{9.84}$$

9.4   *E-M* map

The surjectivity of the trace is crucial for solving the boundary value problem, and it holds on Riemannian manifolds and also on Lipschitz domains.

Lemma 9.12. *Let Ω be a domain with boundary S. Then, there exists an operator of extension* $\text{Ext} \in \mathbf{B}(H_{tan}^{-1/2}(\text{div}_S; S); H(\text{curl}; \Omega))$ *such that*
$$\text{Ext} f = f \quad \text{on} \quad S, \quad \forall f \in H^{-1/2}(\text{div}_S; S).$$

See [34], [81], [5], [12] for the proof.

Let us consider the interior boundary value problem
$$\begin{cases} (\mathcal{H} - \lambda)u = 0 & \text{in} \quad \Omega_{int}, \\ \nu \times u_E = f_E \in H_{tan}^{-1/2}(\text{div}_S; S), \\ \nu \times u_E = 0 & \text{on} \quad S_{int}, \end{cases} \tag{9.85}$$
where $u = (u_E, u_H)$. By virtue of Lemma 9.12, there exists $\widetilde{f}_E \in H(\text{curl}; \Omega_{int})$ such that $\widetilde{f}_E = f_E$ on S and $\widetilde{f}_E = 0$ near S_{int}. We then seek the solution in the form $u = v + f$ with $f = (\widetilde{f}_E \times \nu, 0)$, and are led to the equation
$$\begin{cases} (\mathcal{H}_{int} - \lambda)v = -(\mathcal{H} - \lambda)f & \text{in} \quad \Omega_{int}, \\ \nu \times v_E = 0 & \text{on} \quad S \cup S_{int}, \end{cases} \tag{9.86}$$
where $\mathcal{H}_{int}$ is the Maxwell operator on Ω_{int} with boundary condition $\nu \times u_E = 0$ on $S \cup S_{int}$. By Theorem 7.12, if $\lambda \notin \sigma_p(\mathcal{H}_{int})$, which is a discrete set in $\mathbf{R}$, there exists a unique solution $v \in D(\mathcal{H}_{int})$ to (9.86), hence there is a unique solution u_{int} to (9.85) satisfying $\mathcal{H}u_{int} \in L^2(\Omega_{int})$. Then, $\text{div}_S (u_{int})_H \in H^{-1/2}(\text{div}_S; S)$. We define the *interior E-M (electromagnetic) map* by
$$\Lambda_{int}(\lambda) : H_{tan}^{-1/2}(\text{div}_S; S) \ni f_E \to \nu \times (u_{int})_H \in H_{tan}^{-1/2}(\text{div}_S; S). \tag{9.87}$$

The exterior boundary value problem
$$\begin{cases} (\mathcal{H} - \lambda)u = 0 & \text{in} \quad \Omega_{ext}, \\ \nu \times u_E = f_E \in H_{tan}^{-1/2}(\text{div}_S; S), \\ u \text{ satisfies the outgoing radiation condition,} \end{cases} \tag{9.88}$$
can be solved in the same way as above. We take an extension of $f_E \in H_{tan}^{-1/2}(\text{div}_S ; S)$ such that $\widetilde{f}_E$ is in $H(\text{curl}; \Omega_{ext})$ and compactly supported. Then, there is a unique solution u_{ext} to (9.86) with $\mathcal{H}_{int}$ and Ω_{int} replaced by $\mathcal{H}_{ext}$ and Ω_{ext} satisfyting the outgoing radiation condition. We define the *exterior E-M map* by
$$\Lambda_{ext}(\lambda) : H_{tan}^{-1/2}(\text{div}_S; S) \ni f_E \to \nu \times (u_{ext})_H \in H_{tan}^{-1/2}(\text{div}_S; S). \tag{9.89}$$

9.5 S-matrix determines E-M map

We put

$$\mathcal{F}^{(\pm)}(\lambda)^*\phi = (\mathcal{F}_E^{(\pm)}(\lambda)^*\phi, \mathcal{F}_H^{(\pm)}(\lambda)^*\phi), \quad \phi \in \mathbf{h}_\pm. \tag{9.90}$$

First let us note that

$$\mathcal{F}^{(\pm)}(\lambda)^*\phi \in C^\infty, \quad \text{for } |x| > r_0 - 1. \tag{9.91}$$

In fact, noting that $u = (u_E, u_H) = \mathcal{F}^{(\pm)}(\lambda)^*\phi$ satisfies $(\mathcal{H} - \lambda)u = 0$, and $\mathcal{E}(x) = \mathcal{E}_0$ for $|x| > r_0 - 1$, we have div $u_E =$ div $u_H = 0$ and $(-\Delta - \lambda^2)u_E = (-\Delta - \lambda^2)u_H = 0$ for $|x| > r_0 - 1$. This implies (9.91).

Lemma 9.13. *Assume that $\lambda \notin \sigma(\mathcal{H}_{int})$. If $f_E \in H_{tan}^{-1/2}(S)$ satisfies*

$$(f_E, \nu \times \mathcal{F}_E^{(+)}(\lambda)^*\phi)_{L^2(S)} = 0, \quad \forall\phi \in \mathbf{h}_+,$$

then $\nu \times f_E = 0$.

Proof. Let $f = (\nu \times f_E, 0)$ and put $v = (v_E, v_H) = R(\lambda + i0)\delta_S f = \mathcal{G}(\lambda)f$. Then, (9.84) implies

$$v \simeq \sqrt{2\pi}\frac{e^{i\lambda r}}{r}\mathcal{F}^{(+)}(\lambda)\delta_S f.$$

By the assumption of the lemma, we have

$$0 = (\nu \times f_E, \mathcal{F}_E^{(+)}(\lambda)^*\phi) = (\nu \times f_E, \mathcal{F}_E^{(+)}(\lambda)^*\phi) + (0, \mathcal{F}_H^{(+)}(\lambda)^*\phi)$$
$$= (f, \mathcal{F}^{(+)}(\lambda)^*\phi) = (\mathcal{F}^{(+)}(\lambda)\delta_S f, \phi).$$

Therefore $\mathcal{F}^{(+)}(\lambda)\delta_S f = 0$, hence $v \in \mathcal{B}_0^*$. Since v_E and v_H satisfy the Helmholtz equation outside S, the Rellich-Vekua theorem implies that $v = 0$ outside S. By Lemma 9.11, $\mathcal{G}(\lambda)f$ and $\mathcal{G}_0(\lambda)f$ have the same jump property. By Lemma 9.10, the tangential component of v_E is continuous across S. Hence v satisfies $(\mathcal{H} - \lambda)v = 0$ in Ω_{int} with boundary condition $\nu \times v_E = 0$ on S and also on S_{int}. Therefore, $v = 0$ in Ω_{int}. Lemma 9.10 shows that the jump across S of v_H is $-\frac{i}{\mu_0}\nu \times f_E$, hence it vanishes. This proves the lemma. $\qquad\square$

Suppose we have two Maxwell opearators $\mathcal{H}^{(i)}$ ($i = 1, 2$) satisfying the conditions (C-2), (C-3). Let $S^{(i)}(\lambda)$ and $\Lambda_{int}^{(i)}(\lambda)$ be the associated S-matrix and the interior E-M map.

Theorem 9.3. *Let $\lambda \in \mathbf{R} \setminus \left(\sigma(\mathcal{H}_{int}^{(1)}) \cup \sigma(\mathcal{H}_{int}^{(2)})\right)$. Then, $\widehat{S}^{(1)}(\lambda) = \widehat{S}^{(2)}(\lambda)$ if and only if $\Lambda_{int}^{(1)}(\lambda) = \Lambda_{int}^{(2)}(\lambda)$.*

Proof. Assume that $\Lambda_{int}^{(1)}(\lambda) = \Lambda_{int}^{(2)}(\lambda)$. Let $u^{(i)} = \mathcal{F}_{-}^{(i)}(\lambda)^*\phi$, where $\mathcal{F}_{-}^{(i)}(\lambda)$ is $\mathcal{F}^{(-)}(\lambda)$ for $\mathcal{H}^{(i)}$, and u_{int} be the solution to the interior problem

$$\begin{cases} (\mathcal{H}^{(2)} - \lambda)u_{int} = 0, & \text{in} \quad \Omega_{int}, \\ \nu \times u_{int,E} = \nu \times u_E^{(1)}, & \text{on} \quad S, \\ \nu \times u_{int,E} = 0, & \text{on} \quad S_{int}. \end{cases} \tag{9.92}$$

Note that

$$\nu \times u_E^{(1)} = \nu \times u_E^{(2)} = 0, \quad \text{on} \quad S_{int}. \tag{9.93}$$

We define $u^{(3)}$ on Ω by $u^{(3)} = u_{int}$ on Ω_{int} and $u^{(3)} = u^{(1)}$ on Ω_{ext}. The trace of $\nu \times u_H^{(3)}$ computed from outside of S is $\nu \times u_H^{(3)}\big|_S = \nu \times u_H^{(1)}\big|_S = \Lambda_{int}^{(1)}(\lambda)\nu \times u_E^{(1)}$, since $u^{(1)}$ satisfies $(\mathcal{H}^{(1)} - \lambda)u^{(1)} = 0$ in Ω, hence in Ω_{int}.

On the other hand, the trace computed from inside of S is

$$\nu \times u_H^{(3)}\big|_S = \Lambda_{int}^{(2)}(\lambda)\nu \times u_{int,E} = \Lambda_{int}^{(2)}(\lambda)\nu \times u_E^{(1)} = \Lambda_{int}^{(1)}(\lambda)\nu \times u_E^{(1)}.$$

Therefore, $\nu \times u^{(3)}$ is continuous across S. Hence $(\mathcal{H}^{(2)} - \lambda)u^{(3)} = 0$ on Ω.

Let $u^{(0)} = \mathcal{F}_{-}^{(0)}(\lambda)^*\phi$, where $\mathcal{F}_{-}^{(0)}(\lambda)$ is defined by (8.15). In view of (8.45), we see that $\mathcal{F}_{-}^{(i)}(\lambda)^*\phi - \mathcal{F}_{-}^{(0)}\phi$ satisfies the incoming radiation condition. Then $u^{(3)} - u^{(0)}$ satisfies the incoming radiation condition, since so does $u^{(1)} - u^{(0)}$. Therefore, $v = u^{(3)} - u^{(2)} = (u^{(3)} - u^{(0)}) - (u^{(2)} - u^{(0)})$ is the solution to the equation $(\mathcal{H}^{(2)} - \lambda)v = 0$ satisfying the radiation condition. Then by the Rellich-Vekua theorem, $v = 0$ outside S. Observing the behavior of $u^{(1)} = u^{(2)}$ near infinity (see Theorem 8.14), we get $\hat{S}^{(1)}(\lambda) = \hat{S}^{(2)}(\lambda)$.

Suppose $\hat{S}^{(1)}(\lambda) = \hat{S}^{(2)}(\lambda)$. Let $u^{(i)}$ be as above, and put $w = u^{(1)} - u^{(2)}$. Then, $(H^{(1)} - \lambda)w = 0$ in Ω_{ext}. Since $\hat{S}^{(1)}(\lambda) = \hat{S}^{(2)}(\lambda)$, $w \simeq 0$. Consequently, $w = 0$ outside S, therefore, $u_E^{(1)} = u_E^{(2)}$ and $u_H^{(1)} = u_H^{(2)}$ outside S, hence on S. We then have

$$\begin{aligned} \Lambda_{int}^{(1)}(\lambda)\nu \times \mathcal{F}^{(1)}(\lambda)^*\phi &= \Lambda_{int}^{(2)}(\lambda)\nu \times \mathcal{F}^{(2)}(\lambda)^*\phi \\ &= \lambda_{int}^{(2)}(\lambda)\nu \times \mathcal{F}^{(1)}(\lambda)^*\phi. \end{aligned} \tag{9.94}$$

By Lemma 9.13, $\Lambda_{int}^{(2)}(\lambda)\nu \times \mathcal{F}^{(1)}(\lambda)^*\mathbf{h}_-$ is dense in $H^{1/2}(S)$, which proves the theorem. $\square$

In the last step of the above proof, we have used the following Hahn-Banach theorem.

Theorem 9.4. *Let $\mathcal{X}_0$ be a subspace of a Banach space $\mathcal{X}$. Then, $\mathcal{X}_0$ is dense in $\mathcal{X}$ if and only if the following holds:*

$$f \in \mathcal{X}^*, \quad f = 0 \quad \text{on} \quad \mathcal{X}_0 \Longrightarrow f = 0 \quad \text{on} \quad \mathcal{X}.$$

Let us finally note that in the proof of Theorem 9.3, the assertion "$\widehat{S}^{(1)}(\lambda) = \widehat{S}^{(2)}(\lambda) \implies \Lambda^{(1)}(\lambda) = \Lambda^{(2)}(\lambda)$" is proven without using the individual structure of the interior domains. Therefore, it holds also for the case in which the topology of compact part of the domain is not assumed to coincide. This is important for the inverse scattering to be discussed in the next chapter.

Chapter 10

Inverse scattering

Although the S-matrix deals with the behavior of scattering solutions only at infinity, it contains an abundance of information of the physical system. In this chapter, we study various inverse scattering problems for the Maxwell operator $\mathcal{H} = \mathcal{M}(x)^{-1}\mathcal{L}(-i\nabla_x)$, the topics such as

- Recovery of scalar $\epsilon(x), \mu(x)$,
- Recovery of perturbation domain

$$D = \{x \,;\, \epsilon(x) \neq \epsilon_0, \ \mu(x) \neq \mu_0\},$$

- Recovery of Betti number of the boundary.

The methods are based on the ideas discovered in the study of inverse scattering or inverse boundary value problems for the Schrödinger equation

$$(-\Delta + V(x) - \lambda)u = 0, \tag{10.1}$$

or the acoustic equation

$$\left(-\sum_{i,j=1}^{n} \frac{\partial}{\partial x_i} a_{ij}(x) \frac{\partial}{\partial x_j} - \lambda n(x)\right)u = 0. \tag{10.2}$$

In particular, we review

- Faddeev's theory of inverse scattering,
- Sylvester-Uhlmann's exponentially growing solutions,
- Use of harmonic functions and reflection principle,
- Transmission eigenvalues,
- Linear sampling method,
- Boundary control method.

We focus on the exposition of wide overview of the theory and key facts, and omit the detailed computation.

10.1 Green operators and scattering amplitudes

The inverse scattering problems and inverse boundary value problems for the one-dimensional Schrödinger operator were solved around 1950s by the works of Marchenko [69], Gel'fand-Levitan [36], Levitan-Gasymov [66], Marchenko-Ostrovski [71]. There are also important papers which should be but cannot be cited here. See e.g. Chadan-Sabatier [21], Marchenko [70] and also a good review article Chadan-Colton-Pävärinta-Rundell [20]. The key idea for solving the one-dimensional inverse problem is the integral equation due to Gel'fand and Levitan. Faddeev developed a multi-dimensional analogue of this Gel'fand-Levitan theory from the middle of 1960s until the middle of 1970s ([30], [31], [32]), which elucidated a reconstruction procedure of the potential and the characterization of the S-matrix for the Schrödinger operator, although some parts remain formal. His main idea consists in the new Green function of the Helmholtz equation, which grows up exponentially in some direction and the introduction of the non-physical scattering amplitude constructed by this Green function. This exponentially growing solution of the Helmholtz equation also has a different origin. Calderón [17], independently of Faddeev's idea of inverse scattering, considered this solution to study linearized inverse boundary value problems. Sylvester and Uhlmann [89] implemented this machinery and solved the full inverse boundary value problem which includes e.g. the scalar electric conductivity problem appearing in electrical impedance tomography. The exponentially growing solution of the Schrödinger equation is now often called the complex geometrical optics solution. Faddeev's new Green function was also studied from the view point of complex analysis, and the $\bar{\partial}$-theories for the inverse boundary value problems and inverse scattering problems were developped by Nachman [75] and Novikov and Khenkin [79]. In this section, we briefly review Faddeev's Green function and use it to the inverse scattering and inverse boundary value problems for the Maxwell equation in the next section.

10.1.1 *Abstract theory*

Let $P(x, \partial_x) = \sum_{|\alpha| \leq m} a_\alpha(x) \partial_x^\alpha$ be a differential operator on $\mathbf{R}^n$. By definition, the Green function of $P(x, \partial_x)$ is a solution $g(x, y)$ to the equation

$$P(x, \partial_x)g(x, y) = \delta(x - y),$$

and the Green operator is an integral operator with kernel $g(x, y)$. The Green function is not unique, since $g(x, y) + w(x)$, where $w(x)$ is any solution to the equation $P(x, \partial_x)w(x) = 0$, again becomes a Green function. In physical problems, the Green function is specified by the boundary condition or the behavior at infinity when the domain is unbounded. In Chapter 8, we have used the standard Green operators $(\mathcal{H}_0 - \lambda \mp i0)^{-1}$ and $(\mathcal{H} - \lambda \mp i0)^{-1}$ to derive the physical scattering amplitude. However, adopting different Green operators, we can argue in the same way as in the previous chapters to obtain different, non-physical, scattering amplitudes. As is shown below, there is always a simple relation between these physical scattering amplitudes and non-physical ones.

Let us consider a model case. Let H_0 be a self-adjoint operator in a Hilbert space $\mathbf{H}$ with essential spectrum $\sigma_e(H_0)$. Suppose there are Banach spaces $\mathbf{X}_\pm$ such that

$$\mathbf{X}_+ \subset \mathbf{H} \subset \mathbf{X}_-,$$

and the inclusions $\mathbf{X}_+ \subset \mathbf{H}$, $\mathbf{H} \subset \mathbf{X}_-$ are dense and continuous. We also assume that the inner product $(,)$ of $\mathbf{H}$ is extended to a sesqui-linear map $\mathbf{X}_+ \times \mathbf{X}_- \to \mathbf{C}$ satisfying

$$|(f, u)| \leq C\|f\|_{\mathbf{X}_+}\|u\|_{\mathbf{X}_-}.$$

Let I be a compact set in $\sigma_e(\mathcal{H}_0)$, and assume that for $\lambda \in I$, there exists a strong limit

$$R_0(\lambda \pm i0) = \mathrm{s} - \lim_{\epsilon \to 0} (H_0 - \lambda \mp i\epsilon)^{-1} \in \mathbf{B}(\mathbf{X}_+; \mathbf{X}_-), \qquad (10.3)$$

and $R_0(\lambda \pm i0)$ is a $\mathbf{B}(\mathbf{X}_+; \mathbf{X}_-)$-valued continuous function of $\lambda \in I$. We also assume that there is an auxiliary Hilbert space $\mathbf{h}$ such that for any $\lambda \in I$, there exists an operator $\mathcal{F}_0(\lambda) \in \mathbf{B}(\mathbf{X}_+; \mathbf{h})$ satisfying

$$\mathcal{F}_0(\lambda)H_0 u = \lambda \mathcal{F}_0(\lambda)u, \quad u \in D(H_0) \cap \mathbf{H}_+.$$

We finally assume that the operator

$$(\mathcal{F}_0 u)(\lambda) = \mathcal{F}_0(\lambda)u, \quad \lambda \in I$$

defined on $\mathbf{X}_+$ is uniquely extended to a unitary operator from $E_0(I)\mathbf{H}$ to $L^2(I, \mathbf{h}; d\lambda)$, where $E_0(\cdot)$ is the spectral decomposition of H_0. Let V be a bounded self-adjoint operator on $\mathbf{H}$ such that $V \in \mathbf{B}(\mathbf{X}_-; \mathbf{X}_+)$, and let $H = H_0 + V$.

Let $\lambda \in I$. An operator $G^{(0)} \in \mathbf{B}(\mathbf{X}_+; \mathbf{X}_-)$ is said to be a *Green operator* of $H_0 - \lambda$ if

$$(H_0 - \lambda)G^{(0)} = I_d, \quad \text{on} \quad \mathbf{X}_+, \qquad (10.4)$$

I_d being the identity operator on $\mathbf{X}_+$. An operator $G \in \mathbf{B}(\mathbf{X}_+; \mathbf{X}_-)$ is called a *perturbed Green operator* associated with $G^{(0)}$ if it satisfies

$$G = G^{(0)} - G^{(0)}VG = G^{(0)} - GVG^{(0)}. \tag{10.5}$$

By (10.4) and (10.5), it satisfies $(H - \lambda)G = I_d$. The *scattering amplitude* associated with G is then defined by

$$A = \mathcal{F}_0(\lambda)(V - VGV)\mathcal{F}_0(\lambda)^*. \tag{10.6}$$

Suppose we are given two Green operators $G_1^{(0)}, G_2^{(0)}$ for $H_0 - \lambda$. Let $G^{(1)}$, $G^{(2)}$ be the associated perturbed Green operators, and A_1, A_2 the scattering amplitudes associated with $G^{(1)}, G^{(2)}$. One can then prove the following theorems (see [46]).

Theorem 10.1. *Suppose there exists $M \in \mathbf{B}(\mathbf{h}; \mathbf{h})$ such that*

$$G_2^{(0)} - G_1^{(1)} = \mathcal{F}_0(\lambda)^* M \mathcal{F}_0(\lambda). \tag{10.7}$$

Then we have

$$A_2 = A_1 - A_1 M A_2. \tag{10.8}$$

We add a new assumption.

$$G_1^{(0)}V, G_2^{(0)}V : \mathbf{X}_- \to \mathbf{X}_- \text{ are compact,}$$
$$A_1 : \mathbf{h} \to \mathbf{h} \text{ is compact.}$$

Theorem 10.2. *Under the above assumptions, equation (10.8) is uniquely solvable with respect to A_2:*

$$A_2 = (1 + A_1 M)^{-1} A_1.$$

Theorem 10.2 asserts that, by changing the Green operator, we can get a linear equation between scattering amplitudes, which is always solvable. Using this fact, Faddeev ([30], [32]) found a breakthrough of the inverse scattering theory for the Schrödinger operator $-\Delta + V(x)$ in $\mathbf{R}^3$. The idea consists in introducing a new Green operator of $-\Delta - \lambda$ depending on an arbitrarily chosen direction $\gamma \in S^2$.

10.1.2   Faddeev's Green operator

We first recall the case of $-\Delta$. Let $r_0(z) = (-\Delta - z)^{-1}$ be the resolvent of $-\Delta$ in $\mathbf{R}^3$. For $\lambda > 0$, it satisfies

$$(-\Delta - \lambda)r_0(\lambda \pm i0) = I.$$

Therefore, any convex combination of $r_0(\lambda+i0)$ and $r_0(\lambda-i0)$ has the same property. Take $\gamma \in S^2$, $t \in \mathbf{R}$ arbitrarily, and define

$$m_\gamma^{(\pm)}(t) = (F_{x\to\xi})^{-1}\chi(\pm\gamma \cdot (\xi - t\gamma) \geq 0)F_{x\to\xi}, \qquad (10.9)$$

where $\chi(\pm\gamma \cdot (\xi - t\gamma) \geq 0)$ is the characteristic function of the set $\{\xi \in \mathbf{R}^3\,;\, \pm\gamma \cdot (\xi - t\gamma) \geq 0\}$. For $t \in \mathbf{R}$ and $k > 0$, we define the operator $r_0(k,t) \in \mathbf{B}(\mathcal{H}_+; \mathcal{H}_-)$ by

$$r_{\gamma,0}(k,t) = r_0(\lambda - i0)m_\gamma^{(+)}(t) + r_0(\lambda + i0)m_\gamma^{(-)}(t), \qquad (10.10)$$

where $\lambda = k^2 + t^2$. This is the *Green operator of Faddeev*. Passing to the Fourier transform, it is outgoing in the half-space $\{\gamma \cdot \xi \leq t\}$, and incoming in the other half-space $\{\gamma \cdot \xi \geq t\}$. Note that by the formula (10.10), $r_{\gamma,0}(k,t)$ has the following formal expression

$$r_{\gamma,0}(k,t)f = (2\pi)^{-3/2} \int_{\mathbf{R}^3} \frac{e^{ix\cdot\xi}}{\xi^2 + 2i0\gamma \cdot (\xi - t\gamma) - \lambda}\widehat{f}(\xi)\,d\xi.$$

The formula (10.10) also implies

$$r_{\gamma,0}(k,t) = r_0(\lambda + i0) - 2\pi i F_0(\sqrt{\lambda})^* m_\gamma(\lambda,t)F_0(\sqrt{\lambda}), \qquad (10.11)$$

where $m_\gamma(\lambda,t)$ is the operator of multiplication by the function $\chi(\gamma\cdot(\sqrt{\lambda}\omega - t\gamma) \geq 0)$ on $L^2(S^2)$,

$$F_0(k)f(\omega) = k\widehat{f}(k\omega),$$

and we have used the fact that (see Lemma 8.10)

$$\frac{1}{2\pi i}\left(r_0(\lambda + i0) - r_0(\lambda - i0)\right) = F_0(\sqrt{\lambda})^* F_0(\sqrt{\lambda}). \qquad (10.12)$$

To construct $r_{\gamma,0}(k,t)$, it is convenient to start from the operator

$$\left(g_{\gamma,0}(k,z)f\right)(x) = (2\pi)^{-3/2} \int_{\mathbf{R}^3} \frac{e^{ix\cdot\xi}}{\xi^2 + 2z\gamma \cdot \xi - k^2}\widehat{f}(\xi)\,d\xi. \qquad (10.13)$$

If $\mathrm{Im}\, z > 0$ and $k > 0$, one can make the change of variables $y_1 = \xi^2 + 2\mathrm{Re}\,z\,\gamma \cdot \xi - k^2$, $y_2 = 2\mathrm{Im}\,z\,\gamma \cdot \xi$ to see that

$$\frac{1}{\xi^2 + 2z\gamma \cdot \xi - k^2} \in L^1_{loc}(\mathbf{R}^3).$$

Therefore, the integral in (10.13) is absolutely convergent for $f \in \mathcal{S}(\mathbf{R}^3)$.

The operator (10.13) was also used by Sylvester-Uhlmann [89] to show the uniqueness in the inverse boundary value problem for electric conductivity.

The basic properties of Faddeev's Green operator are summarized in the following lemma (see e.g. Weder [96] for the proof).

Lemma 10.1. *Let $s > 1/2$, and $\mathbf{C}_+ = \{z \in \mathbf{C}\,;\, \mathrm{Im}\, z > 0\}$.*
(1) As a $\mathbf{B}(L^{2,s}; L^{2,-s})$-valued function, $g_{\gamma,0}(k, z)$ is continuous with respect to $k \geq 0, \gamma \in S^2$, $z \in \overline{\mathbf{C}_+}$ except for $(k, z) = (0, 0)$.
(2) $g_{\gamma,0}(k, z)$ is a $\mathbf{B}(L^{2,s}; L^{2,-s})$-valued analytic function of $z \in \mathbf{C}_+$.
(3) For any $\epsilon_0 > 0$, there exists $C > 0$ such that for $k + |z| \geq \epsilon_0$, and $|\alpha| \leq 2$,

$$\|\partial_x^\alpha g_{\gamma,0}(k, z)\|_{\mathbf{B}(L^{2,s}; L^{2,-s})} \leq C(k + |z|)^{-1+|\alpha|}.$$

(4) For $t \in \mathbf{R}$, let $r_{\gamma,0}(k, t) = e^{it\gamma \cdot x} g_{\gamma,0}(k, t) e^{-it\gamma \cdot x}$. Then

$$(-\Delta - k^2 - t^2)r_{\gamma,0}(k, t) = I.$$

The operator $r_{\gamma,0}(k, t)$ defined in (4) coincides with Faddeev's Green operator. In fact, one can show that $r_{\gamma,0}(k, t)$ is outgoing for $\gamma \cdot \xi < t$, and incoming for $\gamma \cdot \xi > t$, hence satisfies (10.10). See [44] for the details.

There is a subtle problem in the analytic continuation of Faddeev's Green operator. In the application to inverse scattering, k corresponds to the energy. We represent it as $k = \sqrt{\lambda - t^2}$ and consider the analytic continuation with respect to t. The integral kernel of $g_{\gamma,0}(k, t) = g_{\gamma,0}(\sqrt{\lambda - t^2}, t)$ is then formally written as

$$(2\pi)^{-3} \int_{\mathbf{R}^3} \frac{e^{i(x-y)\cdot\xi}}{\xi^2 + 2(t + i0)\gamma \cdot \xi + t^2 - \lambda} d\xi. \tag{10.14}$$

Apparently, its analytic continuation with respect to t seems to be

$$(2\pi)^{-3} \int_{\mathbf{R}^3} \frac{e^{i(x-y)\cdot\xi}}{\xi^2 + 2z\gamma \cdot \xi + z^2 - \lambda} d\xi.$$

However, this is not analytic in z. In fact, one can prove the following lemma by a direct computation.

Lemma 10.2. *Let D be an open set in $\mathbf{C}$. Let $p(\xi, z)$ be a $\mathbf{C}$-valued function which is smooth in $\xi \in \mathbf{R}^n$ and analytic in $z \in D$. Let $M_z = \{\xi \in \mathbf{R}^n\,;\, p(\xi, z) = 0\}$ and assume that for $z \in D$, $\nabla_\xi \mathrm{Re}\, p(\xi, z)$ and $\nabla_\xi \mathrm{Im}\, p(\xi, z)$ are linearly independent on M_z. Then, the distribution $S(z)$ defined by*

$$S(z)f = \int_{\mathbf{R}^n} \frac{f(\xi)}{p(\xi, z)} d\xi, \quad f \in C_0^\infty(\mathbf{R}^n)$$

satisfies

$$\overline{\partial_z} S(z)f = \pi \int_{\mathbf{R}^n} f(\xi) \overline{\partial_z p(\xi, z)} \delta(p(\xi, z)) d\xi,$$

where

$$\int_{\mathbf{R}^n} g(\xi) \delta(p(\xi, z)) d\xi = \int_{M_z} g(\xi) dM_z,$$

and dM_z is the induced measure on M_z.

10.1.3 Green operator of Eskin-Raston

The analytic continuation of (10.14) was found by Eskin-Ralston [28] in the ξ-space representation, and then transferred to the x-space representation by [45] with an improvement of estimates. For $c \in \mathbf{R}$, let

$$L_{e^{cr}}^2 = \{f \, ; \, \int_{\mathbf{R}^3} e^{2c|x|} |f(x)|^2 dx < \infty\}. \tag{10.15}$$

Lemma 10.3. *Given any* $\lambda > 0$ *and* $\delta > 0$, *there exist* $\epsilon > 0$ *and an operator* $u_{\gamma,0}(\lambda, z)$ *defined for* $z \in D_\epsilon = \{z \in \mathbf{C}_+ \, ; \, |\mathrm{Re}\, z| < \epsilon/2\}$ *having the following properties.*

(1) $u_{\gamma,0}(\lambda, z)$ *is a* $\mathbf{B}(L_{e^{\delta r}}^2; L_{e^{-\delta r}}^2)$-*valued analytic function with respect to* $z \in D_\epsilon$.

(2) It satisfies

$$(-\Delta - 2iz\gamma \cdot \nabla + z^2 - \lambda)u_{\gamma,0}(\lambda, z) = I.$$

(3) $u_{\gamma,0}(\lambda, z)$ *has a continuous boundary value as* $z \to t \in (\epsilon/2, \epsilon/2)$, *and for* $t \in (-\epsilon/2, \epsilon/2)$

$$u_{\gamma,0}(\lambda, t) = g_{\gamma,0}(\sqrt{\lambda - t^2}, t).$$

(4) For $\tau > 0$

$$u_{\gamma,0}(\lambda, i\tau) = g_{\gamma,0}(\sqrt{\lambda + \tau^2}, i\tau).$$

(5) For $0 < s < 1$ *and* $|\alpha| \le 2$,

$$\|\partial_x^\alpha u_{\gamma,0}(\lambda, i\tau)\|_{\mathbf{B}(L^{2,s}; L^{2,s-1})} \le C_s \tau^{-1+|\alpha|}, \quad \tau > 1.$$

In particular, by (3) we have

$$e^{it\gamma \cdot x} u_{\gamma,0}(\lambda, t) e^{-it\gamma \cdot x} = r_{\gamma,0}(\sqrt{\lambda - t^2}, t). \tag{10.16}$$

Lemma 10.3 (2) shows that $u_{\gamma,0}(\lambda, z)$ is a right inverse of $-\Delta - 2iz\gamma \cdot \nabla + z^2 - \lambda$. The following lemma shows that it is also a left inverse when $z = i\tau, \tau > 0$.

Lemma 10.4. *Let* $0 < s < 1$ *and* $\tau > 0$. *Suppose* $u \in L^{2,s-1}$ *satisfies* $(-\Delta + 2\tau\gamma \cdot \nabla - \tau^2 - \lambda)u \in L^{2,s}$. *Then*

$$u_{\gamma,0}(\lambda, i\tau)(-\Delta + 2\tau\gamma \cdot \nabla - \tau^2 - \lambda)u = u.$$

Proof. We put $w = u_{\gamma,0}(\lambda, i\tau)(-\Delta + 2\tau\gamma \cdot \nabla - \tau^2 - \lambda)u - u$. Since $(-\Delta + 2\tau\gamma \cdot \nabla - \tau^2 - \lambda)w = 0$, we have

$$(\xi^2 + 2i\tau\gamma \cdot \xi - \tau^2 - \lambda)\widehat{w} = 0.$$

Then $\operatorname{supp}\widehat{w}$ is contained in a circle $C_\tau = \{|\xi|^2 = \tau^2 + \lambda\} \cap \{\gamma \cdot \xi = 0\}$, which is a submanifold of codimension 2 in $\mathbf{R}^3$. By the theorem of Agmon-Hörmander ([4] Th 2.2, or [42], Vol. 1, Th. 7.1.27), $\widehat{w}$ is an L^2-function on C_τ, and

$$\int_{C_\tau} |\widehat{w}|^2 dC_\tau \leq C \limsup_{R \to \infty} R^{-2} \int_{|x| < R} |w(x)|^2 dx.$$

Since $w \in L^{2,s-1}$, and $0 < s < 1$, the right-hand side vanishes. This proves the lemma. □

Henceforth, the standard Green operator will be called *physical* Green operator, and the Green operators of Faddeev or Eskin-Ralston are called *direction dependent* Green operators.

By virtue of Lemma 10.3 (3) and (4), $g_{\gamma,0}(\sqrt{\lambda - t^2}, t)$ for t in an open set uniquely determines $g_{\gamma,0}(\sqrt{\lambda + \tau^2}, i\tau)$ for all $\tau > 0$ because of the analyticity of $u_{\gamma,0}(\lambda, z)$. This analytic continuation procedure and the *Faddeev scattering amplitude* constructed in Theorem 10.2 constitute the main ideas of the inverse scattering theory. The main aim of Faddeev [32] is to establish a multi-dimensional analogue of Gel'fand-Levitan theory and to solve the characterization problem of the S-matrix. The following theorem is a remarkable byproduct of Faddeev's Green function.

Theorem 10.3. *An exponentially decreasing potential of the Schrödinger operator* $-\Delta + V(x)$ *is uniquely reconstructed from the S-matrix of one fixed energy.*

In fact, by using direction depedent Green operators, one can construct the Faddeev scattering amplitude $A_F(\lambda, z)$ from the physical scattering amplitude. The potential $V(x)$ is then reconstructed by observing the behavior of $A_F(\lambda, i\tau)$ as $\tau \to \infty$. A remarkable fact in this method is that by passing to a suitable complex manifold, one can obtain a representation formula of the potential in terms of the Faddeev scattering amplitude. By this reason, it is often called a $\overline{\partial}$-*approach* to inverse scattering. See [75], [79], [78].

10.1.4 *Inverse boundary value problem*

The inverse scattering problem can be reduced to the inverse boundary value problem. Suppose that the potential is compactly supported. We take a bounded domain Ω_{int} containing supp V and let $\Omega_{ext} = \mathbf{R}^3 \setminus \overline{\Omega_{int}}$. Solving the Dirichlet problem for $-\Delta + V(x)$ in Ω_{int}, we define the Dirichlet-to-Neumann map, D-N map for short, to be the mapping assigning the Neumann data to the Dirichlet data of the solution. Similary to Theorem 9.3, the S-matrix $S(\lambda)$ determines the D-N map if λ is not a Dirichlet eigenvalue in Ω_{int}. To reconstruct $V(x)$ from the D-N map, we need to use the boundary data which is determined only by the S-matrix.

Let us note that the physical generalized eigenfunction for $-\Delta + V(x)$ is constructed as

$$\varphi(x, \lambda, \omega) = e^{i\sqrt{\lambda}\omega \cdot x} - r(\lambda + i0)(V(x)e^{i\sqrt{\lambda}\omega \cdot x}), \qquad (10.17)$$

where $r(z) = (-\Delta + V - z)^{-1}$ is the perturbed physical Green operator obtained by solving the resolvent equation

$$r(z) = r_0(z) - r_0(z)Vr(z).$$

The physical scattering amplitude appears in the asymptotic expansion of the resolvent term:

$$\varphi(x, \lambda, \omega) - e^{i\sqrt{\lambda}\omega \cdot x} \simeq C(\lambda)\frac{e^{i\sqrt{\lambda}r}}{r}A(\lambda, \theta, \omega), \qquad \theta = x/r. \qquad (10.18)$$

Similarly, the perturbed Green operator of Eskin-Ralston is constructed by solving the resolvent equation

$$u_\gamma(\lambda, t) = u_{\gamma,0}(\lambda, t) - u_{\gamma,0}(\lambda, z)Vu_\gamma(\lambda, t). \qquad (10.19)$$

Letting $g_\gamma(\lambda, z) = e^{-it\gamma \cdot x}u_\gamma(\lambda, t)e^{it\gamma \cdot x}$, we can define a non-physical generalized eigenfunction

$$\psi(x, \lambda, t, \omega) = e^{i\sqrt{\lambda}\omega \cdot x} - g_\gamma(\lambda, t)(V(x)e^{i\sqrt{\lambda}\omega \cdot x}). \qquad (10.20)$$

By virtue of (10.11), it has the asymptotic expansion

$$\psi(x, \lambda, t, \omega) - e^{i\sqrt{\lambda}\omega \cdot x} \simeq C(\lambda)\frac{e^{i\sqrt{\lambda}\omega \cdot x}}{r}a_+(\lambda, t, \theta, \omega)$$
$$+ C(\lambda)\frac{e^{-i\sqrt{\lambda}\omega \cdot x}}{r}a_-(\lambda, t, \theta, \omega), \qquad (10.21)$$

where

$$a_\pm(\lambda, t, \theta, \omega) = m_\gamma^{(\mp)}(t)\big(1 - Vu_\gamma(\lambda, t)\big)\left(V(x)e^{i\sqrt{\lambda}\omega \cdot x}\right).$$

Here, we use Theorem C.3. Observing each term of the right-hand side of (10.21), we obatin the Faddeev scattering amplitude $A_F(\lambda, t, \theta, \omega)$.

Now, suppose for two compactly supported potentials $V_1(x)$ and $V_2(x)$, we have the same physical scattering amplitude. Then, their Faddeev scattering amplitudes coincide. Let $\phi(x) = \psi_1(x, \lambda, t, \omega) - \psi_2(x, \lambda, t, \omega)$ be the difference of the associated non-physical generalized eigenfunctions. Then,

$$(-\Delta - \lambda)\phi(x) = 0, \quad \text{in} \quad \Omega_{ext}, \tag{10.22}$$

and also by (10.21)

$$\frac{1}{R}\int_{|x|<R}|\phi(x)|^2 dx \to 0.$$

By Rellich's theorem, $\phi(x) = 0$ in Ω_{ext}. Since $\psi(x, \lambda, t, \omega)$ is constructed by Eskin-Ralston's Green operator, it has the analytic continuation with respect t. We have thus seen the following lemma.

Lemma 10.5. *The physical S-matrix $S(\lambda)$ determines the non-physical generalized eigenfunction $\psi(x, \lambda, z, \omega)$ in Ω_{ext}, in particular its boundary value on $\partial\Omega_{ext}$.*

The function $\psi(x, \lambda, z, \omega)$ for complex z is then used as the boundary data for the D-N map, which makes it possible to reconstruct the potential from the D-N map. See [75], [78].

10.2 Direction dependent Green operators for the free Maxwell equation

We apply the above procedure to the Maxwell equation. Ola, Päivärinta and Somersalo [80] solved the inverse boundary value problem for the Maxwell equation by using the Faddeev Green function. Therefore, we can solve the inverse scattering problem by reducing it to the inverse boundary value problem. We need to start from the construction of Eskin-Ralston Green operator for the free Maxwell equation. We put

$$\mathcal{H}_0(\xi) = \mathcal{M}_0^{-1}\mathcal{L}(\xi) = \begin{pmatrix} 0 & -\epsilon_0^{-1}\xi\times \\ \mu_0^{-1}\xi\times & 0 \end{pmatrix}. \tag{10.23}$$

Lemma 8.3 shows that $(\mathcal{H}_0 - \lambda \mp i0)^{-1}$ has the form

$$(F_{x\to\xi})^{-1}\left((\xi^2 - (\lambda \pm i0)^2)^{-1}(\mathcal{H}_0(\xi) + \lambda)\mathcal{P}^{(0)}(\xi) - \frac{1}{\lambda}\mathcal{P}_0^{(0)}(\xi)\right)F_{x\to\xi}.$$

Note that $\mathcal{P}_0^{(0)}(\xi)$ is rewritten as

$$\mathcal{P}_0^{(0)}(\xi) = |\xi|^{-2}\begin{pmatrix} |\xi\rangle\langle\xi| & 0 \\ 0 & |\xi\rangle\langle\xi| \end{pmatrix}, \tag{10.24}$$

where $|\xi\rangle\langle\xi|$ is the 3×3 matrix defined by

$$|\xi\rangle\langle\xi| = (\xi_i\xi_j).$$

Therefore, $(\mathcal{H}_0 - \lambda \mp i0)^{-1}$ has the form

$$(F_{x\to\xi})^{-1}(\xi^2 - (\lambda \pm i0)^2)^{-1}\left(\mathcal{H}_0(\xi) + \lambda - \frac{1}{\lambda}\mathcal{K}_0(\xi)\right)F_{x\to\xi}, \qquad (10.25)$$

where

$$\mathcal{K}_0(\xi) = \begin{pmatrix} |\xi\rangle\langle\xi| & 0 \\ 0 & |\xi\rangle\langle\xi| \end{pmatrix}. \qquad (10.26)$$

We have thus arrived at the *formal* formula

$$e^{-it\gamma\cdot x}(\mathcal{H}_0 - \lambda \mp i0)^{-1}e^{it\gamma\cdot x}$$

$$= (F_{x\to\xi})^{-1}\left((\xi + t\gamma)^2 - (\lambda \pm i0)^2\right)^{-1}\left(\mathcal{H}_0(\xi + t\gamma) + \lambda - \frac{1}{\lambda}\mathcal{K}_0(\xi + t\gamma)\right)F_{x\to\xi}.$$

For the case of $-\Delta$, the formal formula

$$(F_{x\to\xi})^{-1}\left((\xi + t\gamma)^2 - (\lambda \pm i0)^2\right)^{-1}F_{x\to\xi}$$

is realized as a direction dependent Green operator. It suggests us to define the *Green operator of Eskin-Ralston for the Maxwell equation* by

$$\mathcal{U}_{\gamma,0}(\lambda, z) = u_{\gamma,0}(\lambda^2, z)\mathcal{V}_{\gamma,0}(\lambda, z), \qquad (10.27)$$

where $u_{\gamma,0}(\lambda, z)$ is introduced in Lemma 10.3, and

$$\mathcal{V}_{\gamma,0}(\lambda, z) = (F_{x\to\xi})^{-1}\left(\mathcal{H}_0(\xi + z\gamma) + \lambda - \frac{1}{\lambda}\mathcal{K}_0(\xi + z\gamma)\right)F_{x\to\xi}. \qquad (10.28)$$

Let us introduce the notation

$$\mathrm{Div}_{z\gamma} = (F_{x\to\xi})^{-1}\langle\xi + z\gamma|\, F_{x\to\xi} = -ie^{-iz\gamma\cdot x}\,\mathrm{div}\,e^{iz\gamma\cdot x}, \qquad (10.29)$$

$$\mathrm{Grad}_{z\gamma} = (F_{x\to\xi})^{-1}\,|\xi + z\gamma\rangle\, F_{x\to\xi} = -ie^{-iz\gamma\cdot x}\,\mathrm{grad}\,e^{iz\gamma\cdot x}, \qquad (10.30)$$

$$\mathrm{Curl}_{z\gamma} = (F_{x\to\xi})^{-1}(\xi + z\gamma) \times\, F_{x\to\xi} = -ie^{-iz\gamma\cdot x}\,\mathrm{curl}\,e^{iz\gamma\cdot x}. \qquad (10.31)$$

Lemma 10.6. *We have*

$$\left(\mathcal{H}_0(-i\nabla_x + z\gamma) - \lambda\right)\mathcal{U}_{\gamma,0}(\lambda, z) = I. \qquad (10.32)$$

Moreover, letting $\mathcal{U}_{\gamma,0}(\lambda, z) = (\mathcal{U}^E_{\gamma,0}(\lambda, z), \mathcal{U}^H_{\gamma,0}(\lambda, z))$, *we also have*

$$\mathrm{Div}_{z\gamma}\,\mathcal{U}^E_{\gamma,0}(\lambda, z) = -\frac{1}{\lambda}\mathrm{Div}_{z\gamma}\,I_E, \qquad (10.33)$$

$$\mathrm{Div}_{z\gamma}\,\mathcal{U}^H_{\gamma,0}(\lambda, z) = -\frac{1}{\lambda}\mathrm{Div}_{z\gamma}\,I_H, \qquad (10.34)$$

where I_E, I_H *are the projections:*

$$I_E : f = (f_E, f_H) \to f_E, \quad I_H : f = (f_E, f_H) \to f_H.$$

Proof.   Letting $f = (f_E, f_H)$ and $u = \mathcal{U}_{\gamma,0}(\lambda, z)f = (u_E, u_H)$, we have

$$u_E = u_{\gamma,0}(\lambda^2, z)\Big(-\epsilon_0^{-1}\,\mathrm{Curl}_{z\gamma}f_H + \lambda f_E - \frac{1}{\lambda}\,\mathrm{Grad}_{z\gamma}\,\mathrm{Div}_{z\gamma}\,f_E\Big), \quad (10.35)$$

$$u_H = u_{\gamma,0}(\lambda^2, z)\Big(\mu_0^{-1}\,\mathrm{Curl}_{z\gamma}f_E + \lambda f_H - \frac{1}{\lambda}\,\mathrm{Grad}_{z\gamma}\,\mathrm{Div}_{z\gamma}\,f_H\Big). \quad (10.36)$$

Using the formulas

$$\begin{aligned}
\mathrm{Div}_{z\gamma}\mathrm{Curl}_{z\gamma} &= 0, \\
\mathrm{Div}_{z\gamma}\mathrm{Grad}_{z\gamma} &= (-i\nabla_x + z\gamma)^2, \\
\mathrm{Curl}_{z\gamma}\mathrm{Curl}_{z\gamma} &= \mathrm{Grad}_{z\gamma}\mathrm{Div}_{z\gamma} - (-i\nabla_x + z\gamma)^2,
\end{aligned} \quad (10.37)$$

we have

$$\mathrm{Div}_{z\gamma}u_E = -\frac{1}{\lambda}\mathrm{Div}_{z\gamma}f_E, \quad \mathrm{Div}_{z\gamma}u_H = -\frac{1}{\lambda}\mathrm{Div}_{z\gamma}f_H,$$

$$-\epsilon_0^{-1}\mathrm{Curl}_{z\gamma}u_H = \lambda u_E + f_E, \quad \mu_0^{-1}\mathrm{Curl}_{z\gamma}u_E = \lambda u_H + f_H,$$

which prove the lemma.   □

We define *Faddeev's Green operator for the free Maxwell equation* by

$$R_{\gamma,0}(\lambda, t) = e^{it\gamma \cdot x}\mathcal{U}_{\gamma,0}(\lambda, t)e^{-it\gamma \cdot x}. \quad (10.38)$$

It satisfies

$$\big(\mathcal{H}_0 - \lambda\big)R_{\gamma,0}(\lambda, t) = I. \quad (10.39)$$

Moreover, we have by (10.11),

$$R_{\gamma,0}(t) = R_0(\lambda + i0) - 2\pi i\,\mathcal{F}_0(\lambda)^* M_\gamma(\lambda, t)\mathcal{F}_0(\lambda), \quad (10.40)$$

where $M_\gamma(\lambda, t)$ is the operator of multiplication by the function

$$\Big(\mathcal{H}_0(\lambda\omega + t\gamma) + t - \frac{1}{t}\mathcal{K}_0(\lambda\omega + t\gamma)\Big)\chi\big(\gamma \cdot (|\lambda|\omega - t\gamma) \geq 0\big). \quad (10.41)$$

10.3   Perturbed direction dependent Green operators

In Sections 10.3, 10.4 and 10.5, we assume that

$$\Omega = \mathbf{R}^3, \quad \text{and} \quad \epsilon(x),\ \mu(x) \text{ are scalars} \in C^\infty(\mathbf{R}^3). \quad (10.42)$$

We put $R_0(z) = (\mathcal{H}_0 - z)^{-1}$, $R(z) = (\mathcal{H} - z)^{-1}$ and

$$V(z) = z(1 - \mathcal{M}_0^{-1}\mathcal{M}). \quad (10.43)$$

Then, the following resolvent equation holds:

$$R(z) = R_0(z)\mathcal{M}_0^{-1}\mathcal{M} - R_0(z)V(z)R(z). \tag{10.44}$$

In the standard perturbation theory, the perturbed Green operators are constructed by solving the resolvent equation. The procedure is well-known for the physical Green operator, which we explain here for the sake of comparison with the case of Faddeev's Green operator. We put

$$\tilde{\mathbf{C}}_+ = \mathbf{C}_+ \cup (\mathbf{R} \setminus \sigma_p(\mathcal{H})). \tag{10.45}$$

Lemma 10.7. *Let $s > 1/2$. Then, as an operator on $L^{2,-s}(\mathbf{R}^3)$, $-1 \notin \sigma(R_0(z)V(z))$ for any $z \in \tilde{\mathbf{C}}_+$. Hence the equation (10.44) is uniquely solvable:*

$$R(z) = \left(1 + R_0(z)V(z)\right)^{-1}R_0(z)\mathcal{M}_0^{-1}\mathcal{M}, \quad z \in \tilde{\mathbf{C}}_+.$$

Proof. We put

$$T_1(z) = \mathcal{P}_0^{(0)}\mathcal{M}_0^{-1}\mathcal{M}R(z), \tag{10.46}$$

$$T_2(z) = \mathcal{P}^{(0)}\mathcal{M}_0^{-1}\mathcal{M}R(z), \tag{10.47}$$

$$\mathcal{N} = \mathcal{M}_0^{-1}\mathcal{M}, \tag{10.48}$$

and derive the equations for $T_1(z)$ and $T_2(z)$. Mutiplying the equation (10.44) by $\mathcal{H}_0 - z$, we obtain

$$(\mathcal{H}_0 - z)R(z) = \mathcal{N} - z(1 - \mathcal{N})R(z).$$

Multiplying by $\mathcal{P}_0^{(0)}$, we have

$$T_1(z) = -\frac{1}{z}\mathcal{P}_0^{(0)}\mathcal{N}. \tag{10.49}$$

Mutiplying the equation (10.44) by $\mathcal{P}^{(0)}\mathcal{N}$, we obtain

$$T_2(z) = \mathcal{P}^{(0)}\mathcal{N}R_0(z)\mathcal{N} - \mathcal{P}^{(0)}\mathcal{N}R_0(z)z(\mathcal{N}^{-1} - 1)(T_1(z) + T_2(z)).$$

Using (10.49), we have

$$(1 + \mathcal{P}^{(0)}\mathcal{N}R_0(z)z(\mathcal{N}^{-1} - 1))T_2(z)$$
$$= \mathcal{P}^{(0)}\mathcal{N}R_0(z)(\mathcal{N} + (\mathcal{N}^{-1} - 1)\mathcal{P}_0^{(0)}\mathcal{N}). \tag{10.50}$$

Let

$$K(z) = \mathcal{P}^{(0)}\mathcal{N}R_0(z)z(\mathcal{N}^{-1} - 1), \quad z \in \mathbf{C} \setminus \mathbf{R},$$

and take $s > 1/2$. We show that $K(z)$ is compact on $L^{2,-s}$ and is bounded when $z \to \lambda \in \mathbf{R} \setminus \sigma_p(\mathcal{H})$. Since $\epsilon(x)$ and $\mu(x)$ are scalar functions, we have

$$[\mathcal{H}_0, \mathcal{N}] = \begin{pmatrix} 0 & -(\operatorname{curl}\mu)\times \\ (\operatorname{curl}\epsilon)\times & 0 \end{pmatrix}.$$

This implies

$$K(z) = \mathcal{P}^{(0)} R_0(z) z(1 - \mathcal{N}) + \mathcal{P}^{(0)} R_0(z) [\mathcal{H}_0, \mathcal{N}] R_0(z) z(\mathcal{N}^{-1} - 1).$$

Since $1 - \mathcal{N}$ and $[\mathcal{H}_0, \mathcal{N}]$ decay at infinity, $K(z)$ is compact on $L^{2,-s}$.

For $\lambda \in \mathbf{R} \setminus \sigma_p(\mathcal{H})$, we take open intervals I_1, I_2, I_3 such that $\lambda \in I_1 \subset\subset I_2 \subset\subset I_3 \subset\subset \mathbf{R} \setminus \sigma_p(\mathcal{H})$. Take $\varphi_i \in C(\mathbf{R})$ such that $\varphi_1(\lambda) = 1$, $\varphi = 0$ $\varphi_3 = 1$ on I_2, $\varphi_3 = 0$ outside I_3. Split $K(z) = K_1(z) + K_2(z)$, where

$$K_i(z) = \mathcal{P}^{(0)} \mathcal{N} \varphi_i(\mathcal{H}_0) R_0(z) z(\mathcal{N}^{-1} - 1) \quad i = 1, 2.$$

Then $z \to \lambda$, $K_2(z)$ is uniformly bounded on $L^{2,-s}$. Using $\varphi_1 \varphi_3 = \varphi_1$, we have

$$K_1(z) = \varphi_3(\mathcal{H}_0) \mathcal{P}^{(0)} \mathcal{N} \varphi_1(\mathcal{H}_0) R_0(z) z(\mathcal{N}^{-1} - 1)$$
$$+ \mathcal{P}^{(0)} [\mathcal{N}, \varphi_3(\mathcal{H}_0)] \varphi_1(\mathcal{H}_0) R_0(z) z(\mathcal{N}^{-1} - 1).$$

Then, the symbol of $\varphi(\mathcal{H}_0) \mathcal{P}^{(0)}$ is smooth. Moreover, by the formula of Helffer-Sjöstrand (Lemma 7.18), $[\mathcal{N}, \varphi_3(\mathcal{H}_0)] \langle x \rangle^s$ is a bounded operator on L^2. Therefore, $K_1(z)$ is uniformly bounded in $\mathbf{B}(L^{2,-s}; L^{2,-s})$ when $z \to \lambda$. Therefore, $K(z)$ has the same property.

We show that $-1 \notin \sigma_p(K(z))$. Let $u_2 = -K(z) u_2$, and put

$$u = -(\mathcal{H} - z) R_0(z) z(\mathcal{N}^{-1} - 1) u_2.$$

Using $\mathcal{H} = \mathcal{N}^{-1} \mathcal{H}_0$, we have

$$u = -\mathcal{N}^{-1}(\mathcal{H}_0 - z\mathcal{N}) R_0(z) z(\mathcal{N}^{-1} - 1) u_2$$
$$= -z\mathcal{N}^{-1}(\mathcal{N}^{-1} - 1)(u_2 + \mathcal{N} R_0(z) z(\mathcal{N}^{-1} - 1) u_2). \tag{10.51}$$

Using the equation $u_2 + K(z) u_2 = 0$, we have

$$u_2 + \mathcal{N} R_0(z) z(\mathcal{N}^{-1} - 1) u_2 = \mathcal{P}_0^{(0)} u_2 = 0.$$

Therefore, $w = R_0(z) z(\mathcal{N}^{-1} - 1) u_2$ satisfies $(\mathcal{H} - z) w = 0$. If $z \notin \mathbf{R}$, $w \in L^2$. If $z \in \mathbf{R}$, w satsifies the radiation condition. Therefore, $w = 0$, which implies $u_2 = 0$. The lemma now follows from the Fredholm theory for compact operators. $\qquad \square$

Let $R_{\gamma,0}(\lambda, t)$ be the Faddeev Green operator for the free Maxwell equation defined by (10.38). Replacing $R_0(z)$ by $R_{\gamma,0}(\lambda, t)$ in (10.44), we obtain the following operator equation on $L^{2,-s}$, by which the perturbed Green operator of Faddeev should be constructed:

$$R_\gamma(\lambda, t) = R_{\gamma,0}(\lambda, t) \mathcal{M}_0^{-1} \mathcal{M} - R_{\gamma,0}(\lambda, t) V(\lambda) R_\gamma(\lambda, t). \tag{10.52}$$

Instead of this equation, however, we consider the equation for the perturbed Green operator of Eskin-Ralston

$$\mathcal{U}_\gamma(\lambda, z) = \mathcal{U}_{\gamma,0}(\lambda, z)\mathcal{M}_0^{-1}\mathcal{M} - \mathcal{U}_{\gamma,0}(\lambda, z)V(\lambda)\mathcal{U}_\gamma(\lambda, z). \tag{10.53}$$

The idea consists in applying the Fredholm theorem to the resolvent equation, and the key step is the uniqueness of the solution. For the case of physical Green operator, it is guaranteed by the self-adjointness of $\mathcal{H}_0$ ($R_0(z)$ is a bounded operator on $L^2(\mathbf{R}^3)$ for $z \notin \mathbf{R}$) or the radiation condition (for $z = t \in \mathbf{R}$). This argument cannot be applied to (10.53), since $\mathcal{U}_{\gamma,0}(\lambda, z)$ is a non-physical Green operator. The remedy comes from the analyticity of $\mathcal{U}_{\gamma,0}(\lambda, z)$ and the Fredholm theorem for compact operator-valued analytic functions. To explain it, we need to recall the following fact.

Let $D = \{z \in \mathbf{C} \, ; \, |z| < 1\}$ be the unit disc in $\mathbf{C}$. The *Hardy class* H^1 is the set of analytic functions $f(z)$ in D such that

$$\sup_{0<r<1} \int_0^{2\pi} |f(re^{i\theta})|d\theta < \infty.$$

Theorem 10.4. *Let $f \in H^1$.*
(1) For a.e. $\theta \in [0, 2\pi]$, there exists a radial limit

$$\lim_{r \to 1} f(re^{i\theta}) =: \tilde{f}(\theta),$$

and $\tilde{f} \in L^1(S^1)$.
(2) If $\tilde{f}(\theta) = 0$ on a set of positive measure in S^1, then $f = 0$ on D.

For the proof see e.g. [40], pp. 38 and 52.

Now, we can prove the following analytic Fredholm theorem.

Theorem 10.5. *Let $\mathbf{H}$ be a Hilbert space, and $D = \{z \in \mathbf{C} \, ; \, |z| < 1\}$. Let $K(z)$ be a $\mathbf{B}(\mathbf{H})$-valued continuous function on $\overline{D}$ such that $K(z)$ is analytic in D, and compact for each $z \in \overline{D}$. Put*

$$S = \{z \in \overline{D} \, ; \, 1 \in \sigma_p(K(z))\}.$$

Then, S is a closed set in $\overline{D}$, and the following alternatives (1) or (2) hold:
(1) $S = \overline{D}$.
(2) S is discrete in D, and $S \cap \partial D$ is of measure 0. In this case, $(1 - K(z))^{-1}$ is continuous on $\overline{D} \setminus S$, meromorphic in D, and analytic in $D \setminus S$.

Proof. We prove that for any $z_0 \in \overline{D}$, there exists a neighborhood $V \subset \overline{D}$ of z_0 such that (1) or (2) holds with D replaced by V. The theorem then follows from the well-known connectedness argument. Given $z_0 \in \overline{D}$, choose

$r > 0$ such that $\|K(z) - K(z_0)\| < 1/4$ on $V = \{z \in \overline{D}; |z - z_0| < r\}$, moreover, there exists a finite rank operator L such that $\|K(z_0) - L\| < 1/4$. Then, $\|K(z) - L\| < 1/2$, hence $(I - K(z) + L)^{-1}$ exists and is analytic in $V \cap D$, continuous in $V \cap \overline{D}$. Since L has finite rank, there exist $\varphi_i, \psi_i \in \mathbf{H}$ such that $Lf = \sum_{i=1}^{n}(f, \varphi_i)\psi_i$. Letting

$$\varphi_i(z) = ((I - K(z) + L)^{-1})^*\varphi_i, \quad M(z) = L(I - K(z) + L)^{-1},$$

we then have

$$M(z)f = \sum_{i=1}^{n}(f, \varphi_i(z))\psi_i.$$

Since $I - K(z) = (I - M(z))(I - K(z) + L)$, we see that $(I - K(z))^{-1}$ exists if and only if $(I - M(z))^{-1}$ exists. Letting $d(z)$ be the determinant of the linear map $I - M(z)$, the latter occurs if and only if $d(z) \neq 0$ for $z \in V$. Let $S_0 = \{z \in V; d(z) = 0\}$. Since $d(z)$ is analytic in $V \cap D$, we have either $S_0 \cap D = V \cap D$ or $S_0 \cap D$ is discrete in D. In the former case, $d(z) = 0$ on $V \cap \partial D$. In the latter case, $S_0 \cap \partial D$ is a 1-dimensional null set, which follows from Theorem 10.4. $\square$

We return to (10.53). Note that $\mathcal{U}_{\gamma,0}(\lambda, z) = u_{\gamma,0}(\lambda^2, z)\mathcal{V}_{\gamma,0}(\lambda, z)$ is not compact due to the 2nd order differential operator term in $\mathcal{V}_{\gamma,0}(\lambda, z)$. Our first task is to remove it. We put

$$Q_1(\lambda, z) = (\mathcal{H}_0(-i\nabla + z\gamma) + \lambda)\, V(\lambda) - \text{Grad}_{z\gamma}[\text{Div}_{z\gamma}, \mathcal{M}^{-1}]\mathcal{M}, \quad (10.54)$$

$$Q_2(\lambda, z) = -\frac{1}{\lambda}\text{Grad}_{z\gamma}[\text{Div}_{z\gamma}, \mathcal{M}^{-1}]\mathcal{M} + \frac{1}{\lambda}\mathcal{K}_0(-i\nabla + z\gamma)(1 - \mathcal{M}_0^{-1}\mathcal{M}).$$
$$(10.55)$$

Lemma 10.8. *Let u be a solution to the equation*

$$u = \mathcal{U}_{\gamma,0}(\lambda, z)\mathcal{M}_0^{-1}\mathcal{M}f - \mathcal{U}_{\gamma,0}V(\lambda)u. \quad (10.56)$$

Then u satisfies

$$\left(1 + u_{\gamma,0}(\lambda^2, z)Q_1(\lambda, z)\right)u = \mathcal{U}_{\gamma,0}(\lambda, z)\mathcal{M}_0^{-1}\mathcal{M}f - u_{\gamma,0}(\lambda^2, z)Q_2(\lambda, z)f.$$
$$(10.57)$$

Proof. We use the notation

$$\mathcal{K}_{0,z\gamma} = \text{Grad}_{z\gamma}\,\text{Div}_{z\gamma}, \quad P_{z\gamma}^2 = \left(\text{Grad}_{z\gamma}\right)^2, \quad u_{\gamma,0} = u_{\gamma,0}(\lambda^2, z).$$

We also put

$$\mathcal{L}_{z\gamma} = \mathcal{L}(-i\nabla + z\gamma), \quad \mathcal{H}_{0,z\gamma} = \mathcal{H}_0(-i\nabla + z\gamma).$$

For $u = (u_E, u_H) \in (L^2_{loc}(\mathbf{R}^3))^3 \times (L^2_{loc}(\mathbf{R}^3))^3$, we define

$$\mathrm{Div}_{z\gamma}\, u = (\mathrm{Div}_{z\gamma}\, u_E, \mathrm{Div}_{z\gamma}\, u_H).$$

We often use the abbreviation

$$\mathrm{Div} = \mathrm{Div}_{z\gamma}, \quad \mathrm{Grad} = \mathrm{Grad}_{z\gamma}, \quad \mathrm{Curl} = \mathrm{Curl}_{z\gamma},$$

until the end of this section.

By the formula (10.37), we have

$$P^2_{z\gamma} - \lambda^2 = (\mathcal{H}_{0,z\gamma} + \lambda)(\mathcal{H}_{0,z\gamma} - \lambda) + \mathcal{K}_{0,z\gamma}. \tag{10.58}$$

Note that

$$(P^2_{z\gamma} - \lambda^2)u_{\gamma,0} = I.$$

Now, from the equation (10.56), we have

$$\mathcal{L}_{z\gamma}u - \lambda \mathcal{M}u = \mathcal{M}f. \tag{10.59}$$

We then have

$$V(\lambda)u = (\mathcal{M}^{-1} - \mathcal{M}_0^{-1})\mathcal{L}_{z\gamma}u - (1 - \mathcal{M}_0^{-1}\mathcal{M})f,$$

which implies

$$\mathrm{Div}\, V(\lambda)u = [\mathrm{Div}, \mathcal{M}^{-1}]\mathcal{M}(\lambda u + f) - \mathrm{Div}\,(1 - \mathcal{M}_0^{-1}\mathcal{M})f, \tag{10.60}$$

$$[\mathrm{Div}, \mathcal{M}^{-1}] = \begin{pmatrix} i\epsilon^{-2}(\nabla\epsilon) & 0 \\ 0 & i\mu^{-2}(\nabla\mu) \end{pmatrix}. \tag{10.61}$$

Therefore, we have

$$\mathcal{K}_{0,z\gamma}V(\lambda)u = \mathrm{Grad}\,[\mathrm{Div}, \mathcal{M}^{-1}]\mathcal{M}(\lambda u + f) - \mathcal{K}_{0,z\gamma}(1 - \mathcal{M}_0^{-1}\mathcal{M})f, \tag{10.62}$$

hence by (10.28)

$$\mathcal{V}_{\gamma,0}(\lambda, z)V(\lambda)u = Q_1(\lambda, z)u + Q_2(\lambda, z)f. \tag{10.63}$$

Using (10.27), we arrive at the equation (10.57). $\qquad \square$

The converse of Lemma 10.8 is proven under the assumption of the analyticity with respect to z.

Lemma 10.9. *Let u be a solution to the equation (10.57) which is meromorphic on D_ϵ and holomorphic on $D_\epsilon \cap \{\mathrm{Im}\, z > T\}$ for some $T > 0$. Then, u satisfies (10.56).*

Proof.   We mulitply (10.57) by $P_{z\gamma}^2 - \lambda^2$. Using (10.58) and

$$\text{Grad}\,[\text{Div}, \mathcal{M}^{-1}]\mathcal{M} = \mathcal{K}_{0,z\gamma} - \text{Grad}\,\mathcal{M}^{-1}\text{Div}\mathcal{M},$$

we have

$$\left(\mathcal{H}_{0,z\gamma} + \lambda\right)\left(\mathcal{H}_{0,z\gamma} - \lambda\mathcal{M}_0^{-1}\mathcal{M}\right)u + \text{Grad}\,\mathcal{M}^{-1}\text{Div}\mathcal{M}u$$

$$= \left(\mathcal{H}_{0,z\gamma} + \lambda\right)\mathcal{M}_0^{-1}\mathcal{M}f - \frac{1}{\lambda}\text{Grad}\,\mathcal{M}^{-1}\text{Div}\mathcal{M}f. \tag{10.64}$$

Mulitiplying this equation by Div and noting that Div commutes with $\mathcal{M}_0^{-1}$, we have

$$\lambda^2\mathcal{M}_0^{-1}\left(\text{Div}\,\mathcal{M}u + \frac{1}{\lambda}\text{Div}\,\mathcal{M}f\right) = P_{z\gamma}^2\mathcal{M}^{-1}\left(\text{Div}\,\mathcal{M}u + \frac{1}{\lambda}\text{Div}\,\mathcal{M}f\right). \tag{10.65}$$

This implies

$$\lambda^2 u_{\gamma,0}\mathcal{M}_0^{-1}\left(\text{Div}\,\mathcal{M}u + \frac{1}{\lambda}\text{Div}\,\mathcal{M}f\right) = \mathcal{M}^{-1}\left(\text{Div}\,\mathcal{M}u + \frac{1}{\lambda}\text{Div}\,\mathcal{M}f\right).$$

By Lemma 10.4, $u_{\gamma,0}$ is a left inverse of $P_{z\gamma}^2 - \lambda^2$, hence we have

$$\lambda^2 u_{\gamma,0}(\mathcal{M}_0^{-1} - \mathcal{M})\left(\text{Div}\,\mathcal{M}u + \frac{1}{\lambda}\text{Div}\,\mathcal{M}f\right) = \mathcal{M}^{-1}\left(\text{Div}\,\mathcal{M}u + \frac{1}{\lambda}\text{Div}\,\mathcal{M}f\right).$$

Since $u_{\gamma,0} = u_{\gamma,0}(\lambda^2, i\tau)$ decays like $O(\tau^{-1})$, we have

$$\left\|\text{Div}\,\mathcal{M}u + \frac{1}{\lambda}\text{Div}\,\mathcal{M}f\right\|_{-s} \leq \frac{C}{\tau}\left\|(\mathcal{M}_0^{-1} - \mathcal{M})\text{Div}\,\mathcal{M}u + \frac{1}{\lambda}\text{Div}\,\mathcal{M}f\right\|_s$$

$$\leq \frac{C}{\tau}\left\|\text{Div}\,\mathcal{M}u + \frac{1}{\lambda}\text{Div}\,\mathcal{M}f\right\|_{-s}.$$

Then, there exists $T > 0$ such that $\text{Div}\,\mathcal{M}u + \frac{1}{\lambda}\text{Div}\,\mathcal{M}f = 0$ for $z = i\tau$, $\tau > T$. Since u is meromorphic with respect to z, we obtain

$$\text{Div}\,\mathcal{M}u + \frac{1}{\lambda}\text{Div}\,\mathcal{M}f = 0, \quad \text{in} \quad D_\epsilon. \tag{10.66}$$

Mutiplying $\mathcal{M}_0^{-1}$ and Grad, we then have

$$\mathcal{K}_{0,z\gamma}\mathcal{M}_0^{-1}\mathcal{M}u + \frac{1}{\lambda}\mathcal{K}_{0,z\gamma}\mathcal{M}_0^{-1}\mathcal{M}f = 0. \tag{10.67}$$

Adding (10.64) and (10.67), and invoking (10.66), we have

$$\left(\mathcal{H}_{0,z\gamma} + \lambda - \frac{1}{\lambda}\mathcal{K}_{0,z\gamma}\right)\left(\mathcal{H}_{0,z\gamma} - \lambda\mathcal{M}_0^{-1}\mathcal{M}\right)u = \left(\mathcal{H}_{0,z\gamma} + \lambda - \frac{1}{\lambda}\mathcal{K}_{0,z\gamma}\right)\mathcal{M}_0^{-1}\mathcal{M}f.$$

We mutiply this equation by $u_{\gamma,0}(\lambda^2, z)$ to obtain the Green operator of Eskin-Ralston. Therefore, we obtain

$$u + \lambda\mathcal{U}_{\gamma,0}V(\lambda)u = \mathcal{U}_{\gamma,0}\mathcal{M}_0^{-1}\mathcal{M}f.$$

We have thus arrived at the equation (10.56). $\square$

Note that by multiplying (10.56) by $\mathcal{H}_{0,z\gamma} - \lambda$, we get the equation

$$(\mathcal{L}_{z\gamma} - \lambda\mathcal{M})u = \mathcal{M}f. \tag{10.68}$$

We discuss the solvability of the equation (10.57).

Lemma 10.10. *Take $\lambda \in \mathbf{R} \setminus \{0\}$ and $\delta > 0$ arbitrarily. Then, there exists a closed subset $\Sigma_\lambda \subset \overline{D}_\epsilon$ having the following properties.*
(1) Σ_λ is discrete in D_ϵ, and $\Sigma_\lambda \cap \{t \, ; \, -\epsilon/2 < t < \epsilon/2\}$ is a 1-dimensional null set.
(2) For any $z, \in \overline{C}_+ \setminus \Sigma_\lambda$, and $f \in L^2_{e^{\delta r}}$, there exists a unique solution $u = u(\lambda, z, f) \in L^2_{e^{-\delta r}}$ of the equation (10.57). The operator: $L^2_{e^{\delta r}} \ni f \to u(\lambda, z, f) \in L^2_{e^{-\delta r}}$ is bounded.
(3) As an $L^2_{e^{-\delta r}}$-valued function of z, $u(\lambda, z, f)$ is continuous on $\overline{D}_\epsilon \setminus \Sigma_\lambda$, analytic on $D_\epsilon \setminus \Sigma_\lambda$ and meromorphic on D_ϵ.

Proof. Since $u_{\gamma,0}(\lambda, z)$ is analytic with respect to $z \in D_\epsilon$, so is $u_{\gamma,0}(\lambda^2, z)Q_1(\lambda, z)$. Moreover, it is a compact operator on $L^2_{e^{-\delta r}}$. By the Fredholm theorem, to construct the inverse $(1 + u_{\gamma,0}(\lambda^2, z)Q_1(\lambda, z))^{-1}$, we have only to show $-1 \notin \sigma_p(u_{\gamma,0}(\lambda^2, z)Q_1(\lambda, z))$, i.e. the uniqueness of the solution to the equation (10.57). It will be easy, if the operator norm of $u_{\gamma,0}(\lambda^2, z)Q_1(\lambda, z)$ decays as $\text{Im } z \to \infty$. However, it does not hold due to the 1st order differential operator term in $Q_1(\lambda, z)$, to remove which is our 2nd task.

Step 1: Transformation of the equation. Let u be a solution to (10.57) satisfying the assumption in Lemma 10.9. We show that u satisfies (10.78) to be defined below. Letting $u = (u_E, u_H), f = (f_E, f_H)$, and taking Div of (10.68), we have

$$\text{Div}\,(\epsilon u_E) = -\frac{1}{\lambda}\text{Div}\,(\epsilon f_E), \tag{10.69}$$

$$\text{Div}\,(\mu u_H) = -\frac{1}{\lambda}\text{Div}\,(\mu f_H), \tag{10.70}$$

which imply

$$\text{Div}\, u_E = \frac{i\nabla\epsilon}{\epsilon} \cdot u_E - \frac{1}{\lambda\epsilon}\text{Div}\,(\epsilon f_E), \tag{10.71}$$

$$\text{Div}\, u_H = \frac{i\nabla\mu}{\mu} \cdot u_H - \frac{1}{\lambda\mu}\text{Div}\,(\mu f_H). \tag{10.72}$$

Letting

$$\nabla \mathcal{M} = \begin{pmatrix} \nabla\epsilon & 0 \\ 0 & \nabla\mu \end{pmatrix}, \tag{10.73}$$

we have

$$[\mathrm{Div}, \mathcal{M}^{-1}]\mathcal{M} = \mathcal{M}^{-1}(i\nabla\mathcal{M}). \tag{10.74}$$

We then have

$$\mathcal{V}_{\gamma,0}(\lambda, z)V(\lambda)u = \mathcal{V}(\lambda, z)u + \mathcal{W}(\lambda, z)f, \tag{10.75}$$

where

$$\mathcal{V}(\lambda, z)u = \left(\mathcal{H}_{0,z\gamma} + \lambda\right)V(\lambda)u - \begin{pmatrix} \mathrm{Grad}\left(\frac{i\nabla\epsilon}{\epsilon} \cdot u_E\right) \\ \mathrm{Grad}\left(\frac{i\nabla\mu}{\mu} \cdot u_H\right) \end{pmatrix}, \tag{10.76}$$

$$\mathcal{W}(\lambda, z)f = \begin{pmatrix} \mathrm{Grad}\left(\left(\frac{1}{\lambda\epsilon} - \frac{1}{\lambda\epsilon_0}\right)\mathrm{Div}\left(\epsilon f_E\right)\right) \\ \mathrm{Grad}\left(\left(\frac{1}{\lambda\mu} - \frac{1}{\lambda\mu_0}\right)\mathrm{Div}\left(\mu f_H\right)\right) \end{pmatrix}. \tag{10.77}$$

The equation (10.57) is thus rewritten as

$$u = u_{\gamma,0}\mathcal{V}(\lambda, z)u + u_{\gamma,0}\mathcal{W}(\lambda, z)f + \mathcal{U}_{\gamma,0}(\lambda, z)\mathcal{M}_0^{-1}\mathcal{M}f. \tag{10.78}$$

Step 2: 2nd transformation. We further transform the equation (10.78) to (10.88) below. Taking note of

$$\mathrm{Grad}\left(\frac{i\nabla\epsilon}{\epsilon} \cdot u_E\right) = e^{-iz\gamma \cdot x}(-i\nabla_x)\left(\frac{i\nabla\epsilon}{\epsilon} \cdot e^{iz\gamma \cdot x}u_E\right),$$

and using (A.22) in the Appendix, we have

$$\mathrm{Grad}\left(\frac{i\nabla\epsilon}{\epsilon} \cdot u_E\right) = \left(\frac{i\nabla\epsilon}{\epsilon} \cdot \mathrm{Grad}\right)u_E + \frac{i\nabla\epsilon}{\epsilon} \times \mathrm{Curl}\, u_E$$
$$- i(u_E \cdot \nabla)\frac{i\nabla\epsilon}{\epsilon} - iu_E \times \mathrm{curl}\frac{i\nabla\epsilon}{\epsilon}$$
$$= \left(\frac{i\nabla\epsilon}{\epsilon} \cdot \mathrm{Grad}\right)u_E + \lambda\frac{i\nabla\epsilon}{\epsilon} \times (\mu u_H) + \frac{i\nabla\epsilon}{\epsilon} \times (\mu f_H)$$
$$- i(u_E \cdot \nabla)\frac{i\nabla\epsilon}{\epsilon}, \tag{10.79}$$

where we have used $\operatorname{curl}\dfrac{\nabla\epsilon}{\epsilon} = \operatorname{curl}\operatorname{grad}\log\epsilon = 0$. Similarly, we have

$$
\begin{aligned}
\operatorname{Grad}\left(\frac{i\nabla\mu}{\mu}\cdot u_H\right) &= \left(\frac{i\nabla\mu}{\mu}\cdot\operatorname{Grad}\right)u_H + \frac{i\nabla\mu}{\mu}\times\operatorname{Curl}u_H \\
&\quad - i(u_H\cdot\nabla)\frac{i\nabla\mu}{\mu} - iu_H\times\operatorname{curl}\frac{i\nabla\mu}{\mu} \\
&= \left(\frac{i\nabla\mu}{\mu}\cdot\operatorname{Grad}\right)u_H - \lambda\frac{i\nabla\mu}{\mu}\times(\epsilon u_E) - \frac{i\nabla\mu}{\mu}\times(\epsilon f_E) \\
&\quad - i(u_H\cdot\nabla)\frac{i\nabla\mu}{\mu}.
\end{aligned}
$$

$$(10.80)$$

By a direct computation

$$
\mathcal{H}_{0,z\gamma}\mathcal{M}_0^{-1}\mathcal{M} = \mathcal{M}_1 + \begin{pmatrix} \mu & 0 \\ 0 & \epsilon \end{pmatrix}\mathcal{L}_{z\gamma}. \tag{10.81}
$$

where

$$
\mathcal{M}_1 = \begin{pmatrix} 0 & i(\nabla\mu)\times \\ -i(\nabla\epsilon)\times & 0 \end{pmatrix}. \tag{10.82}
$$

We have, therefore,

$$
\mathcal{H}_{0,z\gamma}\mathcal{M}_0^{-1}\mathcal{M}u = \mathcal{M}_1 u + \epsilon\mu(\lambda u + f),
$$

which yields

$$
(\mathcal{H}_{0,z\gamma} + \lambda)V(\lambda)u = \lambda^2(1 - \epsilon\mu)u - \lambda\mathcal{M}_1 u + \lambda(\mathcal{M}_0^{-1}\mathcal{M} - \epsilon\mu)f. \tag{10.83}
$$

We have thus obtained

$$
\begin{aligned}
\mathcal{V}(\lambda,z)u &= (\mathcal{M}^{-1}i\nabla\mathcal{M})\operatorname{Grad}u + \lambda^2(1 - \epsilon\mu)u - \lambda(M_1 + M_2)u + M_3 u \\
&\quad + \lambda(\mathcal{M}_0^{-1}\mathcal{M} + M_2\mathcal{M} - \epsilon\mu)f,
\end{aligned}
$$

$$(10.84)$$

where

$$
(\mathcal{M}^{-1}i\nabla\mathcal{M})\operatorname{Grad} = \begin{pmatrix} \dfrac{i\nabla\epsilon}{\epsilon}\cdot\operatorname{Grad} & 0 \\ 0 & \dfrac{i\nabla\mu}{\mu}\cdot\operatorname{Grad} \end{pmatrix}, \tag{10.85}
$$

$$
\mathcal{M}_2 = \begin{pmatrix} 0 & -\dfrac{\nabla\epsilon}{\epsilon}\times \\ \dfrac{\nabla\mu}{\mu}\times & 0 \end{pmatrix}, \tag{10.86}
$$

$$M_3 u = \begin{pmatrix} (u_E \cdot \nabla) \dfrac{\nabla \epsilon}{\epsilon} \\[2mm] (u_H \cdot \nabla) \dfrac{\nabla \mu}{\mu} \end{pmatrix}. \tag{10.87}$$

Therefore, the equation (10.78) is rewritten as

$$u + u_{\gamma,0}(M^{-1}i\nabla M)\operatorname{Grad} u = u_{\gamma,0}Q_1(\lambda)u + u_{\gamma,0}Q_2(\lambda, z)f, \tag{10.88}$$

where

$$Q_1(\lambda) = \lambda^2(1 - \epsilon\mu) - \lambda(M_1 + M_2) + M_3, \tag{10.89}$$

$$Q_2(\lambda, z) = W(\lambda, z) + \lambda(M_0^{-1}M + M_2 M - \epsilon\mu) + (\mathcal{H}_{0,z\gamma} + \lambda - \frac{1}{\lambda}\mathcal{K}_{0,z\gamma})M_0^{-1}M. \tag{10.90}$$

Step 3: Gauge transformation. We remove the term $(M^{-1}i\nabla M)\operatorname{Grad}$ in (10.88) by a gauge transformation. Multiplying (10.88) by $-\Delta - 2iz\gamma \cdot \nabla + z^2 - \lambda^2 = (\operatorname{Grad})^2 - \lambda^2$, we have

$$\left((\operatorname{Grad})^2 - \lambda^2 + (M^{-1}i\nabla M)\operatorname{Grad}\right)u = Q_1(\lambda)u + Q_2(\lambda, z)f. \tag{10.91}$$

Here, we note that for a positive function a

$$a^{-1}\operatorname{Grad} a = \operatorname{Grad} - \frac{i\nabla a}{a},$$

$$a^{-1}(\operatorname{Grad})^2 a = (\operatorname{Grad})^2 - 2\frac{i\nabla a}{a} \cdot \operatorname{Grad} - (\frac{\nabla a}{a})^2 - \nabla \cdot (\frac{\nabla a}{a}).$$

Therefore, letting

$$\sqrt{M} = \begin{pmatrix} \sqrt{\epsilon} & 0 \\ 0 & \sqrt{\mu} \end{pmatrix}, \tag{10.92}$$

$$(\nabla\sqrt{M})^2 = \begin{pmatrix} (\nabla\sqrt{\epsilon})^2 & 0 \\ 0 & (\nabla\sqrt{\mu})^2 \end{pmatrix}, \tag{10.93}$$

$$\nabla \cdot (\sqrt{M}^{-1}\nabla\sqrt{M}) = \begin{pmatrix} \nabla \cdot (\nabla\sqrt{\epsilon}/\sqrt{\epsilon}) & 0 \\ 0 & \nabla \cdot (\nabla\sqrt{\mu}/\sqrt{\mu}) \end{pmatrix}, \tag{10.94}$$

we have

$$\sqrt{M}^{-1}\left((\operatorname{Grad})^2 - \lambda^2 + (M^{-1}i\nabla M)\operatorname{Grad}\right)\sqrt{M}$$
$$= (\operatorname{Grad})^2 - \lambda^2 - M^{-1}(\nabla\sqrt{M})^2 - \nabla \cdot (\sqrt{M}^{-1}\nabla\sqrt{M}). \tag{10.95}$$

We then see that $\widetilde{u}$ defined by

$$\widetilde{u} = \sqrt{\mathcal{M}}^{-1}u, \tag{10.96}$$

satisfies the equation

$$\left((\mathrm{Grad})^2 - \lambda^2 - \mathcal{M}^{-1}(\nabla\sqrt{\mathcal{M}})^2 - \nabla \cdot (\sqrt{\mathcal{M}}^{-1}\nabla\sqrt{\mathcal{M}})\right)\widetilde{u}$$
$$= \sqrt{\mathcal{M}}^{-1}Q_1(\lambda)\sqrt{\mathcal{M}}\,\widetilde{u} + \sqrt{\mathcal{M}}^{-1}Q_2(\lambda, z)f. \tag{10.97}$$

Here, we assume that $z = i\tau$, $\tau > 0$. Then, since u is written as $u = u_{\gamma,0}F$ for some $F \in L^2_{e^{\delta r}}$, by Lemma 10.3 (5), we have $u \in L^{2,s-1}$ for some $0 < s < 1$. Thanks to Lemma 9.4, $\widetilde{u}$ satisfies the equation

$$\widetilde{u} = A(\lambda, i\tau)\widetilde{u} + B(\lambda, i\tau)f, \tag{10.98}$$

where

$$A(\lambda, z) = u_{\gamma,0}(\lambda^2, z)\widetilde{Q}_1(\lambda), \tag{10.99}$$

$$\widetilde{Q}_1(\lambda) = \mathcal{M}^{-1}(\nabla\sqrt{\mathcal{M}})^2 + \nabla \cdot (\sqrt{\mathcal{M}}^{-1}\nabla\sqrt{\mathcal{M}}) + \sqrt{\mathcal{M}}^{-1}Q_1(\lambda)\sqrt{\mathcal{M}}, \tag{10.100}$$

$$B(\lambda, z) = u_{\gamma,0}(\lambda^2, z)\widetilde{Q}_2(\lambda, z), \tag{10.101}$$

$$\widetilde{Q}_2(\lambda, z) = \sqrt{\mathcal{M}}^{-1}Q_2(\lambda, z). \tag{10.102}$$

Let us note that $\widetilde{Q}_1(\lambda)$ is an operator of multiplication by a compactly supported function, hence is bounded from $L^2_{e^{-\delta r}}$ to $L^2_{e^{\delta r}}$.

Suppose u satisfies (10.57) with $f = 0$. Then, $\widetilde{u}$ satisfies (10.98) with $f = 0$. In view of Lemma 10.3 (5), we have $\|\widetilde{A}(\lambda, i\tau)\| \leq C/\tau$, therefore

$$\|u\| \leq \frac{C}{\tau}\|u\|.$$

Hence $u = 0$ for large $\tau > 0$.

Step 4: Proof of Lemma 10.10. In Step 3, we have proven that there exists a constant $T > 0$ such that $1 \notin \sigma_p(u_{\gamma,0}Q_1(\lambda, i\tau))$ for $\tau > T$. The assertions in Lemma 10.10 then follows from Theorem 10.5. □

Let $z = t \in \mathbf{R}$ in (10.53) and conjugate it by $e^{-it\gamma \cdot x}$. Using (10.38), we obtain the equation for the perturbed Green operator of Faddeev (10.52). Thanks to Lemma 10.10, it is also solvable.

Lemma 10.11. *For $t \in (-\epsilon/2, \epsilon/2) \setminus \Sigma_\lambda$, as a solution of (10.52), the perturbed Green operator of Faddeev $R_\gamma(\lambda, t)$ exists and*

$$R_\gamma(\lambda, t) = e^{it\gamma \cdot x}\mathcal{U}_\gamma(\lambda, t)e^{-it\gamma \cdot x}.$$

10.4   From physical scattering amplitude to Faddeev scattering amplitude

We fix $\lambda \notin \sigma_p(\mathcal{H})$ arbitrarily. By (8.45), we have

$$\mathcal{F}_+(\lambda) = \mathcal{F}_+^{(0)}(\lambda)\left(\mathcal{M}_0^{-1}\mathcal{M} - V(\lambda)R(\lambda + i0)\right).$$

The scattering amplitude is then rewritten as

$$A(\lambda) = \mathcal{F}_+^{(0)}(\lambda)\left(V(\lambda) - V(\lambda)R(\lambda + i0)V(\lambda)^*\right)\mathcal{F}_+^{(0)}(\lambda)^*. \qquad (10.103)$$

Let Σ_λ be the set specified in Lemma 10.10. By Lemma 10.11, the perturbed Green operator of Faddeev $R_\gamma(\lambda, t)$ exists for $t \in (-\epsilon/2, \epsilon/2) \setminus \Sigma_\lambda$. Define the Faddeev scattering amplitude $A_F(\lambda, t)$ by

$$A_F(\lambda, t) = \mathcal{F}_+^{(0)}(\lambda)\left(V(\lambda) - V(\lambda)R_\gamma(\lambda, t)V(\lambda)^*\right)\mathcal{F}_+^{(0)}(\lambda)^*. \qquad (10.104)$$

Lemma 10.12. *The following equation holds:*

$$A_F(\lambda, t) - A(\lambda) = 2\pi i\, A(\lambda)M_\gamma(\lambda, t)A_F(\lambda, t). \qquad (10.105)$$

Proof.   We let

$$R_\gamma = R_\gamma(\lambda, t), \quad R = R(\lambda + i0), \quad R_{\gamma,0} = R_{\gamma,0}(\lambda, t), \quad R_0 = R_0(\lambda + i0),$$

$$T = R_{\gamma,0} - R_0, \quad V = V(\lambda).$$

Subtracting the resolvent equations

$$R_\gamma = R_{\gamma,0}\mathcal{M}_0^{-1}\mathcal{M} - R_{\gamma,0}VR_\gamma, \quad R = R_0\mathcal{M}_0^{-1}\mathcal{M} - R_0VR,$$

we have

$$(1 + R_0V)(R_\gamma - R) = T(\mathcal{M}_0^{-1}\mathcal{M} - VR_\gamma).$$

Using

$$V^*(z) = z(\mathcal{M}^{-1}\mathcal{M}_0 - 1), \qquad (10.106)$$

and

$$R(z) = R_0(z)\mathcal{M}_0^{-1}\mathcal{M} - R(z)V^*(z)R_0(z)\mathcal{M}_0^{-1}\mathcal{M}, \qquad (10.107)$$

we have

$$-RV^* + R_0V - RV^*R_0V = 0,$$

hence

$$\left(1 - RV^*\right)\left(1 + R_0V\right) = 1. \qquad (10.108)$$

We then obtain

$$R_\gamma - R = (1 - RV^*)T(\mathcal{M}_0^{-1}\mathcal{M} - VR_\gamma).$$

By (10.103) and (10.104), we then have

$$A_F - A = -\mathcal{F}_+^{(0)}(\lambda)V(1 - RV^*)T(\mathcal{M}_0^{-1}\mathcal{M} - VR_\gamma)V^*\mathcal{F}_+^{(0)}(\lambda)^*.$$

In view of (10.40), we have proven (10.105).   □

Lemma 10.13. *For $t \in (-\epsilon/2, \epsilon/2) \setminus \Sigma_\lambda$, the equation (10.105) is solvable with respect to $A_F(\lambda, t)$:*

$$A_F(\lambda, t) = \left(1 - 2\pi i A(\lambda) M_\gamma(\lambda, t)\right)^{-1} A(\lambda).$$

Proof. Letting

$$S_1 = 2\pi i \, \mathcal{F}_+^{(0)}(\lambda) V,$$

$$S_2 = \left(1 - RV^*\right) \mathcal{F}_+^{(0)}(\lambda)^* M_\gamma(\lambda, t),$$

we have

$$2\pi i \, A(\lambda) M_\gamma(\lambda, t) = S_1 S_2.$$

Let $\widetilde{K} = (1 - RV^*)TV$. In view of (10.40), we have

$$S_2 S_1 = -\widetilde{K}. \tag{10.109}$$

Since S_1 is compact, $S_1 S_2$ and $S_2 S_1$ are also compact. In Lemma 10.12, we have shown the existence of $(1 + R_0 V)^{-1}$. By (10.108), we then have $1 - RV^* = (1 + R_0 V)^{-1}$. Therefore,

$$(1 + R_0 V)(1 + \widetilde{K}) = 1 + R_{\gamma,0} V, \tag{10.110}$$

since $T = R_{\gamma,0} - R_0$. Lemma 10.11 implies the existence of $(1 + R_{\gamma,0} V)^{-1}$. Hence $-1 \notin \sigma_p(\widetilde{K})$. By (10.109), we then have

$$1 \notin \sigma_p(S_2 S_1), \tag{10.111}$$

which yields

$$1 \notin \sigma_p(S_1 S_2). \tag{10.112}$$

In fact, if $S_1 S_2 u = u$, we have $S_2 S_1 S_2 u = S_2 u$. By (10.111), $S_2 u = 0$, hence $u = 0$. The compactness of $S_1 S_2$ then implies the existence of $(1 - S_1 S_2)^{-1}$. This proves the lemma. $\qquad \square$

For the case of Schrödinger or Dirac equations, one can reconstruct the potential from the Faddeev scattering matrix, hence from the physical S-matrix. However, it seems to be difficult for the case of the Maxwell equation. Therefore, we reduce the issue to the inverse boundary value problem in a bounded domain.

10.5 Determination of medium

Take $\eta \in \mathbf{R}^2$ such that $\eta \cdot \gamma = 0$, $|\eta| = \sqrt{\lambda^2 + \tau^2}$, and put

$$\zeta = \eta + i\tau\gamma. \tag{10.113}$$

Then, we have

$$\zeta^2 = \lambda. \tag{10.114}$$

Conversely, any ζ satisfying (10.114) is rewritten as (10.113). We consider a new Green operator

$$g_\zeta^{(0)} = e^{-i\eta \cdot x} g_{\gamma,0}(\sqrt{\lambda + \tau^2}, i\tau) e^{i\eta \cdot x}, \tag{10.115}$$

having the integral kernel

$$g_\zeta^{(0)}(x - y) = (2\pi)^{-3} \int_{\mathbf{R}^3} \frac{e^{i(x-y)\cdot\xi}}{\xi^2 + 2\zeta \cdot \xi} d\xi. \tag{10.116}$$

As can be checked easily $g^{(0)} = e^{i\zeta \cdot x} g_\zeta^{(0)} e^{-i\zeta \cdot x}$ satisfies

$$(-\Delta - \lambda) g^{(0)} = I. \tag{10.117}$$

In [80], this Green operator $g^{(0)}$ was used to constuct the Faddeev Green operator for the Maxwell equation. Invoking the construction in the previous section, this means that their perturbed Green operator $\mathcal{G}$ is given by

$$\mathcal{G} = e^{i\zeta \cdot x} \mathcal{G}_\zeta(\lambda, i\tau) e^{-i\zeta \cdot x}, \tag{10.118}$$

$$\mathcal{G}_\zeta = e^{-i\eta \cdot x} \mathcal{U}_\gamma(\lambda, i\tau) e^{i\eta \cdot x}. \tag{10.119}$$

Then,

$$(\mathcal{H} - \lambda)\mathcal{G} = I. \tag{10.120}$$

Now, suppose we are given a Maxwell operator $\mathcal{H}$ on $\mathbf{R}^3$ such that $\epsilon(x) = \epsilon_0$ and $\mu(x) = \mu_0$ in an exterior domain Ω_{ext}. Letting

$$u^{(in)} = (E_0, M_0) = e^{i\zeta \cdot x}(\eta_E, \eta_M), \tag{10.121}$$

where $\eta_E, \eta_M \in \mathbf{C}^3$, we put

$$u = (E, M) = u^{(in)} - \mathcal{G}_\zeta u^{(in)}. \tag{10.122}$$

As has been discussed for the Schrödinger operator, given $u^{(in)}$, the boundary value of u at S is uniquely determined by the S-matrix $S(\lambda)$ (see Lemma 10.5).

By virtue of Theorem 9.3, from the S-matrix $S(\lambda)$, one can uniquely determine the E-M map $\Lambda(\lambda)$ in the interior domain $\Omega_{int} = \mathbf{R}^3 \setminus \overline{\Omega_{ext}}$. We are now in a position to following the arguments of [80]. We put

$$I = (\nu \times E, M_0^*)_{L^2(S)} + (\Lambda_{int}(\lambda)\nu \times E, E_0^*)_{L^2(S)}, \tag{10.123}$$

where (E_0^*, M_0^*) is a solution to the free Maxwell equation which we now specify. The choice is as follows. Let $R > 0$, $a = (|a|, 0, 0)$ and

$$\zeta = \left(|a|/2, i(|a|^2/4 + R^2)^{1/2}, (R^2 + \lambda)^{1/2}\right),$$

$$\eta_E = (1, 1, -(\zeta_1 + \zeta_2)/\zeta_3), \quad u\eta_M = (\lambda\mu_0)^{-1}\zeta \times u_E,$$

$$E_0^* = e^{i\zeta \cdot x}\eta_E^*, \quad M_0^* = e^{i\zeta \cdot x}\eta_M^*,$$

$$\eta^* = a - \zeta, \quad \eta_E^* = (1, -1, (\zeta_1 + \zeta_2)/\zeta_3), \quad \eta_M^* = (\lambda\mu_0)^{-1}\zeta^* \times \eta_E^*.$$

Then, by the same computataion as in [80], letting $R \to \infty$ in (10.123), one obtains

$$\lim_{R \to \infty} I = \frac{1}{\lambda\mu_0}\tilde{J}(a), \tag{10.124}$$

where $\tilde{J}$ is the Fourier transform of

$$J = \Delta f + F(f, g)/f, \tag{10.125}$$

with

$$f = (\mu/\mu_0)^{1/2}, \quad g = (\epsilon/\epsilon_0)^{1/2},$$

$$F(z_1, z_2) = \lambda z_1 (1 - z_1^2 z_2^2).$$

Therefore, the reconstruction of $\epsilon(x), \mu(x)$ is reduced to the equation

$$\begin{cases} \Delta f + F(f, g) = pf, \\ \Delta_g + F(g, f) = qg, \end{cases} \tag{10.126}$$

where p, q are computed from the data. Ola-Päivärinta-Somersalo [80] proved that the semi-linear elliptic equation (10.126) has a unique solution. Therefore, we arrive at the following conclusion.

Theorem 10.6. *For the scattering problem in* $\mathbf{R}^3$, *if* $\epsilon(x) - \epsilon_0$ *and* $\mu(x) - \mu_0$ *are in* $C_0^\infty(\mathbf{R}^3; \mathbf{R})$, *they are uniquely reconstructed from the S-matrix* $S(\lambda)$ *of one fixed energy* λ.

For the Schrödinger equation, one can reconstruct exponentially decreasing potentials from the S-matrix of one fixed energy. For the Maxwell equation, such results are not yet known.

10.6   Anisotropic medium

When $\epsilon(x)$ and $\mu(x)$ are matrices, the inverse problem becomes much harder. Kurylev-Lassas-Sommersalo [62], and Kenig-Salo-Uhlmann [52] studied the boundary value problem in which $\epsilon(x) = \kappa(x)\mu(x)$ for some $\kappa(x) > 0$. Using Theorem 9.3, one can discuss the inverse scattering for this case.

Ola-Päivärinta-Somersalo [80] took into account, in addition to the electric permittivity $\epsilon(x)$ and the magnetic permeability $\mu(x)$, the electric conductivity $\sigma(x)$ of the medium, and studied the following boundary value problem

$$\mathcal{M}(\lambda)^{-1}\mathcal{L}u = \lambda u, \quad \mathcal{M}(\lambda) = \begin{pmatrix} \epsilon + \dfrac{i\sigma}{\lambda} & 0 \\ 0 & \mu \end{pmatrix}. \tag{10.127}$$

The *chiral medium* was studied by Joshi-McDowall [48]. It is formulated as the boundary value problem

$$\mathcal{M}^{-1}(x)\mathcal{L}u = \lambda u, \quad \mathcal{M} = \begin{pmatrix} a & c \\ -c & b \end{pmatrix}. \tag{10.128}$$

The method of complex geometrical optics solutions also works for these non-selfadjoint boundary value problems (10.127), (10.128).

It is an interestng problem to extend inverse scattering theory to these non-self adjoint problems. Petkov [82] developed a scattering theory for non-selfadjoint hyperbolic equation, and discussed some inverse problems different from the ones treated here. There is also a long history of the inverse problem for the wave equation related to the reflection of light by obstacles. The recent results are summarized in Petkov-Stoyanov [83].

10.7   Identifiability of the domain

For the exterior boundary value problem in uniform media, the main issue of the inverse problem is the identification of the domain. Let us begin with the Helmholtz equation

$$(-\Delta - \lambda)u(x) = 0, \quad \text{in} \quad \Omega, \tag{10.129}$$

where Ω is an exterior domain in $\mathbf{R}^n$, and the Dirichlet (or Neumann) condition is imposed on $\partial\Omega$. We assume that $\partial\Omega$ is compact and sufficiently smooth. One can the find the solution $u(x)$ to (10.129) having the asymptotic expansion

$$u(x) \simeq e^{i\sqrt{\lambda}\omega \cdot x} + \frac{e^{i\sqrt{\lambda}r}}{r^{(n-1)/2}} a_\infty(\lambda; \theta, \omega)$$

as $r = |x| \to \infty$, where $\theta = x/r$. The 1st term $e^{i\sqrt{\lambda}\omega \cdot x}$ represents an incident plane wave with propagation direction $\omega \in S^{n-1}$ and frequency $\sqrt{\lambda}$, and the *far field pattern* $a_\infty(\lambda; \theta, \omega)$ is the integral kernel of the scattering amplitude up to a constant depending on λ. The first inverse result is attributed to Schiffer ([65]): Ω is uniquely determined from the far field pattern of all frequency $\lambda > 0$. The complete proof was given by Colton and Kress [22]. This result is further extended to the case of a fixed frequency and also to the Maxwell equation. See [23], Theorems 5.1 and 7.1.

Theorem 10.7. *Given two exterior domains* Ω_1, Ω_2 *in* $\mathbf{R}^n$, *and an arbitrarily fixed frequency* $\sqrt{\lambda} > 0$, *suppose that the associated S-matrices for the Helmholtz equation coincide. Then,* $\Omega_1 = \Omega_2$.

Theorem 10.8. *Given two exterior domains* Ω_1, Ω_2 *with compact* C^∞- *boundaries, consider the Maxwell operator* $\mathcal{M}_0^{-1}\mathcal{L}$ *with* $\mathcal{M}_0 = \begin{pmatrix} \epsilon_0 I_3 & 0 \\ 0 & \mu_0 I_3 \end{pmatrix}$ *where* ϵ_0, μ_0 *are positive constants, and the boundary condition* $\nu \times E = 0$ *on* $\partial\Omega_1$, $\partial\Omega_2$. *Suppose their S-matrices coincide for an arbitrarily fixed frequency* $\lambda > 0$. *Then,* $\Omega_1 = \Omega_2$.

One can also deal with impedance boundary conditions. It is interesting that for polyhedral domains, only a finite number of incident waves with a fixed frequency is sufficient to determine the domain (see the works of Liu, Yamamoto and Zou, Liu and Zou [68], [67]).

The factorization method due to Kirsch [54], [55] is based on the deep connection between the scattering and the D-N map in the interior domain. The basic idea is as follows. Suppose that the measurement is done by the operator A defined on the measurement domain. Then A is factorized as

$$A = FBF^*, \tag{10.130}$$

where B is an operator on the boundary of the unknown domain, and F is an operator with range in the measurement of the domain. Once we have established (10.130), we can prove that the knowledge of A determines the range of F which then determines the obstacle. For the scattering of acoustic waves in an exterior domain, in which case the operator is $-\Delta - k^2 c(x)$, A is the scattering amplitude (with suitable modification), B

is the D-N map and F is the operator which assigns the far-field pattern of the solution to the reduced wave equation in the exterior domain with prescribed data on the boundary. Nachman-Päivärinta-Terilä [76] extended Kirsch's argument as well as its functional analytic back-ground.

The inverse scattering problem in the anisotropic medium

$$\left(-\sum_{i,j=1}^{n} \frac{\partial}{\partial x_i}\left(a_{ij}(x)\frac{\partial}{\partial x_j}\right) - \lambda n(x)\right)u(x) = 0 \qquad (10.131)$$

is much harder, and in general the coefficents are not determined uniquely from the far-field pattern. A more accesible problem is that of determining the shape of the penetrable medium D on which $(a_{ij}(x)) \neq I$ or $n(x) \neq n_0$, n_0 being a constant, more precisely, the smallest region D such that $(a_{ij}(x)) = I$ and $n(x) = n_0$ on $\mathbf{R}^n \setminus D$. Hähner [38] proved that the domain D is determined from the far-field pattern. This results was extended to the Maxwell equation by Cakoni and Colton [13]. The idea of Hähner is based on two ingredients, the linear sampling method and the interior transmission problem, which we explain in the next sections.

10.8   Linear sampling method

Let us return to the Helmholtz equation in an exterior domain $\Omega_{ext} \subset \mathbf{R}^n$:

$$\begin{cases} (-\Delta - \lambda)u = 0, & \text{in} \quad \Omega_{ext}, \\ u = f \in H^{-1/2}(\partial\Omega), & \text{on} \quad \partial\Omega_{ext}, \\ \left(\dfrac{\partial}{\partial r} - i\sqrt{\lambda}\right)u \in \mathcal{B}_0^*. \end{cases} \qquad (10.132)$$

Then, there is a unique solution u of this equation, which has the following asymptotic expansion:

$$u - \frac{e^{i\sqrt{\lambda}r}}{r^{(n-1)/2}}a_\infty(\theta; f) \in \mathcal{B}_0^*, \qquad \theta = x/r. \qquad (10.133)$$

The linear sampling method is a way for the reconstruction of the interior domain $\Omega_{int} = \mathbf{R}^n \setminus \overline{\Omega_{ext}}$ from the far-field pattern. It consists of the following obeservation. For $z \in \Omega_{int}$, let $\Phi(x, z)$ be the fundamental solution to the Helmholtz equation satisfying

$$\left(-\Delta - \lambda\right)\Phi(x, z) = \delta(x - z), \qquad (10.134)$$

and put

$$\phi(\theta, z) = a_\infty(\theta; \Phi(\cdot, z)). \qquad (10.135)$$

Let $z_n \in \Omega_{int}$ tend to $z \in \partial\Omega_{int}$. Then, since $\Phi(x,z)$ is singular at $x = z$, we will see that $\phi(\theta, z_n) \to \infty$.

Take two bounded domains $\Omega_{int}^{(i)}$, $i = 1, 2$, such that the far-field patterns for $\Omega_{int}^{(1)}$ and $\Omega_{int}^{(2)}$ coincide for a fixed frequency $\sqrt{\lambda}$. Suppose $\Omega_{int}^{(1)} \neq \Omega_{int}^{(2)}$ and, for the sake of simplicity, consider the case in which $\Omega_{int}^{(1)} \cap \Omega_{int}^{(2)} \neq \emptyset$. For $z_n \in \Omega_{int}^{(1)} \cap \Omega_{int}^{(2)}$, put

$$\phi^{(i)}(\theta, z_n) = a_\infty^{(i)}(\theta; \Phi(\cdot, z_n)), \quad i = 1, 2,$$

where $a_\infty^{(i)}$ is the far-field pattern for the domain $\Omega_{int}^{(i)}$. Let $u^{(i)}(x, z_n)$ be the solution of (10.132) for $\Omega = \Omega_{ext}^{(i)}$ and $f = \Phi(x, z_n)$. Let $\Omega_{ext}^{(3)} = \mathbf{R}^n \setminus \overline{\Omega_{int}^{(1)} \cap \Omega_{int}^{(2)}}$ and assume that $\Omega_{ext}^{(3)}$ is connected. Since the far-field patterns for $u^{(i)}(x, z_n)$ coincide for $i = 1, 2$, the difference $u^{(1)}(x, z_n) - u^{(2)}(x, z_n)$ satisfies the Helmholtz equation in $\Omega_{ext}^{(3)}$ and belongs to $\mathcal{B}_0^*$. Therefore, by Rellich's theorem, $u^{(1)}(x, z_n) = u^{(2)}(x, z_n)$ on $\Omega_{ext}^{(3)}$. Then, there is a sequence $\{z_n\}$ in $\Omega_{int}^{(1)} \cap \Omega_{int}^{(2)}$ such that $z_n \to z \in \partial\Omega_{int}^{(1)}$, however, $z \in \Omega_{int}^{(2)}$. Then, we will see that $\phi^{(1)}(\theta, z_n) \to \infty$, while $\phi^{(2)}(\theta, z_n) \to \phi^{(2)}(\theta, z)$, a contradiction. In this way, we can determine the domain Ω_{int} from the far-field pattern. More precise assertion is as follows.

Theorem 10.9. *Let $A(\lambda)$ be the scattering amplitude for $-\Delta$ in an exterior domain Ω_{ext}, and $\Phi_\infty(\theta, z) = e^{-i\lambda\theta\cdot z}$ the far-field pattern of the fundamental solution $\Phi(x, z)$ of the Helmholtz equation. Assume that λ is not a Dirichlet eigenvalue of $-\Delta$ in Ω_{int} with Dirichlet boundary condition.*
(1) For $z \in \Omega_{int}$ and a given $\epsilon > 0$, there exists $g_z^\epsilon \in L^2(S^{n-1})$ such that

$$\|A(\lambda)g_z^\epsilon - \Phi_\infty(\cdot, z)\|_{L^2(S^{n-1})} < \epsilon, \tag{10.136}$$

and v_z^ϵ defined by

$$v_z^\epsilon = C(\lambda) \int_{S^{n-1}} e^{i\sqrt{\lambda}\theta\cdot x} g_z^\epsilon(\theta) d\theta, \tag{10.137}$$

$C(\lambda)$ being a constant depending only on λ, converges to $w \in H^1(\Omega_{int})$ to the solution $w \in H^1(\Omega_{int})$

$$\begin{cases} (-\Delta - \lambda)w = 0, & in \quad \Omega_{int}, \\ w = -\Phi_\infty(\cdot, z) & on \quad \partial\Omega_{int}, \end{cases} \tag{10.138}$$

as $\epsilon \to 0$.
(2) For $z \notin \Omega_{int}$, any $g_z^\epsilon \in L^2(S^{n-1})$ satisfying (10.136) also satisfies

$$\|v_z^\epsilon\|_{H^1(\Omega_{int})} \to \infty, \quad as \quad \epsilon \to 0,$$

where v_z^ϵ is defined by (10.137).

For the proof, see Theorem 5.34 of [23].

10.9 Transmission eigenvalue

Let us recall Theorem 8.14. Given an incoming data $\varphi_-(x) \in \mathbf{h}_\lambda$, there exist a unique $u \in \mathcal{N}(\lambda)$ and $\varphi_+ = S_J(\lambda)\varphi_- \in \mathbf{h}$ satisfying

$$(\mathcal{H} - \lambda)u = 0, \quad \text{in} \quad \mathbf{R}^3, \tag{10.139}$$

$$u \simeq -i(2\pi c_0)^{-1/2}r^{-1}\left(e^{i\frac{\lambda}{c_0}r}\varphi_+(\widehat{x}) - e^{-i\frac{\lambda}{c_0}r}\varphi_-(\widehat{x})\right). \tag{10.140}$$

If $\epsilon(x) = \epsilon_0$ and $\mu(x) = \mu_0$, this u is written by u_0, where by (8.15),

$$u_0 = \mathcal{F}_-^{(0)}(\lambda)^*\varphi(x) = \frac{\lambda}{(2\pi c_0)^{3/2}}\int_{S^2} e^{-i\frac{\lambda}{c_0}\omega\cdot x}\mathcal{P}_-^{(0)}(\omega)\varphi(\omega)d\omega. \tag{10.141}$$

Using this, one can rewrite (10.140) as

$$u - u_0 \simeq \frac{e^{i\frac{\lambda}{c_0}r}}{r}a_\infty(\lambda, \widehat{x}). \tag{10.142}$$

The operator

$$F_\infty(\lambda) : \varphi_- \to a_\infty \tag{10.143}$$

is often called the *far-field operator*. It coincides with $A(\lambda)J$ up to a constant, where $A(\lambda)$ is the scattering amplitude (8.61). This is directly related with the scattering experiment. To make the explanation simpler, we change the definition slightly. Let us call

$$\widehat{A}(\lambda) = \widehat{S}(\lambda) - I \tag{10.144}$$

the scattering amplitude. Since $\widehat{A}(\lambda)$ is compact (see Theorem 8.11), and $\widehat{S}(\lambda)$ is unitary, the non-zero eigenvalues of $\widehat{A}(\lambda)$ is of the form

$$e^{i\lambda_n} - 1, \quad n = 1, 2, \cdots,$$

where $\lambda_n \in (0, 2\pi)$ with accumulation points of only 0 and 2π. If 0 is an eigenvalue for $\widehat{A}(\lambda)$, so is for $F_\infty(\lambda)$ and the associated eigenvector cannot be observed in the scattering experiment.

Theorem 10.10. *Let $\Omega \subset \mathbf{R}^3$ be a bounded domain such that $\epsilon(x) = \epsilon_0$, $\mu(x) = \mu_0$ outside Ω. Then, the far-field operator $F_\infty(\lambda)$ has 0 as an eigenvalue if and only if there is a non-trivial solution $v, w \in C^2(\Omega) \cap C^1(\overline{\Omega})$ of the following problem*

$$\begin{cases} (\mathcal{H} - \lambda)w = 0 & \text{in} \quad \Omega, \\ (\mathcal{H}_0 - \lambda)v = 0 & \text{in} \quad \Omega, \\ \nu \times (v - w) = 0 & \text{on} \quad \partial\Omega, \end{cases} \tag{10.145}$$

where v is given by (10.141) with non-zero $\varphi \in L^2(S^2)$.

Proof. Assume $F_\infty(\lambda)\varphi_- = 0$ for some $0 \neq \varphi_- \in L^2(S^2)$, and further that there are no non-trivial solutions to (10.145). Then u in (10.139) satisfies $(\mathcal{H}_0 - \lambda)(u - u_0) = 0$ outside Ω and $u - u_0 \in \mathcal{B}_0^*$. By Rellich's theorem $u = u_0$ outside Ω. We then see that $(v, w) = (u_0, u)$ satisfies (10.145). By the assumption, we have $u = 0$ in $\mathbf{R}^3$, hence $w = 0$ in $\mathbf{R}^3$. This yields $\varphi_- = 0$ and the contradiction.

Conversely, suppose there exists a non-trivial solution (v, w) of (10.145) with non-zero φ. Then, letting u to be u_0 with $\varphi_- = \varphi$ outside Ω and $u = w$ on Ω, we see that u is the one specified by (10.139) and (10.140). We then have $u = u_0$ outside Ω, hence $F_\infty(\lambda)\varphi = 0$. $\qquad \square$

We are thus led to the boundary value problem (10.145) in the interior domain Ω. The value of λ for which (10.145) has a non-trivial solution is called the *transmision eigenvalue*. Sometimes we pick up only the electric part, and consider the following boundary value problem

$$\begin{cases} \operatorname{curl}\operatorname{curl} E = \lambda^2 \epsilon \mu_0 E, & \text{in} \quad \Omega, \\ \operatorname{curl}\operatorname{curl} E_0 = \lambda^2 \epsilon_0 \mu_0 E_0, & \text{in} \quad \Omega, \\ \nu \times (E - E_0) = G, & \text{on} \quad \partial\Omega, \\ \nu \times \operatorname{curl}(E - E_0) = H, & \text{on} \quad \partial\Omega. \end{cases} \qquad (10.146)$$

As can be easily inferred, the same problem is formulated for the acoustic equation

$$\begin{cases} \Delta u + k^2 c(x)u = 0 & \text{in} \quad \mathbf{R}^3, \\ u = u^{(in)} + u^{(out)}, \\ \left(\dfrac{\partial}{\partial r} - ik\right)u^{(out)} = o(r^{-1}). \end{cases} \qquad (10.147)$$

The transmission eigenvalues do exist, and are shown to be discrete in general. The knowledge of transmission eigenvalues determines the sound speed $c(x)$. The study of transmission eigenvalues is now rapidly growing, and a lot of new ideas and techniques has been found. See the review articles of Colton, Päivärinta and Sylvester [24], and Cakoni-Haddar [15]. For the Maxwell equation, see e.g. Cossonière and Hadder [26]. In the recent works of Petkov-Vodev [84], [93], the Weyl asymptotics of transmission eigenvalues for the acoustic equation is studied.

10.10   Determination of Betti number

We are also interested in the geometrical inverse problem for the Maxwell equation in an exterior domain Ω. The issue is to obtain the geometric information of the domain Ω or of the boundary $\partial\Omega$ from the knowledge of the scattering data. The assumption is minimal. We assume that $\epsilon(x)$, $\mu(x)$ are positive definite, and that there exist constants $R, \epsilon_0, \mu_0 > 0$ such that

$$\epsilon(x) = \epsilon_0, \quad \mu(x) = \mu_0, \quad \text{for} \quad |x| > R. \tag{10.148}$$

We put

$$D = \Omega \cap \{x \in \mathbf{R}^3 \,;\, |x| < R\}. \tag{10.149}$$

Letting $\Omega_{out} = \Omega \setminus \overline{D}$ and $\Omega_{int} = D$, we have seen that the S-matrix for the energy for λ uniquely determines the E-M map on $S = \{|x| = R\}$. We use the following property of $\Lambda_{int}(\lambda)$.

Lemma 10.14. *For $f \in H^{3/2}(\partial\Omega)$, $\Lambda_{int}(z)f$ is an $H^{-1/2}(S)$-valued meromorphic function of $z \in \mathbf{C}$ with poles at $\sigma(\mathcal{H}_{int})$.*

Proof. By Theorem 7.2, $f \in H^{3/2}(S)$ is extended to $F \in H^2(\Omega_{int})$ vanishing near $\partial\Omega$. Then,

$$u(z) = F - (\mathcal{H}_{int} - z)^{-1}(\mathcal{H} - z)F$$

is the unique solution to the boundary value problem

$$\begin{cases} (\mathcal{H} - z)u = 0, & \text{in} \quad \Omega_{int}, \\ \nu \times E = f, & \text{on} \quad S, \\ \nu \times E = 0, & \text{on} \quad \partial\Omega. \end{cases}$$

Then, for $\lambda \in \mathbf{R} \setminus \sigma(\mathcal{H}_{int})$, $\Lambda_{int}(z)f = \nu \times H$ on S. Theorem 7.4 shows that it has the desired property. $\qquad\square$

Now, in the interior domain D, consider the time-dependent boundary value problem

$$\begin{cases} \mu(x)^{-1}\operatorname{curl} E = -\partial_t H, & \text{in} \quad D, \\ \epsilon(x)^{-1}\operatorname{curl} H = \partial_t E, & \text{in} \quad D \end{cases} \tag{10.150}$$

with initial data

$$E(x,t)\big|_{t=-\tau_f} = 0, \quad H(x,t)\big|_{t=-\tau_f} = 0, \tag{10.151}$$

and boundary data

$$\nu \times E(x,t)\big|_{\partial D \times \mathbf{R}_-} = \begin{cases} f(x,t) & \text{on} \quad S \times \mathbf{R}_-, \\ 0 & \text{on} \quad \partial\Omega \times \mathbf{R}_-, \end{cases} \tag{10.152}$$

and τ_f is chosen so that $f(x,t) = 0$ if $t < -\tau_f$. The *response operator* is then defined by

$$R_S : \nu \times E\big|_{S \times \mathbf{R}_-} \to \nu \times H\big|_{S \times \mathbf{R}_-}. \tag{10.153}$$

Krupchyk-Kurylev-Lassas [61] proved the following theorem.

Theorem 10.11. *The knowledge of the response operator R_S determines the 1st Betti number of D.*

We now assume that we know the S-matrix $\widehat{S}(\lambda)$ for all frequency $\lambda \in \mathbf{R} \setminus \sigma_p(\mathcal{H})$. Then, it determines the $E - M$ map $\Lambda_{int}(\lambda)$. We show that it determines the Response operator. Taking the Fourier transform

$$\widehat{u}(z) = \int_{-\infty}^{\infty} e^{izt} u(t) dt,$$

we see that $\widehat{u}(z)$ is analytic in the upper-half plane $\{\text{Im } z > 0\}$ and satisfies

$$\begin{cases} (\mathcal{H} - z)\widehat{u}(z) = 0, & \text{in} \quad D, \\ \nu \times \widehat{u}(z) = \widehat{f}(z), & \text{on} \quad S, \\ \nu \times \widehat{u}(z) = 0, & \text{on} \quad \partial\Omega. \end{cases} \tag{10.154}$$

Then, $\widehat{u}(z) = \Lambda_{int}(z)\widehat{f}(z)$ for z in the upper-half plane, which determines $u(t)$ by virtue of Lemma 10.14. We have thus shown that the knowledge of S-matrix determines the response operator. Therefore, as the inverse scattering problem, we have proven the following theorem.

Theorem 10.12. *The 1st Betti number of the boundary $\partial\Omega$ is determined from the knowledge of the S-matrix $\widehat{S}(\lambda)$ for all frequency $\lambda \in \mathbf{R} \setminus \sigma_p(\mathcal{H})$.*

In particular, the number of handles of the boundary $\partial\Omega$ is determind from the S-matrix of the Maxwell operator in the exterior domain Ω.

Remark 10.1. It is not difficult to extend the above theorem on Riemannian manifolds. Consider a non-compact connected 3-dimensional Riemannian manifold $\mathcal{M}$ with the following property. $\mathcal{M}$ is split into two parts

$$\mathcal{M} = \mathcal{K} \cup \mathcal{M}_\infty, \tag{10.155}$$

where $\mathcal{K}, \mathcal{M}$ are smooth Riemannian manifolds. Assume that $\mathcal{M}_\infty$ is isometric to $\{x \in \mathbf{R}^3 \, ; \, |x| > R\}$ endowed with the Euclidean metric $(dx)^2$, and

$\mathcal{K}$ is a relatively compact 3-dimensional manifold with arbitrary topolgy and metric. Moreover, we give two other metrics $\epsilon(x)$ and $\mu(x)$ on $\mathcal{M}_\infty$, which are equal to $\epsilon_0(dx)^2$ and $\mu_0(dx)^2$ with constants $\epsilon_0, \mu_0 > 0$, and define the Maxwell operator on $\mathcal{M}$ having $\epsilon(x)$ and $\mu(x)$ as dielectric permittivity and magnetic permeability. We do not give the precise formulation, since it is the same as [61]. Then, the scattering theory can be developped in the same way as above, and Theorem 10.12 holds also on this manifold $\mathcal{M}$.

The argument in [61] is closely related to the *boundary control method* (BC-method), which has its origin in the works of Krein [58], [59] in 1950s. While the method of Gel'fand-Levitan and Marchenko uses the stationary Schrödinger equation, especially spectral quantities in the complex plane, Krein considered the wave equation and passed the problem to the time domain. This idea was further developped by Blagvestcenskii [10], [11], and Belishev [7] established the BC method on n-dimensional Euclidean space. It was further extended to Riemannian manifolds by Belishev-Kurylev [9], and is now regarded as an only general method for the reconstruction of the manifold itself and its Riemannian metric. See the review article [8]. We also refer the monograph of Katchalov-Kurylev-Lassas[49] for the inverse boundary value problem, and Isozaki-Kurylev [47] for the inverse scattering problem.

Appendix A

Formulae in vector calculus

The vector product of $a = (a^1, a^2, a^3), b = (b^1, b^2, b^3) \in \mathbf{R}^3$ are defined by

$$a \times b = (a^2 b^3 - a^3 b^2, a^3 b^1 - a^1 b^3, a^1 b^2 - a^2 b^1). \tag{A.1}$$

We have only to remember the 1st component, since the remaining ones are obtained by the *cyclic change* of indices: $1 \to 2 \to 3 \to 1$. We put $a \cdot b = \sum_{i=1}^{3} a^i b^i$, and

$$\mathbf{e}_1 = (1, 0, 0), \quad \mathbf{e}_2 = (0, 1, 0), \quad \mathbf{e}_3 = (0, 0, 1).$$

We summarize useful formulas in the following lemmas, which are proven by direct computations.

Lemma A.1.

$$a \times b = -b \times a, \tag{A.2}$$

$$\mathbf{e}_i \times \mathbf{e}_j = \mathbf{e}_k, \quad (i, j, k) = (1, 2, 3), (2, 3, 1), (3, 1, 2), \tag{A.3}$$

$$(a \times b) \cdot c = \det(a, b, c), \tag{A.4}$$

$$(a \times b) \times c + (b \times c) \times a + (c \times a) \times b = 0, \tag{A.5}$$

$$(a \times b) \times c = (c \cdot a)b - (b \cdot c)a, \tag{A.6}$$

$$(a \times b) \cdot (c \times d) = (a \cdot c)(b \cdot d) - (a \cdot d)(b \cdot c), \tag{A.7}$$

$$|a|^2 |b|^2 = |a \times b|^2 + (a \cdot b)^2. \tag{A.8}$$

$$a \times b = \begin{pmatrix} 0 & -a^3 & a^2 \\ a^3 & 0 & -a^1 \\ -a^2 & a^1 & 0 \end{pmatrix} \begin{pmatrix} b^1 \\ b^2 \\ b^3 \end{pmatrix}. \tag{A.9}$$

Lemma A.2. *Let* $\nabla = \partial/\partial x = {}^t(\partial/\partial x^1, \partial/\partial x^2, \partial/\partial x^3)$. *Then, we have*

$$\text{curl grad} = (\nabla \times)\nabla = 0, \tag{A.10}$$

$$\text{div curl} = \nabla \cdot (\nabla \times) = 0, \tag{A.11}$$

$$\text{div grad} = \nabla \cdot \nabla = \Delta, \tag{A.12}$$

$$\text{curl curl} = \nabla \times (\nabla \times) = \text{grad div} - \Delta = \nabla(\nabla \cdot) - \Delta. \tag{A.13}$$

Lemma A.3. *For* $F(x) = {}^t(F^1(x), F^2(x), F^3(x))$, *we put*

$$F'(x) = \left(\frac{\partial F}{\partial x^1}, \frac{\partial F}{\partial x^2}, \frac{\partial F}{\partial x^3} \right). \tag{A.14}$$

Then, for $a(x) = (a^1(x), a^2(x), a^3(x))$ *and* $b(x) = (b^1(x), b^2(x), b^3(x))$,

$$\big((\nabla \times F) \times a\big) \cdot b = \big((a \cdot \nabla)F\big) \cdot b - a \cdot \big((b \cdot \nabla)F\big), \tag{A.15}$$

$$(\nabla \times F) \cdot (a \times b) = (F'a) \cdot b - a \cdot (F'b). \tag{A.16}$$

Proof.   A straight forward computation yields (A.15). Letting $c = \nabla \times F$, we have by (A.4)

$$c \cdot (a \times b) = \det(a, b, c) = \det(c, a, b) = (c \times a) \times b,$$

which implies

$$(\nabla \times F) \cdot (a \times b) = \big((\nabla \times F) \times a\big) \cdot b.$$

We use (A.15) and compute directly to show (A.16).   □

Lemma A.4. *For scalar fields* f, g *and vecrtor fields* F, G, *we have*

$$\nabla \cdot (fF) = \nabla f \cdot F + f \nabla \cdot F, \tag{A.17}$$

$$\nabla \times (fF) = \nabla f \times F + f \nabla \times F, \tag{A.18}$$

$$\nabla \cdot (F \times G) = (\nabla \times F) \cdot G - F \cdot (\nabla \times G), \tag{A.19}$$

$$\nabla(fg) = g\nabla f + f\nabla g, \tag{A.20}$$

$$\nabla \times (F \times G) = (G \cdot \nabla)F - (F \cdot \nabla)G + F\nabla \cdot G - G\nabla \cdot F, \tag{A.21}$$

$$\nabla(F \cdot G) = (F \cdot \nabla)G + (G \cdot \nabla)F + F \times (\nabla \times G) + G \times (\nabla \times F). \tag{A.22}$$

Appendix B

1-forms and vector fields

We start with giving a definite meaning for dx by using algebraic notions. Let U be an open set in $\mathbf{R}^n$, and $\mathcal{R} = C^\infty(U; \mathbf{R})$ the set of all real-valued C^∞-functions on U. It is a *ring*, i.e. it admits the algebraic operations of addition and multiplication. Take $p \in U$ and put

$$\mathcal{I}_p = \{f(x) \in \mathcal{R}\,;\, f(p) = 0\}.$$

This is an *ideal* of $\mathcal{R}$ in the sense that

$$f(x), g(x) \in \mathcal{I}_p \Longrightarrow f(x) + g(x) \in \mathcal{I}_p,$$

$$f(x) \in \mathcal{I}_p,\ h(x) \in \mathcal{R} \Longrightarrow h(x)f(x) \in \mathcal{I}_p.$$

For an ideal $\mathcal{I}_p$, we define $\mathcal{I}_p^2$ to be the set of all finite sums of the product $f_i(x)g_i(x)$, where $f_i(x), g_i(x) \in \mathcal{I}_p$. It is again an ideal. Recall the formula for $f(x) \in \mathcal{R}$

$$
f(x) - f(p) = \sum_{i=1}^n \frac{\partial f}{\partial x^i}(p)(x^i - p^i)
$$
$$
+ \sum_{i,j=1}^n \left(\int_0^1 \frac{\partial^2 f}{\partial x^i \partial x^j}(t(x-p)+p)dt \right)(x^i - p^i)(x^j - p^j). \tag{B.1}
$$

The 1st term of the right-hand side is in $\mathcal{I}_p$, and the 2nd term in $\mathcal{I}_p^2$. Since $\mathcal{I}_p$ is a vector space and $\mathcal{I}_p^2$ is its vector subspace, we can consider the factor space

$$T_p^*(U) := \mathcal{I}_p/\mathcal{I}_p^2. \tag{B.2}$$

It simply means that we regard the term in $\mathcal{I}_p^2$ to be 0. Therefore in $T_p^*(U)$, $f(x) - f(p)$ is equal to the 1st term of the right-hand side of (B.1), which is called the *differential* of f at p, and denoted as $(df)_p$, i.e.

$$(df)_p \equiv \sum_{i=1}^n \frac{\partial f}{\partial x^i}(p)(x^i - p^i). \tag{B.3}$$

261

Here the symbol $\equiv$ means that the difference of both sides is in $\mathcal{I}_p^2$. Note that we have by definition

$$\left(dx^i\right)_p = x^i - p^i, \quad \text{in} \quad T_p^*(U), \tag{B.4}$$

and also

$$\left(df\right)_p = \sum_{i=1}^{n} \frac{\partial f}{\partial x^i}(p)\left(dx^i\right)_p. \tag{B.5}$$

Each element of $T_p^*(U)$ is called a *cotangent vector* at p, or a *1-form* at p.

Lemma B.1. *The differential defined above has the following properties.*
(1) $\left(d(f+g)\right)_p = \left(df\right)_p + \left(dg\right)_p, \quad f,g \in \mathcal{R}.$
(2) $\left(d(fg)\right)_p = \left(df\right)_p g(p) + f(p)\left(dg\right)_p, \quad f,g \in \mathcal{R}.$
(3) $\left(dx^1\right)_p, \cdots, \left(dx^n\right)_p$ *form a basis of* $T_p^*(U).$

Proof. (1) is obvious. Neglecting the terms in $\mathcal{I}_p^2$, we compute

$$f(x)g(x) - f(p)g(p) = (f(x) - f(p))g(x) + f(p)(g(x) - g(p))$$
$$\equiv (f(x) - f(p))g(p) + f(p)(g(x) - g(p)),$$

which proves (2). Since $(dx^i)_p = x^i - p^i$, (B.3) implies that they span $T_p^*(U)$. Suppose $\sum_{i=1}^{n} c_i (dx^i)_p \equiv 0$. Then

$$\sum_{i=1}^{n} c_i(x^i - p^i) = \sum_{j} f_j(x)g_j(x)$$

holds on U, where $f_j(x), g_j(x) \in \mathcal{I}_p$. Differentiating both sides and putting $x = p$, we obtain $c^i = 0$, which proves (3). $\square$

We put

$$T^*(U) = \{(p,\omega)\,;\, \omega \in T_p^*(U)\}. \tag{B.6}$$

Take C^∞-functions $a_i(x) \in \mathcal{R}$, $i = 1, \cdots, n$, and consider the mapping

$$U \ni p \to \sum_{i=1}^{n} a_i(p)\left(dx^i\right)_p \in T^*(U), \tag{B.7}$$

which is called a *1-form* on U and is denoted by

$$\sum_{i=1}^{n} a_i(x)dx^i.$$

Similarly, the mapping $U \ni p \to (df)_p \in T^*(U)$ is denoted by

$$df(x) = \sum_{i=1}^{n} \frac{\partial f}{\partial x^i}(x)dx^i, \tag{B.8}$$

and is also called the differential of f on U.

Since $T_p^*(U)$ is a vector space, we can consider its dual space. Let $T_p(U)$ be the set of all linear mappings from $T_p^*(U)$ to $\mathbf{R}$. Each element of $T_p(U)$ is called a *tangent vector* at p. The value of $v \in T_p(U)$ at $\omega \in T_p^*(U)$ is denoted by $\langle v, \omega \rangle_p$, i.e.

$$v(\omega) = \langle v, \omega \rangle_p, \quad v \in T_p(U), \quad \omega \in T_p^*(U). \tag{B.9}$$

By definition, the dual basis of $(dx^1)_p, \cdots, (dx^n)_p \in T_p^*(U)$ are the vectors $v_1, \cdots, v_n \in T_p(U)$ such that $\langle v_i, (dx^j)_p \rangle_p = \delta_{ij}$. Recalling that $(dx^i)_p = x^i - p^i$, it is convenient to denote v_i by $(\partial/\partial x^i)_p$, i.e.

$$\left\langle \left(\frac{\partial}{\partial x^i}\right)_p, (dx^j)_p \right\rangle_p = \delta_i^j = \begin{cases} 1 & (i = j), \\ 0 & (i \neq j). \end{cases} \tag{B.10}$$

We put

$$T(U) = \{(p, v) \,;\, v \in T_p(U)\}. \tag{B.11}$$

Take C^∞-functions $a^i(x) \in \mathcal{R}$, $i = 1, \cdots, n$, and consider the mapping

$$U \ni p \to \sum_{i=1}^n a^i(p)\left(\frac{\partial}{\partial x^i}\right)_p \in T(U). \tag{B.12}$$

This is called a *vector field* on U and is denoted by

$$\sum_{i=1}^n a^i(x)\frac{\partial}{\partial x^i}.$$

Take a vector $a = (a^1, \cdots, a^n)$ and a curve $c(t)$ in U such that $c(0) = p$, $\dot{c}(0) = a$. Then we have for $f(x) \in \mathcal{R}$

$$\begin{aligned} \frac{d}{dt}f(c(t))\Big|_{t=0} &= \sum_{i=}^n a^i \frac{\partial f}{\partial x^i}(p) \\ &= \left\langle \sum_{i=1}^n a^i \left(\frac{\partial}{\partial x^i}\right)_p, \sum_{j=1}^n \frac{\partial f}{\partial x^j}(p)(dx^j)_p \right\rangle_p. \end{aligned} \tag{B.13}$$

The formula (B.13) shows that vector fields act not only on 1-forms but also on functions.

Let U, V be open sets in $\mathbf{R}^n$, and suppose there is a diffeomorphism $\varphi : U \ni x \to y = \varphi(x) \in V$. For $f(y) \in C^\infty(V ; \mathbf{R})$, we put

$$(\varphi^* f)(x) = f(\varphi(x)).$$

Then we have

$$d(\varphi^* f)(x) = \sum_{j=1}^n \left(\sum_{i=1}^n \frac{\partial f}{\partial y^i}(\varphi(x))\frac{\partial \varphi^i}{\partial x^j}(x) \right) dx^j. \tag{B.14}$$

Let $q = \varphi(p)$, and $f(y) = y^i$. Using the abbreviation $y^i = (\varphi^* y^i)(x) = \varphi^i(x)$, we have

$$(d\varphi^i)(p) = \sum_{j=1}^{n} \frac{\partial y^i}{\partial x^j}(p)(dx^j)_p.$$

We now use $y^1, \cdots, y^n$ to mean the coordinate functions on $\mathbf{R}^n$ to see that

$$(d\varphi^i)(p) = \varphi^i(x) - \varphi^i(p) = y^i - p^i = (dy^i)_q,$$

which implies

$$(dy^i)_q = \sum_{j=1}^{n} \frac{\partial y^i}{\partial x^j}(p)(dx^j)_p. \tag{B.15}$$

Noting that $(\partial x^i/\partial y^j) = (\partial y^i/\partial x^j)^{-1}$ and that $(\partial/\partial y^i)_q$'s are the dual basis of $(dy^i)_q$'s , we have from (B.15)

$$\left(\frac{\partial}{\partial y^i}\right)_q = \sum_{j=1}^{n} \frac{\partial x^j}{\partial y^i}(q)\left(\frac{\partial}{\partial x^j}\right)_p. \tag{B.16}$$

This agrees with the chain rule of differentiation.

Let U be an open set in $\mathbf{R}^m$ and φ a 1 to 1 C^∞-map from U to $\mathbf{R}^N$ satisfying

$$\text{rank}\left(\frac{\partial\varphi}{\partial u^1}, \cdots, \frac{\partial\varphi}{\partial u^m}\right) = m, \quad \text{on} \quad U. \tag{B.17}$$

Then $M = \varphi(U)$ is an m-dimensional submanifold of $\mathbf{R}^N$. Take $u_0 \in U$ arbitrarily, and put $x_0 = \varphi(u_0)$. One then has 1-forms and tangent vectors

$$(dx^1)_{x_0}, \cdots, (dx^N)_{x_0}, \quad \left(\frac{\partial}{\partial x^1}\right)_{x_0}, \cdots, \left(\frac{\partial}{\partial x^N}\right)_{x_0}, \quad \text{at} \quad x_0 \in \mathbf{R}^N,$$

$$(du^1)_{u_0}, \cdots, (du^m)_{u_0}, \quad \left(\frac{\partial}{\partial u^1}\right)_{u_0}, \cdots, \left(\frac{\partial}{\partial u^m}\right)_{u_0}, \quad \text{at} \quad u_0 \in U.$$

Let us consider their relations. Noting the formula

$$(d\varphi^i(u))_{u_0} = \sum_{j=1}^{m} \frac{\partial\varphi^i}{\partial u^j}(u_0)(du^j)_{u_0},$$

we define the *pull-back*: $T^*(\mathbf{R}^N) \to T^*(U)$ by

$$\varphi^* : \sum_{i=1}^{N} a_i(x)dx^i \to \sum_{j=1}^{n}\left(\sum_{i=1}^{N} a_i(\varphi(u))\frac{\partial\varphi^i}{\partial u^j}(u)\right)du^j, \tag{B.18}$$

and the *push-forward*: $T(U) \to T(\mathbf{R}^N)$ by

$$\varphi_* : \sum_{j=1}^{m} b^j(u) \left(\frac{\partial}{\partial u^j} \right)_u \to \sum_{i=1}^{N} \left(\sum_{j=1}^{m} b^j(u) \frac{\partial \varphi^i}{\partial u^j}(u) \right) \left(\frac{\partial}{\partial x^i} \right)_{\varphi(u)}. \qquad \text{(B.19)}$$

The formula (B.18) gives the restriction of the 1-form $\sum_{i=1}^{N} a_i(x) dx^i$ on $\mathbf{R}^N$ to M, and (B.19) gives the imbedding of the tangent vector on U to $\mathbf{R}^N$. Therefore, it is natural to define

$$T_{\varphi(u)}(M) = T_u(U), \quad T^*_{\varphi(u)}(M) = T^*_u(U). \qquad \text{(B.20)}$$

and call them tangent and cotangent space of M at $\varphi(u)$. As a geometrical object, we identify $T_{\varphi(u)}(M)$ with $\varphi_* T_u(U)$. Similarly, we put

$$T(M) = T(U), \quad T^*(M) = T^*(U), \qquad \text{(B.21)}$$

and call them tangent and cotangent *bundle* of M. The elements of $T^*(M)$ and $T(M)$ are called 1-forms and vector fields on M, respectively.

Appendix C

Stationary phase method

We consider the following integral with a large parameter $t > 0$:

$$I(t) = \int_{\mathbf{R}^n} e^{it\phi(x)} a(x) dx. \tag{C.1}$$

Let $\Omega = \{x \in \mathbf{R}^n \, ; \, |x| < 1\}$, and assume the following conditions.

(A-1) $\phi(x)$ is a real-valued C^∞-function on Ω satisfying $C_\alpha = \sup_{x \in \Omega} |\partial_x^\alpha \phi(x)| < \infty$ for all α.

(A-2)   There exists a unique $x^* \in \Omega$ such that $\nabla \phi(x^*) = 0$. Moreover, it satisfies $\det \left(\partial_{x_i} \partial_{x_j} \phi(x^*) \right) \neq 0$.

(A-3) $a(x) \in C_0^\infty(\Omega)$.

Theorem C.1. *Let $H = \left(\partial_{x_i} \partial_{x_j} \phi(x^*) \right)$, and $\sigma = n_+ - n_-$, where $n_+ \, (n_-)$ is a number of positive (negative) eigenvalues of H. Let $N \geq 1$. Then, as $t \to \infty$*

$$I(t) = \left(\frac{2\pi}{t} \right)^{n/2} |\det H|^{-1/2} e^{\sigma \pi i / 4} e^{it\phi(x^*)} \left(\sum_{j=0}^{N-1} t^{-j} a_j(x^*) + t^{-N} R_N(t) \right),$$

where $a_0(x) = a(x)$, $a_j(x) = P_{2j} a(x)$, P_{2j} being a differential operator of order $2j$ with coefficients depending only on derivatives of $\phi(x)$,

$$|R_N(t)| \leq C \sum_{|\alpha| \leq K} \sup_{x \in \Omega} |\partial_x^\alpha a(x)|,$$

K is a constant depending only on N and n, and C is a constant depending only on C_α and $\det H$.

For the proof, see e.g. [42], Vol. 1, Theorem 7.7.7.

Theorem C.2. *For* $\varphi(\omega) \in C^\infty(S^{n-1})$ *and* $\lambda > 0$, *let*

$$I(x) = \int_{S^{n-1}} e^{i\sqrt{\lambda}\omega \cdot x} \varphi(\omega) \, d\omega.$$

Then, we have, letting $r = |x| \to \infty$, $\hat{x} = x/r$,

$$I(x) = C(\lambda) r^{-(n-1)/2} e^{i\sqrt{\lambda}r} \Big(\sum_{j=0}^{N-1} (\sqrt{\lambda}r)^{-j} \varphi_j^{(+)}(\hat{x}) + (\sqrt{\lambda}r)^{-N} R_N^{(+)}(x, \lambda) \Big)$$

$$+ \overline{C(\lambda)} r^{-(n-1)/2} e^{-i\sqrt{\lambda}r} \Big(\sum_{j=0}^{N-1} (\sqrt{\lambda}r)^{-j} \varphi_j^{(-)}(\hat{x}) + (\sqrt{\lambda}r)^{-N} R_N^{(-)}(x, \lambda) \Big),$$

$$C(\lambda) = e^{-(n-1)\pi i/4} (2\pi)^{(n-1)/2} \lambda^{-(n-1)/4},$$

$$\varphi_j^{(\pm)}(\hat{x}) = \sum_{|\alpha| \le 2j} C_{j\alpha}^{(\pm)}(\hat{x}) \varphi^{(\alpha)}(\pm \hat{x}),$$

$$|R_N^{(\pm)}(x, \lambda)| \le C,$$

where $C_{j\alpha}(\hat{x}) \in C^\infty(S^{n-1})$, *and* $\varphi^{(\alpha)}$ *is an* α*-th derivative of* φ. *In particular,*

$$\varphi_0^{(\pm)}(\hat{x}) = \varphi(\pm\hat{x}).$$

Proof.   Without loss of generality, we assume that $U = \operatorname{supp} \varphi$ is sufficiently small, and that we can take $\theta = (\omega_1, \cdots, \omega_{n-1})$ as local coordinates. Then, $\omega_n = \pm\sqrt{1 - \theta^2}$, and

$$\frac{\partial}{\partial \theta_i} \omega \cdot x = x_i \mp \frac{\theta_i}{\sqrt{1 - \theta^2}} x_n.$$

Hence, $\nabla_\theta \omega \cdot x = 0$ if and only if $\omega = \pm\hat{x}$. We can then assume that $\hat{x} = (0, \cdots, 0, 1) =: e_n$. Take $0 < \epsilon < 1$ small enough, and let $\chi_\pm(t) \in C^\infty(\mathbf{R})$ be such that $\chi_+(t) = 1$ ($t > 1 - \epsilon$), $\chi_-(t) = 0$ ($t < 1 - 2\epsilon$), $\chi_-(t) = \chi_+(-t)$. Split $\varphi(\omega)$ as

$$\varphi(\omega) = \chi_+(\omega \cdot e_n)\varphi(\omega) + \chi_-(\omega \cdot x)\varphi(\omega) + \varphi_0(\omega) =: \varphi_+(\omega) + \varphi_-(\omega) + \varphi_0(\omega),$$

and also $I(x)$ as

$$I(x) = I_+(x) + I_-(x) + I_0(x),$$

$$I_k(x) = \int_{S^{n-1}} e^{i\sqrt{\lambda}\omega \cdot x} \varphi_k(\omega) \, d\omega, \quad k = \pm, 0.$$

On supp $\varphi_+(\omega)$, taking $\theta = (\omega_1, \cdots, \omega_{n-1})$ as local coordinates

$$I_+(x) = \int e^{i\sqrt{\lambda}rS(\theta)}\varphi_+(\theta, \sqrt{1-\theta^2})\frac{d\theta}{\sqrt{1-\theta^2}},$$

$$S(\theta) = \omega \cdot \widehat{x} = \sqrt{1-\theta^2}.$$

The stationary phase point, i.e. the zero of $\nabla_\theta S(\theta)$, is $\theta = 0$, at which all eigenvalues of the Hessian are equal to -1. Therefore, one can apply Theorem C.1 to $I_+(x)$. Similary, one can compute $I_-(x)$. On the support of $\varphi_0(\omega)$, taking suitable local coordinates θ, we have $|\nabla_\theta \omega \cdot x| \geq C|x|$ for a constant $C > 0$. Then, by integraion by parts

$$|I_0(x)| \leq C_N(1+r)^{-N}, \quad \forall N > 0.$$

This completes the proof of the theorem. $\qquad\square$

Theorem C.3. *For any* $L^2(S^{n-1})$,

$$(2\pi)^{-(n-1)/2} \int_{S^{n-1}} e^{ik\omega \cdot x}a(\omega)d\omega \simeq \frac{e^{ikr}}{(ikr)^{(n-1)/2}}a(\widehat{x}) + \frac{e^{-ikr}}{(-ikr)^{(n-1)/2}}a(-\widehat{x}),$$

$$(C.2)$$

$(k > 0, \widehat{x} = x/r, r = |x|)$, *where* $f(x) \simeq g(x)$ *means that*

$$\frac{1}{R}\int_{|x|<R}|f(x) - g(x)|^2dx \to 0, \quad R \to \infty. \tag{C.3}$$

Proof. When $\varphi \in C^\infty(S^{n-1})$, the theorem follows from Theorem C.2. If $\varphi \in L^2(S^{n-1})$, we approximate φ by $\phi \in C^\infty(S^{n-1})$. Letting $u = F_0(k)^*(\varphi - \phi)$, we have

$$\limsup_{R\to\infty} \frac{1}{R}\int_{|x|<R}|u(x)|^2dx \leq \|u\|_{\mathcal{B}^*}^2 \leq C\|\varphi - \phi\|_{L^2(S^{n-1})}^2,$$

from which we can prove the theorem immediately. $\qquad\square$

Appendix D

Interpolation theorem

Let us recall Hadamard's three line theorem.

Theorem D.1. *Suppose $f(z)$ is analytic in $\Omega = \{z = x + iy\,;\, 0 < x < 1,\ -\infty < y < \infty\}$, bounded continuous on $\overline{\Omega} = \{z = x + iy\,;\, 0 \le x \le 1,\ -\infty < y < \infty\}$, and*

$$\sup_{-\infty < y < \infty} |f(iy)| \le M_0, \tag{D.1}$$

$$\sup_{-\infty < y < \infty} |f(1 + iy)| \le M_1. \tag{D.2}$$

Then, for any $0 < x < 1$, we have

$$\sup_{-\infty < y < \infty} |f(x + iy)| \le M_0^{1-x} M_1^x. \tag{D.3}$$

Proof. Put $g(z) = f(z) M_0^{z-1} M_1^{-z}$. Then,

$$\sup_y |g(iy)| \le 1, \quad \sup_y |g(1 + iy)| \le 1.$$

Letting $g_n(z) = g(z) \exp(z^2/n)$, we have $|g_n(z)| \le |g(z)| \exp((x^2 - y^2)/n)$, hence $g_n(z)$ is bounded on $\overline{\Omega}$ and $g_n(z) \to 0$ as $|z| \to \infty$. By the maximum principle, $|g_n(z)|$ attains its maximum at $x = 0$ or $x = 1$. Therefore, $|g_n(z)| \le \exp(1/n)$ on $\overline{\Omega}$. Letting $n \to \infty$, we have $|g(z)| \le 1$, which yields

$$|f(z)| \le |M_0^{1-z} M_1^z| = M_0^{1-x} M_1^x,$$

and proves the lemma. □

Theorem D.2. *Let A, B be positive definite bounded self-adjoint operators on a Hilbert space $\mathbf{H}$ and T a bounded operator. If*

$$\|A^{\alpha_0} T B^{\beta_0}\| \le C_0, \quad \|A^{\alpha_1} T B^{\beta_1}\| \le C_1,$$

holds, then T satisfies

$$\|A^{\alpha_\theta} T B^{\beta_\theta}\| \le C_0^{1-\theta} C_1^{\theta}, \quad 0 < \theta < 1$$

for $\alpha_\theta = \alpha_0(1 - \theta) + \alpha_1 \theta$, $\beta_\theta = \beta_0(1 - \theta) + \beta_1 \theta$.

Proof.   Let $E_A(\lambda)$ and $E_B(\lambda)$ be the spectral decomposition of A and B. For a bounded interval I and $u, v \in \mathbf{H}$, put

$$f(z) = (E_A(I)A^{\alpha_z}TB^{\beta_z}E_B(I)u, v).$$

Since $f(z)$ is analytic in $0 < \operatorname{Re} z < 1$, bounded continuous in $0 \leq \operatorname{Re} z \leq 1$, we have for $y \in \mathbf{R}$

$$|f(iy)| \leq C_0\|u\|\|v\|, \quad |f(1+iy)| \leq C_1\|u\|\|v\|.$$

Theorem D.1 implies

$$|f(\theta)| \leq C_0^{1-\theta}C_1^{\theta},$$

hence we have

$$\|E_A(I)A^{\alpha_\theta}TB^{\beta_\theta}E_B(I)\| \leq C_0^{1-\theta}C_1^{\theta}.$$

Letting $I \to \mathbf{R}$, we prove the theorem. $\square$

Appendix E

Theory of quadratic forms

Let $\mathcal{H}$ be a Hilbert space, and D a dense subspace of $\mathcal{H}$. A mapping $a(\cdot,\cdot) : D \times D \to \mathbf{C}$ is said to be a *symmetric quadratic form* with domain D if it satisfies

$$a(\lambda u + \mu v, w) = \lambda a(u, w) + \mu a(v, w) \tag{E.1}$$

for any $\lambda, \mu \in \mathbf{C}$ and $u, v, w \in D$, and

$$\overline{a(u, v)} = a(v, u), \quad u, v \in D. \tag{E.2}$$

It is said to be *positive definite* if there exists a constant $C > 0$ such that

$$a(u, u) \geq C\|u\|^2, \quad \forall u \in D. \tag{E.3}$$

In this case, $a(\cdot,\cdot)$ is an inner product on D. If D is complete with respect to the norm $\|u\|_a = \sqrt{a(u,u)}$, i.e. for any Cauchy sequence $\|u_n - u_m\|_a \to 0$ $(u_n \in D)$ there exists $u \in D$ such that $\|u_n - u\|_a \to 0$, $a(\cdot,\cdot)$ is said to be a *closed form*. A symmetric positive definite quadratic form $a(\cdot,\cdot)$ is said to be *closable* if it satisfies

$$u_n \in D, \ \|u_n\| \to 0, \ \|u_n - u_m\|_a \to 0 \Longrightarrow \|u_n\|_a \to 0. \tag{E.4}$$

For a closable form $a(\cdot,\cdot)$, we define a subspace $\widetilde{D}$ by

$$u \in \widetilde{D} \Longleftrightarrow \exists u_n \in D \text{ s.t. } \|u_n - u\| \to 0, \ \|u_n - u_m\|_a \to 0. \tag{E.5}$$

We define a new quadratic form $\widetilde{a}(\cdot,\cdot)$ with domain $\widetilde{D}$ as follows: For $u, v \in \widetilde{D}$, we take $u_n, v_n \in D$ satisfying $\|u_n - u\| \to 0$, $\|v_n - v\| \to 0$, $\|u_n - u_m\|_a \to 0$, $\|v_n - v_m\|_a \to 0$, and define

$$\widetilde{a}(u, v) = \lim_{m,n \to \infty} a(u_m, v_n). \tag{E.6}$$

One can then show that $\widetilde{a}(u, v)$ does not depend on the choice of sequences $\{u_n\}, \{v_n\}$, and $\widetilde{a}(\cdot,\cdot)$ is a positive definite closed form on $\widetilde{D}$. We call it the *closed extension* of $a(\cdot, \cdot)$. Note that $D \subset \tilde{D}$ and $a(u, v) = \tilde{a}(u, v)$ for $u, v \in D$.

Theorem E.1. *Let $a(\cdot, \cdot)$ be a positive definite closed symmetric form with domain D. Then, there exists uniquely a self-adjoint operator A with domain $D(A)$ such that $D(A) \subset D$ and*

$$a(u, v) = (Au, v), \quad u \in D(A), \quad v \in D.$$

Moreover, $D = D(A^{1/2})$.

For the proof, see [51] Chap. 6, Theorems 2.1 and 2.23, or [85] Vol. 1, Theorem 8.15.

Bibliography

[1] R. Adams, *Sobolev Spaces*, Academic Press, (1975).

[2] S. Agmon, *Lectures on Elliptic Boundary Value Problems*, New York, Van Nostrand (1965).

[3] S. Agmon, *Spectral properties of Schrödinger operators and scattering theory*, Ann. Scoula Norm. Sup. Pisa **2** (1975), 151-218.

[4] S. Agmon and L. Hörmander, *Asymptotic properties of solutions of differential equations with simple characteristics*, J. d'Anal. Math. **30** (1976), 1-38.

[5] A. Alonso and A. Valli, *Some remarks on the characterization of the space of tangential traces of $H(\text{rot}\,;\Omega)$ and the construction of an extension operator*, Manuscripta Math. **89** (1996), 159-178.

[6] H. Ammari and H. Kang, *Reconstruction of Small Inhomogeneities from Boundary Measurements*, Springer-Verlag, Berlin-Heidelberg, (2004).

[7] M. I. Belishev, *An approach to multidimensional inverse problems for the wave equation*, Dokl. Akad. Nauk SSSR **297** (1987), 524-527; English translation, Soviet Math. Dokl. **36** (1988), 481-484.

[8] M. I. Belishev, *Boundary control in constrution of manifolds and metics (the BC method)*, Inverse Problems **13** (1997), R1-R45.

[9] M. I. Belishev and Y. V. Kurylev, *To the reconstruction of a Riemannian manifold via its spectral data (BC method)*, Comm. in P. D. E. **17** (1992), 767-804.

[10] A. S. Blagovestcenskii, *The local method of solution of the non-stationary inverse scattering problem for an inhomogeneous string*, Trudy Mat. Inst. Steklova, **115** (1971), 28-38 (in Russian).

[11] A. S. Blagovestcenskii, *The nonselfadjoint inverse matrix boundary problem for a hyperbolic differential equation*, In : Problems of mathemtical physics, **5**, Spectral Theory, Izdat. Lenigrad Univ., Leningrad (1971), 38-62 (in Russian).

[12] A. Buffa, M. Costabel and D. Sheen, *On traces for $H(\text{curl}\,,\Omega)$ in Lipschitz domains*, J. Math. Anal. Appl. **276** (2002), 845-867.

[13] F. Cakoni and D. Colton, *A uniqueness theorem for an inverse electromagnetic scattering problem in inhomogeneous anisotropic media*, Proc.

Edinburgh Math. Soc. **46** (2003), 293-314.

[14] F. Cakoni and D. Colton, *A Qualitative Approach to Inverse Scattering Theory*, Springer, New York- Heidelberg-Dordrecht-London (2014).

[15] F. Caconi and H. Haddar, *Transmission eigenvalues in inverse scattering theory*, Inside Out II, MSRI Publications, Vol **60**, 2012.

[16] A. P. Calderón and R. Vaillancourt, *A class of bounded pseudo-differential operators*, Proc. Nat. Acad, Sci. USA., **69** (1972), 1185-1187.

[17] A. P. Calderón, *On an in verse boundary value problems, Seminar on Numerical Analysis and its Applications to Continuum Physics*, Soc. Brasileria de Matematica, Rio de Janeiro (1980), 65-73.

[18] J. Cantarella, D. DeTurck and H. Gluck, *Vector calculus and the topology of domains in 3-space*, The American Mathematical Monthly, **109** No. 5 (2002), 409-442.

[19] M. Cessenat, *Mathematical Methods in Electromagnetism, Linear Theory and Applications*, World Scientific, Singapore-New Jersey-London-Hong Kong, (1996).

[20] K. Chadan, D. Colton, L. Päivärinta and W. Rundell, *An Introduction to Inverse Scattering and Inverse Spectral Problems*, SIAM, Philadelphia (1997).

[21] K. Chadan and P. Sabatier, *Inverse Problems in quantum scattering theory*, Springer Verlag, New York (1977).

[22] D. Colton and R. Kress, *Integral Equation Methods in Scattering Theory*, John Wiley and Sons, (1983).

[23] D. Colton and R. Kress, *Inverse Acoustic and Electromagnetic Scattering Theory*, Springer-Verlag, Berlin-Heidelberg, (1992).

[24] D. Colton, L. Päivärinta and J. Sylvester, *The interior transmission problem*, Inv. Prob. Imag. **1** (2007), 13-28.

[25] R. R. Coifman, A. McIntosh and Y. Meyer, *L'intégrale de Cauchy définit un opérateur borné sur L^2 pour les courbes Lipschiziennes*, Ann. of Math. **116** (1982), 361-387.

[26] A. Cossonnière and H. Haddar, *The electromagnetic interior transmission problem for regions with cavities*, SIAM J. Math. Anal. **43** (2011), 1698-1715).

[27] V. Enss, *Asymptotic completeness for quantum mechanical potential scattering I, Shortrange potentials*, Commun. Math. Phys. **61** (1978), 285-291.

[28] G. Eskin and J. Ralston, *Inverse scattering problem for the Schrödinger equation with magnetic potential at a fixed energy*, Commun. Math. Phys. **173** (1995), 199-224.

[29] E. B. Fabes, M. Jodeit and N. M. Rivières, *Potential techniques for boundary value problems on C^1 domains*, Acta Math. **141** (1978), 165-186.

[30] L. D. Faddeev, *Increasing solutions of the Schrödinger equations*, Sov. Phys. Dokl. **10** (1966), 1033-1035.

[31] L. D. Faddeev, *Factorization of the S-matrix for the multi-dimensional Schrödinger operator*, Sov. Phys. Dokl. **11** (1966), 209-211.

[32] L. D. Faddeev, *Inverse problem of quantum scattering theory*, J. Sov. Math. **5** (1976), 334-396.

[33] K. O. Friedrichs, *Mathematical Methods of Electromagnetic Theory*, Courant Inst. of Math. Sciences, New York University, New York ; American Math. Society, Providence, Rhode Island (2014).

[34] V. Georgescu, *Some boundary value problems for differential forms on compact Riemannian manifolds*, Ann. Mat. Pura Appl. **122** (1979), 159-198.

[35] V. Girault and P. A. Raviart, *Finite Element Methods for Navier-Stokes Equations*, Springer Verlag (1986).

[36] I. M. Gel'fand and B. M. Levitan, *On the determination of a differential equuation from its spectral function*, Trans. Amer. Math. Soc. **1** (1951), 253-304.

[37] P. Grisvard, *Elliptic Problems in Nonsmooth Domains*, Pitman (1985).

[38] P. Hähner, *On the uniqueness of the shape of a penetrable, anisotropic obstacle*, Jour. Comp. Applied Math. **116** (2000), 167-180.

[39] B. Helffer and J. Sjöstrand, *Equation de Schrödinger avec champ magnétique et équation de Harper*, Lecture Notes in Phys. **345**, *Schrödinger Operators*, pp. 118-197, eds. H. Holden, A. Jensen, Springer (1989).

[40] K. Hoffman, *Banach Spaces of Analytic Functions*, Prentice-Hall, Englewood Cliffths, N. J. (1962).

[41] L. Hörmander, *Lower bounds at infinity for solutions of differential equations with constant coefficients*, Israel J. Math. **16** (1973), 103-116.

[42] L. Hörmander, *The Analysis of Linear Partial Differential Operators I, II, III*, Springer-Verlag (1983), (1984).

[43] V. Isakov, *Inverse Problem for Partial Differential Equations*, Springer-Verlag, New York, (1998).

[44] H. Isozaki, *Multi-dimensional inverse scattering theory for Schrödinger operators*, Reviews in Math. Phys. **8** (1996), 591-622.

[45] H. Isozaki, *Inverse scatteirng theory for Dirac operators*, Ann. l'I. H. P. Physique Théorique **66** (1997), 237-270.

[46] H. Isozaki, *Inverse spectral theory*, in *Topics in the Theory of Schrödinger Operators*, pp. 93-143, eds. H. Araki and H. Ezawa, World Scientific (2003).

[47] H. Isozaki and Y. Kurylev, *Introduction to Spectral Theory and Inverse Problems on Asymptotically Hyperbolic Manifolds*, MSJ Memoires **32**, Math. Soc. of Japan (2014).

[48] M. S. Joshi and S. R. Mcdowall, *Total determination of material parameteres from electromagnetic boundary oinformation*, Pacific J. of Math. **193** (2000), 107-129.

[49] A. Katchalov, Y. Kurylev and M. Lassas, *Inverse Boundary Spectral Problems*, Pure and Applied Mathematics, vol **123**, Chapman Hall/CRC, (2001).

[50] T. Kato, *Growth properties of solutions of the reduced wave equation with a variable coefficient*, Comm. Pure Appl. Math. **12** (1959), 403-425.

[51] T. Kato, *Perturbation Theory for Linear Operators*, Springer-Verlag, (1966).

[52] C. Kenig, M. Salo and G. Uhlmann, *Inverse problems for the anisotropic Maxwell equations*, Duke Math. J. **157** (2011), 369-419.

[53] A. Kirsch, *Surface gradients and continuity properties for some integral*

operators in classical scattering theory, Math. Meth. in Appl. Sci. **11** (1989), 789-804.

[54] A. Kirsch, *Characterization of the scattering obstacle by the spectral data of the far field operator*, Inverse Problems **14** (1998), 1489-1512.

[55] A. Kirsch, *The factorization method for a class of inverse elliptic problems*, Math. Nachr. **278** (2005), 258-277.

[56] A. Kirsch, *An Introduction to the Mathematical Theory of Inverse Problems*, Springer, New York-Dordrecht-Heidelberg-London, (2011).

[57] H. Kozono and T. Yanagisawa, *L^r-variational inequality for vector fields and the Helmholtz-Weyl decomposition in bounded domains*, Indiana Univ. Journal, **58** (2009), 1853-1920.

[58] M. G. Krein, *Solution of the inverse Sturm-Liouville problem*, Doklady Akad. Nauk SSSR (N.S.) **76** (1951), 21-24 (in Russian).

[59] M. G. Krein, *On the transfer funtion of a one-dimensional boundary value problem of the second order*, Dokl. Akad. Nauk SSSR (N.S.) **88** (1953), 405-408 (in Russian).

[60] R. Kress, *On an exterior boundary-value problem for the time-harmonic Maxwell equations with boundary conditions for the normal components of the electric and magnetic field*, Math. Meth. Appl. Sci. **8** (1986), 77-92.

[61] K. Krupchyk, Y. Kurylev and M. Lassas, *Reconstruction of Betti numbers of manifolds for anisotropic Maxwell and Dirac systems*, Commun. in Anal. and Geom. **18** (2010), 963-985.

[62] Y. Kurylev, M. Lassas and E. Somersalo, *Maxwell's equations with a polarization independent wave velocity : direct and inverse problems*, J. Math. Pures Appl. **86** (2006), 237-270.

[63] P. D. Lax, *Symmetrizable linear transformations*, Comm. Pure Appl. Math. **7** (1954), 246-273.

[64] P. D. Lax and R. S. Phillips, *Local boundary conditions for dissipative symmetric linear differential operators*, Comm. Pure Appl. Math. **13** (1960), 427-455.

[65] P. D. Lax and R. S. Phillips, *Scatering Theory*, Academic Press (1967).

[66] B. M. Levitan and M. G. Gasymov, *Determination of a differential equation by two of its spectra*, Russian Math. Surveys **19**:2 (1964), 1-63.

[67] H. Liu, M. Yamamoto and J. Zou, *Reflection principle for the Maxwell equations and its applications to inverse electromagnetic scattering*, Inverse Problems **23** (2007), 2357-2366.

[68] H. Liu and J. Zou, *Uniqueness in an inverse aoustic obstacle scattering problem for both sound-hard and sound-soft polyhderal scatterers*, Inverse Problems **22** (2006), 515-524.

[69] V. A. Marchenko, *Concerning the theory of a differential operator of second order*, Dokl. Akad. Nauk. SSR, **72** (1950), 457-470.

[70] V. A. Marchenko, *Sturm-Liouville Operators and Applications*, Birkhäuser, Basel (1986).

[71] V. A. Marchenko and I. M. Ostrovski, *A characterization of the spectrum of the Hill operator*, Math. USSR Sbornik **26** (1975), 493-554.

[72] S. Mizohata, *The Theory of Partial Differential Equations*, Cambridge

University Press (1973).

[73] K. Mochizuki, *Spectral and scattering theory for symmetric hyperbolic systems in an exterior domain*, Publ. RIMS, Kyoto Univ. **5** (1969), 219-258.

[74] P. Monk, *Finite Element Methods for Maxwell's Equation*, Clarendon Press (2003)

[75] A. Nachman, *Reconstruction from boundary measurements*, Ann. of Math. **128** (1988), 531-576.

[76] A. Nachman, L. Päivärinta, A. Terilä, *On imaging obstacles inside inhomogeneous media*, J. Funct. Anal. **252** (2007), 490-516.

[77] R. G. Newton, *Inverse Schrödinger Scattering in Three Dimensions*, Springer Verlag, New York (1989).

[78] R. G. Novikov, *Multi-dimensional inverse spectral problem for the equation* $-\Delta\psi + (v(x) - E)\psi = 0$, Funct. Anal. Appl. **22** (1988), 263-278.

[79] R. G. Novikov and G. M. Khenkin, *The $\bar{\partial}$-equation in the multi-dimensional inverse scattering problem*, Russian Math. Survey **42** (1987), 109-180.

[80] P. Ola, L. Päivärinta and E. Somersalo, *An inverse boundary value problem in electrodynamics*, Duke Math. J. **70** (1993), 617-653

[81] L. Paquet, *Problèm mixtes pour le système de Maxwell*, Ann. Fac. Sci. de Toulouse, **4** (1982), 103-141.

[82] V. Petkov, *Scattering Theory for Hyperbolic Operators*, North-Hollnad, Amsterdam-New York - Oxford - Tokyo, (1989).

[83] V. Petkov and L. Stoyanov, *Geomtry of the Generalized Geodesic Flow and Inverse Spectral Problems*, 2nd edition, John Wiley & Sons, Ltd., Chichester, (2017).

[84] V. Petkov and G. Vodev, *Asymptotics of the number of the interior transmission eigenvalues*, J. Spectr. Theory **7** (2017), 1-31.

[85] M. Reed and B. Simon, *Methods of Modern Mathematical Physics, Vol 1* $\sim$ *4*, Academic Press (1980).

[86] F. Rellich, *Über das asymptotische Verhalten der Lösungen von* $\Delta u + \lambda u = 0$ *in unendlichen Gebieten*, Jber. Deutsch. Math. Verein. **53** (1943), 57-65.

[87] G. Schwarz, *Hodge Decomposition - A Method for Solving Boundary Value Problems*, Lecture Notes in Mathematics **1607**, Springer Berkin-Heidelberg-New York (1995).

[88] E. M. Stein and G. Weiss, *Introduction to Fourier Analysis on Euclidean Spaces*, Princeton University Press (1971).

[89] J. Sylvester and G. Uhlmann, *A global uniqueness theorem for an inverse boundary value problem*, Ann. Math. **125** (1987), 153-169.

[90] R. Temam, *Navier-Stokes Equations*, North-Holland (1979).

[91] E. Vekua, *On metaharmonic functions*, Trudy Tbiliss. Mat. Inst. **12** (1943), 105-174.

[92] G. Verchota, *Layer potentials and regularity for the Dirichlet problem for Laplae equation in Lipshitz domain*, J. Funct. Anal. **59** (1984), 572-611.

[93] G. Vodev, *Transmission eigenvalues for strictly convex domains*, Math. Ann. **366** (2016).

[94] G. N. Watson, *A Treatise on The Theory of Bessel Functions*, Cambridge Univ. Press, (1944).

[95] C. Weber, *A local compactness theorem for Maxwell's equations*, Math. Meth. in the Appl. Sci. **2** (1980), 12-15.

[96] R. Weder, *Generalized limiting absorption method and multidimensional inverse scatteirng theory*, Math. Meth. in Appl. Sci. **14** (1991), 509-524.

[97] W. Wendland, *Die Fredholmsche Alternative für Operatoren, die bezülich eines bilinearen Funktionals adjungiert sind*, Math. Zeit. **101** (1967), 61-64.

[98] P. Werner, *On an integral equation in electro magnetic diffaction theory*, J. Math. Anal. Appl. **14** (1966), 445-462.

[99] J. Wloka, *Partial Differential Equations*, Cambridge University Press, Cambridge (1987).

[100] D. R. Yafaev, *Mathematical Scattering Theory : General Theory*, Amer. Math. Soc., Providence, Rhode Island, (1992).

[101] D. R. Yafaev, *Mathematical Scattering theory : Analytic Theory*, Amer. Math. Soc., Providence, Rhode Island, (2010).

[102] K. Yosida, *Functional Analysis*, Springer-Verlag, Berlin (1966).

[103] M. Zworski, *Semiclassical Analysis*, Graduate Studies in Mathematics, **138**, Amer. Math. Soc. Providence, Rhode Island, (2012).

Histoical References

[104] J. J. Cross, *Integral theorems in Cambridge mathematical physics, 1830-1855, Wranglers and physicists: studies on Cambridge mathematical physics in the nineteenth century*, ed. P. M. Harman, Manchester Univ. Press, (1985).

[105] I. Fredholm, *Sur une nouvelle méthode pour la résolution du problème de Dirichlet*, äOfversigt af Kongl. Vetenskaps-Akademiens Fäorhandelingar **1** (1900), 39-46.

[106] C. F. Gauss, *Die Attraction homogener elliptischer Sphaeroiden nach einer neuen Methode*, Mon. Corr. **28**, 37, 125, 385, (1813).

[107] C. F. Gauss, *Allgemeine Lehrsätze in Beziehung auf die im verkehrten Verhältnisse des Quadrats der Entfernung wirkeden Amziehungs- und Abstrossungs- kräfte*, Res. Beob. magn. Vereins **4**. 1, (1840).

[108] G. Green, *An essay on the application of mathematical analysis to the theories of electricity and magnetism (1828) ; J. f. Math. **39**, 73, (1850), **44**, 356, (1852), **47**, 161, (1854)*.

[109] M. V. Ostrogradsky, *Note sur la théorie de la chaleur*, Mém. Acad. Sci. St.-Pétersb. **1**. 129 (1831) ; Deuxième note sur la théorier de la chaleur, ibid. **1**. 123, (1831).

[110] F. Riesz, *Über lineare Funktionalgleichungen*, Acta Math. **41** (1918), 71-98.

[111] J. Schauder, *Über lineare, vollstetige Funktionaloperationen*, Studia Math., **2** (1930), 183-196.

[112] *The correspondence between Sir George Gabriel Stokes and Sir William Thomson, Baron Kelvin of Largs*, 2 vols., Cambridge Univ. Press, (1990).

[113] W. Thomson and P. G. Tate, *Treatise on Natural Philosophy*, Clarendon, (1867).

Index

$H(\text{curl}\,;\Omega)$, 145
$H(\text{div}\,;\Omega)$, 144
$H_0(\text{curl}\,;\Omega)$, 146
$H_0(\text{div}\,;\Omega)$, 146
*-operator, 25
δ-operator, 25
0-form, 23
1-form, 21
2-form, 23

absolutely continuous spectrum, 157
adjoint operator, 66
almost analytic extension, 168
analytic Fredholm theorem, 237
asymptotically complete, 164

Betti number, 113, 114
Biot-Savart's law, 99
boundary control method, 258

capacitance, 107
cavities, 108
closed, 29
cohomology group, 113
compact, 66
conjugate operator, 65
continuous spectrum, 157
Coulomb gauge, 97
cross-sectional surface, 118

dielectric permittivity, 93
diffeomorphism, 1

Dirac measure, 131
Dirichlet condition, 75
discrete spectrum, 156
distribution derivative, 136
divergence, 18
divergence theorem, 18
double layer potential, 47, 59
dual operator, 65
dual space, 65

eigenvalue, 72
electric dipole operator, 213
electromagnetic potential, 96
electrostatic capacity, 107
essential spectrum, 156
Euler characteristics, 114
exterior E-M map, 218
exterior differential, 24
exterior domain, 75
exterior product, 23

final set, 192
Fourier transform, 129, 131
frequency, 139
fundamental solution, 31

gauge, 96
generalized Cauchy formula, 167
genus, 114
gradient, 18

handle, 113

handle body, 114
Hardy class, 237
harmonic, 36
harmonic p-form, 26
harmonic vector fields, 121
Helffer-Sjöstrand formula, 167
Helmholtz decomposition, 116, 147
Helmholtz equation, 31
Hilbert-Schmidt, 66
Hodge theory, 116
Huygens' principle, 42
Hölder continuous, 50

incoming radiation condition, 87
incoming wave, 87
induced metric, 9
initial set, 192
inner boundary, 75
interior E-M (electromagnetic) map,
    218
interior domain, 75

Lagrange's identity, 13
Laplace equation, 31
Laplacian, 31
limiting absorption principle, 177
line element, 16
line integral, 16
linking number, 102
local compactness, 153
local coordinates, 13
logarithmic potential, 49
Lorentz gauge, 97

magnetic dipole operator, 213
magnetic permeability, 93
maximum principle, 37
mean value property, 36
multiplicity, 72

Neumann condition, 75
Neumann harmonic vector field, 128
nullspace, 72

outer boundary, 75
outgoing radiation condition, 87

outgoing wave, 87

Parseval's formula, 130
partial isometry, 192
partition of unity, 15
permeability tensor, 104
permittivity tensor, 104
Plancherel's formula, 130
point spectrum, 72
Poisson equation, 31
polarizations, 103
positively oriented, 19
principal value, 48

radiation condition, 179
range, 72
resolvent, 72
resolvent set, 72
response operator, 257
Riemannian metric, 9

S-matrix, 196
scattering amplitude, 196
scattering operator, 163
Schwartz' space of rapidly decreasing
    functions, 130
self-adjoint, 140
short-range, 164
Silver-Müller radiation condition, 182
simply connected, 105
single layer potential, 47, 59
singular continuous spectrum, 157
solid torus, 115
spectral decomposition, 154
spectral measure, 154
spectrum, 72
Stokes' theorem, 20
Stone's formula, 155
submanifolds, 13
surface curl, 206
surface divergence, 206
surface element, 12
surface gradient, 118, 206
surface with boundary, 20

tangent space, 6

tempered distribution, 131
trace operator, 144
triangulation, 114

unique continuation theorem, 88

volume element, 12
wave length, 139
wave operator, 163
weak derivative, 136
Weingarten map, 56

Printed in the United States
By Bookmasters